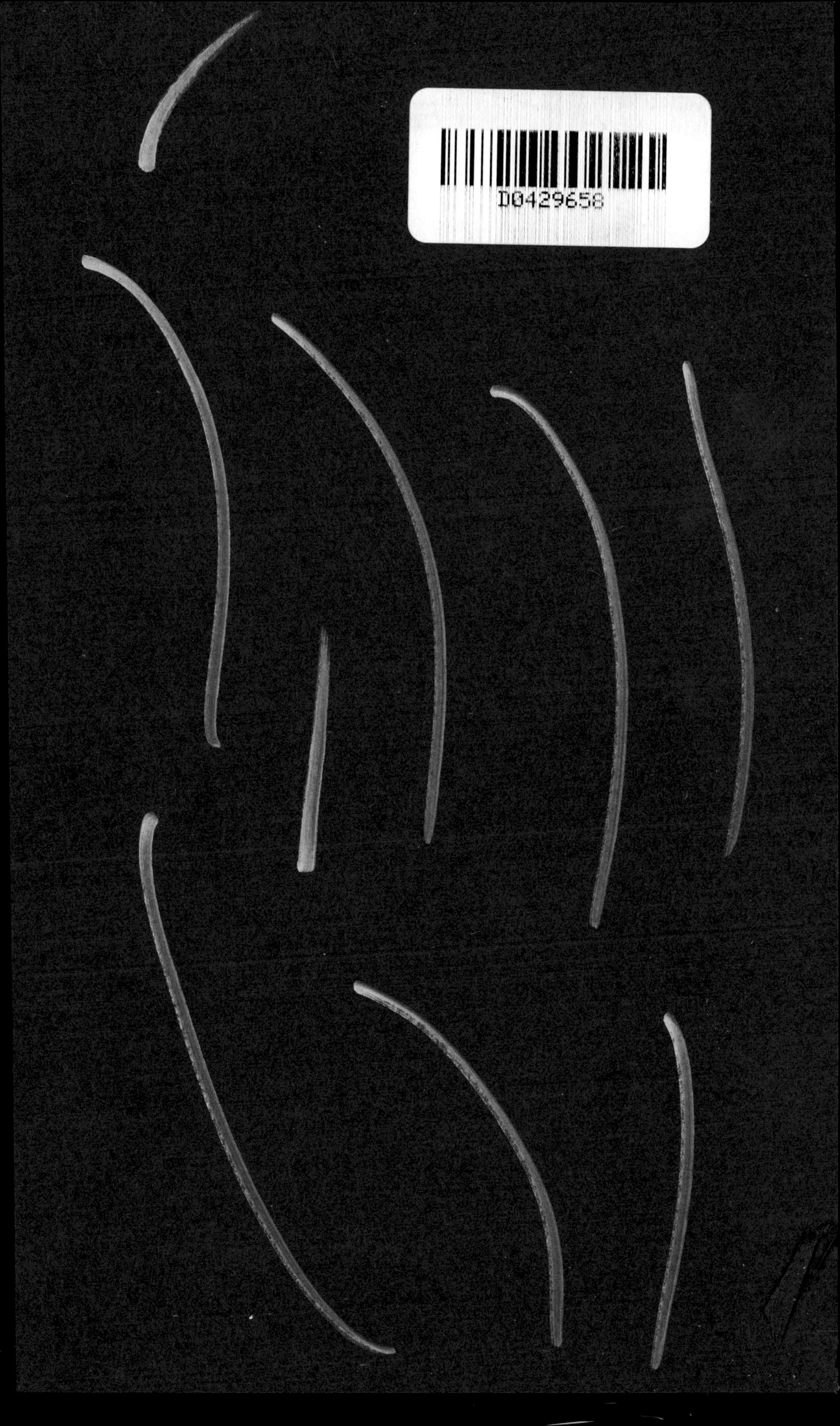

Handbook of Material & Capacity Requirements Planning

Handbook of Material & Capacity Requirements Planning

Howard W. Oden

Gary A. Langenwalter

Raymond A. Lucier

McGraw-Hill, Inc.

New York San Francisco Washington, D.C. Auckland Bogotá
Caracas Lisbon London Madrid Mexico City Milan
Montreal New Delhi San Juan Singapore
Sydney Tokyo Toronto

Library of Congress Cataloging-in-Publication Data

Oden, Howard W.
Handbook of material and capacity requirements planning / Howard W. Oden, Gary A. Langenwalter, Raymond A. Lucier.
p. cm.
Includes index.
ISBN 0-07-047909-7 (acid-free paper)
1. Materials management—Handbooks, manuals, etc. 2. Inventory control—Handbooks, manuals, etc. I. Langenwalter, Gary A. II. Lucier, Raymond A. III. Title.
TS161.033 1993
658.5'03—dc20 93-14719
CIP

1 2 3 4 5 6 7 8 9 0 DOC/DOC 9 9 8 7 6 5 4 3

ISBN 0-07-047909-7

The sponsoring editor for this book was James H. Bessent, Jr., the editing supervisor was Robert C. Walters, and the production supervisor was Pamela A. Pelton. It was set in Baskerville by Carol Woolverton, Lexington, Massachusetts, in cooperation with Warren Publishing Services, Eastport, Maine.

Printed and bound by R. R. Donnelley & Sons Company.

This book is printed on recycled, acid-free paper containing a minimum of 50% recycled de-inked fiber.

Contents

Part 3. Advanced Aspects of M&CRP

Preface

As Dr. Joseph Orlicky stated in the Preface to his landmark work, *Material Requirements Planning, someone* had to write this book. The first edition of Orlicky's text, although a major milestone in the development and spread of MRP throughout manufacturing, was written in 1975, *before* the widespread availability of:

- Master Production Scheduling
- Capacity Requirements Planning
- Just-in-Time and zero inventories
- Lot sizes of 1
- Time-Based Competition
- Bar coding and shop floor data collection
- Total Quality Management
- On-line systems
- Most minicomputers
- PCs and Local Area Networks
- Electronic connection between customers, companies, and suppliers
- Practitioners who understand the requirements for truly successful implementation
- CAD/CAM and CAE

Practitioners are having increasing difficulty relating Orlicky's text to their current environment. Therefore, the time has come for a text written for the practitioner that incorporates the most contemporary practices and techniques in Materials and Capacity Requirements Planning (M&CRP).

The book's primary purpose is to provide materials practitioners, and others who are interested in materials management, with an up-to-date and accurate understanding of:

- The functions of M&CRP and its main components:
  Materials Requirements Planning (MRP)
  Capacity Requirements Planning (CRP)
- How M&CRP can best operate in specific manufacturing environments
- How to utilize M&CRP to benefit "real-world" manufacturing companies, including its direct effects on customer service and communications, inventory levels, and supplier negotiations and communication
- The effects of choices that face the practitioner during system selection, implementation, and operation
- The interrelationship between M&CRP and other functions in an organization

The authors intend this book to assist materials practitioners in passing the Materials and Capacity Requirements certification examination, given by the American Production and Inventory Control Society. However, this is not to be taken as an endorsement or sponsorship of the book by APICS.

Other persons who will directly benefit include:

- Persons who want to increase their knowledge of M&CRP, with the intent to work with practitioners in the field (such as engineers, accountants, marketing professionals, information systems professionals, and the like)
- Persons who want to transfer into the materials management field
- Students who desire a practical background to materials management in manufacturing

The book focuses on the practical application of M&CRP, and the relationship between M&CRP and its interfaces. Its goal is to illuminate M&CRP and the factors that impact M&CRP in a manufacturing company. It does not attempt to serve as a reference in the non-M&CRP areas, such as engineering, information systems management, accounting, purchasing and the like. Nor does this book attempt to provide mathematical rigor in areas such as the theory of order quantities; it leaves those subjects for other texts. Although this book is designed for use in a JIT-oriented world, it does not attempt to explain JIT theory or application outside of M&CRP.

The authors have over 50 years' experience collectively, encompassing all aspects of materials management and manufacturing systems, including:

- Hands-on materials management
- Manufacturing systems consulting
- Manufacturing systems design, development, and implementation (both package and custom code)
- Education at both college and practitioner levels

This book is written from the perspective of the materials manager to facilitate easy understanding of the concepts by the professionals who work in manufacturing. Although the style is deliberately nonacademic, the content will withstand the most rigorous academic scrutiny for accuracy and applicability.

We have incorporated the underlying concepts of JIT (Just-in-Time), including continuous improvement and the elimination of waste, throughout. We assume that computers are widely used for inventory and manufacturing management, order entry and marketing, accounting and finance, and engineering. Additionally, we approach M&CRP from a holistic, or enterprise-wide, viewpoint (as well as from an analytical viewpoint of how it works inside).

The book is organized to create a basic foundation of knowledge in the first chapters and to build on that with more detailed information in the later chapters. We start with an overview of the different manufacturing environments and the materials and information foundations required for each, then cover the basics of inventory and capacity control, M&CRP, and bills of material.

In the central section of the book, we expand on the M&CRP understanding by discussing the various options available to the practitioner, the information systems aspects of M&CRP, and how M&CRP interfaces to other functions.

We conclude by discussing proven M&CRP selection, justification, and implementation approaches, performance measurements, and some topics that will increase in importance in the future. For each chapter, we reference additional books or publications.

With the myriad unique manufacturing environments, the creativity of the practitioners, and hundreds of MRP II software offerings, there are currently many, many contemporary practices. In fact, even the author team had spirited debates about the "best" way to view many different subjects. We have come to accept that this book cannot be all things to all people. As we urge practitioners, so too are we committed to continuous improvement. We sincerely hope that readers will contact us to let us know what they are doing, how the book has helped them, and how the book can be improved. We can be reached at the following addresses:

Howard W. Oden
Professor of Management
Nichols College
Dudley, MA 01571

Gary A. Langenwalter
President
Langenwalter & Associates
20 Seven Star Lane
Stow, MA 01775
508-562-2289

Raymond A. Lucier
Materials Manager
Artel Communications Corporation
22 Kane Industrial Drive
Hudson, MA 01749
508-562-2100

Acknowledgments

Two long years after we uttered the statement, "Somebody ought to write a book," our work became a printed book. During that period, we received substantial assistance from many fellow professionals who were generous with their time, praise, and constructive suggestions. We are grateful to the following:

- The Worcester Chapter of APICS, Peter Langford, Vice President, Region I, APICS, and Andy Nicoll, Vice President, Education Programs and Materials, APICS, for their support and counsel
- The members of the APICS Materials & Capacity Requirements Planning Committee: Paul Rosa, George Adams, Merle Ehlers, Don Frank, Terry Lunn, Merle Thomas, Jr., Gus Vargas, and Nancy Ann Varney, for their insightful comments
- Jim Bessent, our senior sponsoring editor at McGraw-Hill, and Bob Walters, editing supervisor, for their encouragement and willingness to work with us, far above and beyond the call of duty
- Keith Langenwalter, who at the age of 13 transformed semilegible images into professionally drawn figures under very tight deadlines
- The professionals who reviewed and commented on all or parts of the book in draft form: David Caruso of D&B Software, Linda Chamberlayne of Artel Communications, Jeff Davis of Nova Biomedical, Mike Donovan of R. Michael Donovan Inc., Dave Downs of Leaf Systems, Harris Footer of Easy Day Manufacturing, Mark Gordon of Vivid Technologies, Mike Harding of Michael S. Harding and Associates, Dave Hassell of IEC, Al

Lapierre of Albert R. Lapierre and Associates, Bob Lucas of Du Pont NEM Products, Peter MacMurray of Bird-Johnson Company, Bob McInturff of McInturff and Associates, Patricia Moody, Editor of AME *Target,* Amato Prudente of R.R. Donnelley & Sons, Geoff Rezek of Geoffrey Rezek and Associates, Mark Richards of Mark Richards and Associates, Phil Roe of Escom, Paul Routhier of Inframetrics, Al Sjoholm of Fourth Shift New England, and Aaron Wizel of Aaron Wizel Consultants

- Students at Nichols College who reviewed and commented on the book in draft form: Marie Cusanello, Ron LaPointe, Colin Leavitt, Paul Lincoln, Tim Moulton, and Joanne Wood
- R. R. Donnelley & Sons, Hudson, Massachusetts, for printing the first two drafts under extremely tight deadlines
- Jim Kadra and the staff of Nichols College at Southboro, Massachusetts, who generously provided the facilities where we met during the writing and editing
- The late Dr. Joseph Orlicky, whose pioneering efforts and original book on MRP provided the foundation for our understanding and the springboard for our efforts

Most importantly, we wish to thank our wives and children, who put up with the long hours and encouraged us when the tasks seemed almost insurmountable:

- Carmela Oden
- Janet, Karl, and Keith Langenwalter
- Debra, Brian, and Sarah Lucier

To all of the above, plus the many others who have encouraged and influenced our work, we say, "Thank you."

Howard Oden
Gary Langenwalter
Ray Lucier

Handbook of Material & Capacity Requirements Planning

1

Understanding Manufacturing

The Environment of M&CRP

Introduction

A Materials and Capacity Requirements Planning (M&CRP) system is a major element in a manufacturing company and is also the heart of MRP II (Manufacturing Resources Planning). This chapter describes manufacturing capabilities at a basic level, to provide the basis for understanding the environment in which M&CRP systems exist, and the issues they must successfully address.

Although this chapter focuses on the materials planning professional, and others who have experience in manufacturing, the concepts are explained in terms that will be understandable to a person having little manufacturing experience. We do not expect you to become an expert in manufacturing capabilities after reading this one chapter, or even this entire book, but you will understand the manufacturing process well enough to know how and why the M&CRP system works within it. Additional references are listed at the end of the chapter that will provide more complete coverage of manufacturing systems.

We first define manufacturing and then describe how it operates. Next we describe how a manufacturing process is designed and show that the process should be designed hierarchically (in steps from top to bottom) and concurrently with the design of the product.

The strategic design of the manufacturing process requires making decisions in three basic areas:

1. Manufacturing process
2. Demand response
3. Planning and control system

Because they are highly interrelated, the decisions made in each area directly impact the others. Therefore, you must create the design in an iterative (multiple attempts or passes, each learning from the previous), manner. We are most interested in planning and control design, because one of the possible choices in this area is the M&CRP system.

Types of planning and control systems are suggested that are appropriate for each manufacturing process and Demand Response Strategy, emphasizing the manufacturing environments for which M&CRP is most appropriate. The chapter concludes with an overview of MRP II and M&CRP systems.

Basics of Manufacturing

What Is Manufacturing?

As shown in Figure 1-1, manufacturing transforms material, labor, and other inputs into tangible outputs (goods) desired by society. A manufacturing capability consists of four essential components or subsystems: inputs, transformation processes, outputs, and management; all operating in an environment. In the following paragraphs we describe each of these subsystems, and then describe how a more complex version of manufacturing operates.

Inputs include all the incoming resources that are used or retained by the system. They can be divided into consumable (or expense) inputs and retained (or capital) inputs.

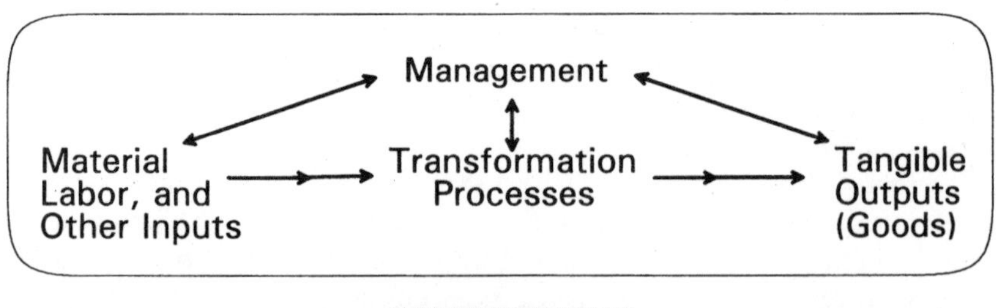

Figure 1-1. Simple schematic of manufacturing.

- *Consumable* inputs include all those resources that are completely consumed in the manufacture of the product such as: raw materials, components, utilities, labor hours, and supplies. Consumable inputs are normally classified as "expense" items by traditional accounting systems.
- *Retained* inputs include all those resources that are not immediately consumed in the production of the product such as: facilities, equipment, personnel, patents, and information. Retained inputs are normally "capitalized" by the accounting system and amortized, or written off, over a number of years.

Outputs of a manufacturing organization are tangible and are technically called *goods.* Outputs of a service system are intangible and are called *services.* Throughout this book, we will refer to the output of a manufacturing organization as the *product,* rather than use the more archaic term of goods. The Standard Industrial Classification (SIC) system classifies manufacturers into 33 groups, depending on their products, with 5 groups in the agriculture, forestry, and fishing division; 5 groups in the mining division; 3 groups in the construction division, and 20 groups in the general manufacturing division.

We are most interested in the general manufacturing division. This division produces: food, tobacco, textiles, apparel, lumber-wood, furniture, paper, printing-publishing, chemicals, petroleum refining, rubber, leather, stone-clay, primary metals, fabricated metals, nonelectrical machinery, electrical-electronic machinery, transportation equipment, instrumentation, and other miscellaneous products.

Transformation Processes include all the activities that must take place to convert the inputs into the desired outputs. The transformation processes performed by a firm are highly dependent upon the desired products and the selected inputs. For example, a metal-fabricating firm may use a number of processes, such as casting, joining, forming, and machining, in order to transform the selected input (metal) into the desired output (customer specified product). These processes may be markedly different for different metals and end products. A firm producing other products, such as food, paper, textiles, or furniture, will utilize a completely different set of processes. Although the primary manufacturing processes vary from one industry to the next, the supporting processes, for example, purchasing, assembling, transporting, distributing, and warehousing, are much more common across industry lines.

The *environment* includes everything that is not a part of the manufacturing organization. Manufacturing is only one of the major elements of a manufacturing firm. The others, top management, sales and marketing, engineering, accounting and finance, MIS, and human resources, are shown in Figure 1-2 as manufacturing's *internal environment.* Other organi-

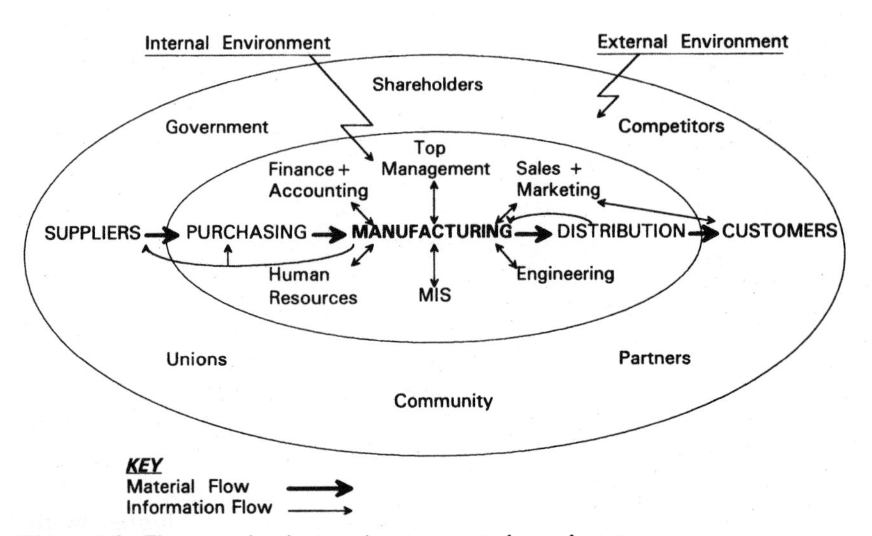

Figure 1-2. The internal and external environment of manufacturing.

zations outside the firm that interact with manufacturing (shareholders, competitors, customers, partners, community, unions, suppliers, and government) are shown in Figure 1-2 as manufacturing's *external environment.*

Management is an important component of a manufacturing organization. In order for the manufacturing organization to function effectively, skilled persons must plan (decisions), and then execute those plans effectively, comparing the feedback of the actual results to the intended results. People at all levels must acquire the inputs, control the transformation or productive process, and ensure that outputs are available at the proper time and place to satisfy demand. This information-processing function can be performed by either humans or computers. The Material and Capacity Requirements Planning system is a subsystem of the management component of the manufacturing organization as shown in Figure 1-1. We will discuss in Chapter 7 how M&CRP fits into an MRP II manufacturing management system.

How Does a Manufacturing Organization Operate?

Figure 1-3 shows the sequence of operation of the major components of the manufacturing organization. The heavy line shows the material flow and the light line shows information flow. Figure 1-3 is intended to show the major functions to provide a framework for future discussion; it cannot begin to accurately show all the interacting operations that occur in manufac-

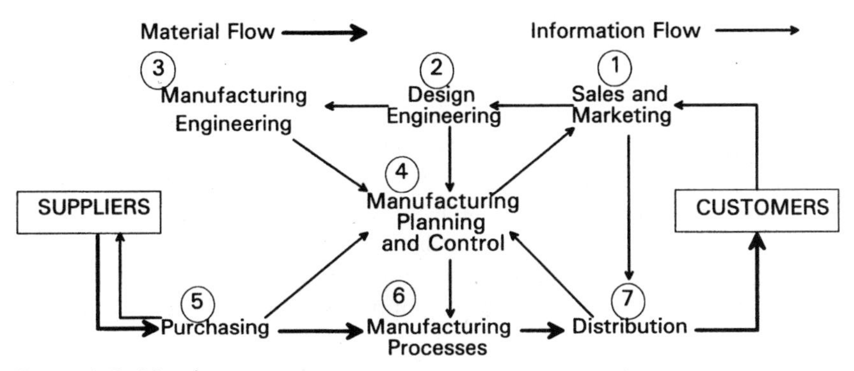

Figure 1-3. Manufacturing operations sequence.

turing. The operational sequence starts with the receipt of a customer order by Sales and Marketing, and follows the sequence indicated by the numbers in Figure 1-3.

In operation 1, when Sales accepts an order for a product that has been made before, the order goes directly from Sales and Marketing to Manufacturing Planning and Control for scheduling; functions 2 and 3 have already been completed. Alternatively, the sales order for a ship from stock product goes directly to Distribution for immediate shipment to the customer. If the product has not been made before, the sales order will go to Design Engineering.

In operation 2, Design Engineering designs the product and documents the design by means of drawings, specifications, and a design Bill of Materials. Manufacturing Engineering performs operation 3, including: advising Design Engineering on the producibility of the product, preparing a process plan or routing for the product, preparing a manufacturing Bill of Materials from the design Bill of Materials, and preparing operating instructions for the human operators. In the past, Design Engineers would pass the information to Manufacturing Engineers without much discussion. The trend toward Concurrent Engineering, which we discuss below, has radically changed this practice.

In operation 4, Manufacturing Planning and Control receives the inputs from the first three operations and schedules the product so that it will be manufactured and delivered in the quantity, quality, and time required by the customer. Manufacturing Planning and Control includes the product in the Master Production Schedule, in conjunction with all other products ordered by other customers. With the Master Production Schedule and the manufacturing Bill of Materials as inputs, the Material and Capacity Requirements Planning (M&CRP) system "explodes" the product to determine the required quantity and timing for the manufacture and/or purchase of the subassemblies, parts, and raw materials needed to build the product on time.

In operation 5, Purchasing uses the output of the M&CRP system to determine which purchased parts are required, and on which dates they are needed. Current trends have created a buyer-planner position to integrate the purchasing and production planning functions more closely. In operation 6, using the output of the same M&CRP system, the manufacturing people manage the entire manufacturing process to support the M&CRP plan. Finally, operation 7 shows Distribution feeding information concerning the levels of inventory at each distribution warehouse directly to Manufacturing Planning and Control.

Manufacturing Product and Process Design

The term *design* denotes the technical planning process that occurs in a manufacturing firm to define its product and process. Design (i.e., technical planning) takes many forms and should not be considered as drawing and engineering analysis only. A myriad of activities, such as orally receiving information from the customer, are part of the design process.

Product and Process Design Concurrency

Product and process design are essentially information-processing efforts that solve problems. Figures 1-4*a* and 1-4*b* depict two contrasting approaches to problem solving in this situation: sequential and concurrent.

In the sequential approach, each step must be fully completed before anything is passed on to the next step. There is little feedback between steps. Each function plays a specific and limited role: engineering designs the product and manufacturing designs the process. For example, product engineering completes the product design, "throws it over the wall" to manufacturing, and essentially washes its hands of the product, letting manufacturing struggle to figure out how to make what the engineers have designed.

The concurrent product and process design approach utilizes a multifunctional team that simultaneously considers all aspects of product and process performance and cost throughout the product's life cycle. This team is organized early in the product's life cycle to allow for more up-front work, when product and process changes cost much less in time and money. Concurrent design results in continuous interaction; product designs are influenced and changed by what we learn as we design manufacturing processes, and manufacturing process designs are influenced and changed by what we learn as we design products. Figure 1-5 shows the per-

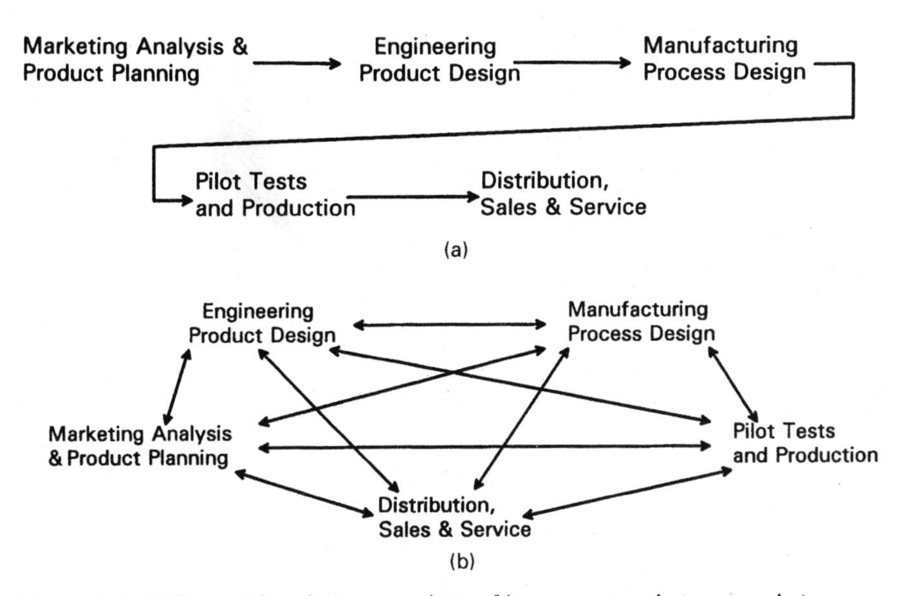

Figure 1-4. (a) Sequential product–process design; (b) concurrent product–process design.

centage of total product cost committed at each stage of the design process. Because 85 percent of the total product cost is determined in the early stages of design, concurrent design can have a dramatic effect on a company's profitability.[1]

Product and Process Design Hierarchy

As indicated in Figure 1-4*b*, all functions that are required to get a new product to market should be represented in the multifunctional effort. However, we will concentrate on the interaction between the two most important elements, product design and the manufacturing process design. The concurrent product-process design procedure, shown in Figure 1-6, is divided into three levels: strategic, tactical, and operational.

Strategic design has a long-range time frame (greater than two years) and broad scope, focusing on the total product line and the entire manufacturing organization. We use the term product line to denote all current and planned products of the company. Strategic product and process design is guided by the firm's overall corporate strategy.

Tactical design has an intermediate time frame (four months to two years),

[1]Hal Mather, "Strategic Logistics—A Total Company Focus," in Patricia E. Moody (ed.), *Strategic Manufacturing*, Business One Irwin/APICS, 1990.

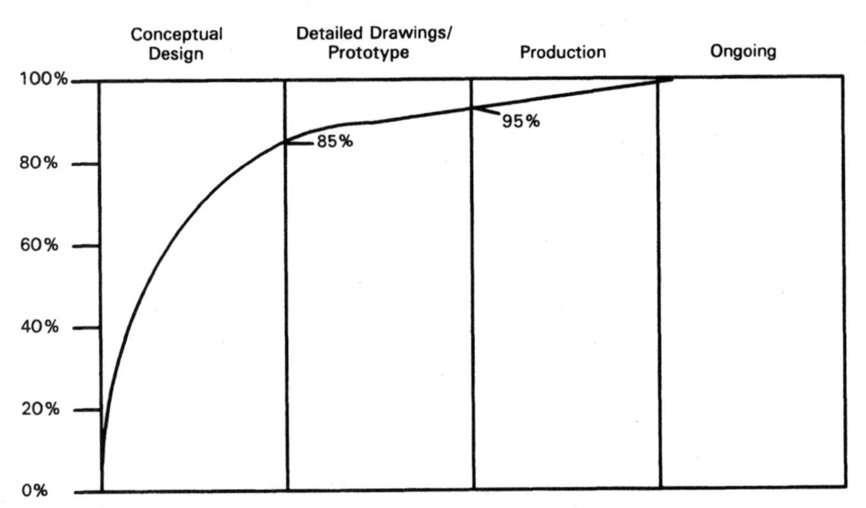

Figure 1-5. Percentage of product cost committed at each stage of development.

and medium scope, focusing on specific sections of both the product line and the manufacturing organization. The design may consider certain product groups or families, and certain manufacturing plants, lines, or cells.

Operational design has a short time frame (less than four months) and narrow scope, often concentrating on one product and that portion of the manufacturing organization that produces that product. Operational de-

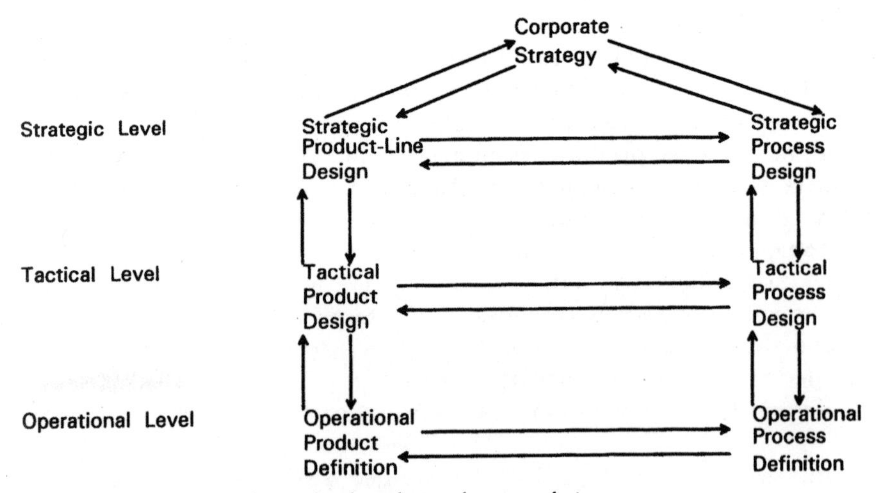

Figure 1-6. Concurrent hierarchical product and process design.

sign is the routine product and process design that is part of the normal sequence of manufacturing operations depicted in Figure 1-3.

These three levels of product and process design proceed in a hierarchical fashion, with each level taking guidance and constraints from the level above and, in turn, passing on guidance and constraints to the level below. However, each level also provides feedback to the level above. The time frames defined for strategic, tactical, and operational design are typical for most manufacturing firms. They may differ, however, by industry or by companies in the same industry.

Strategic Product-Line Design

Strategic product-line design encompasses all the activities leading up to the introduction, revision, or phase-out of products. Although our coverage of the subject must be brief, an understanding of the subject makes the strategic manufacturing process design more meaningful.

Product-line design is an ongoing process; it is never completed. Intense competition, expiring patents, and rapid technological innovations all challenge an organization's ability to produce a product that delights the customer, while providing value, service, and on-time delivery. Corporate strategy guides product-line design, ensuring compatibility with the firm's overall goals.

A firm that neglects to introduce new products periodically will eventually decline. Sales and profits from any given product will normally decrease over time, so the firm must plan to introduce new products before existing ones peak. This concept is shown graphically in Figure 1-7. A prod-

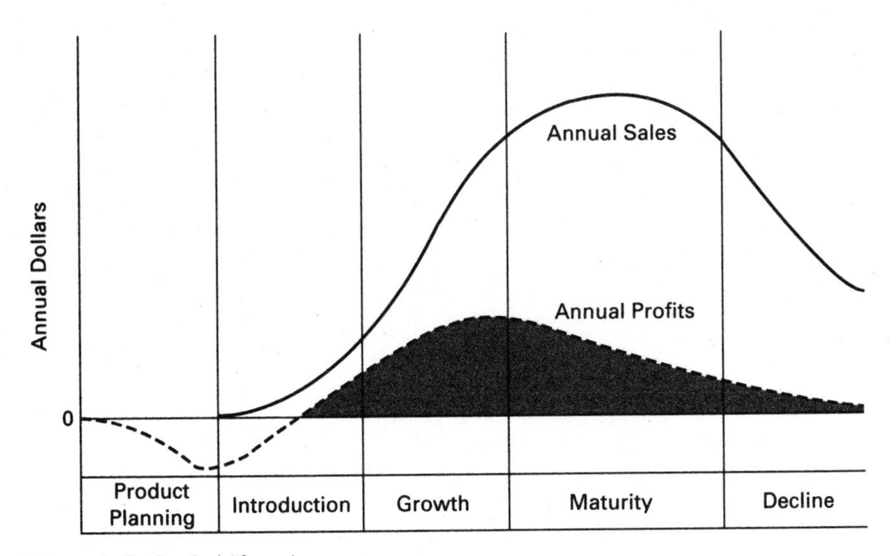

Figure 1-7. Product life cycle.

uct life cycle consists of five stages through which a product passes: product planning, introduction, growth, maturity, and decline. Figure 1-7 shows the change in annual sales and profits as the product passes through the five stages.

The length of the life cycle varies greatly from product to product. The life cycles of such commodities as salt and flour are essentially infinite. The life cycle of automobile models is three to five years. In the high-tech computer and microchip industry, products can become obsolete in months. The length of the product's life cycle determines the nature of the product and process design procedures. At one time, long life cycles permitted leisurely sequential product and process design procedures. Currently, short product life cycles require rapid, concurrent product and process designs. With the life cycles of most products continuing to decrease, a company that can design and produce products more quickly has a competitive edge.

The life cycle of a product in a specific company can be quite different from its life cycle within an industry. As indicated in Figure 1-8, an entrance-exit strategy outlines a firm's choice of when to enter a market and when to leave it. Choosing one of the three basic strategies has important implications for manufacturing and product design.

Early-Early. Small, product-innovative firms often choose to stay in the low-volume, custom product business by entering the market early and exiting early. This strategy requires no painful transition. When the product reaches the maturity stage and profit margins begin to shrink, the firm sells the product line (sometimes with the plant and personnel) and focuses on new products. Throughout the product life cycle, manufacturing maintains a small, flexible production system that can readily be adapted to changing products.

Strategy	Stage to enter	Stage to exit	Implication for manufacturing
Early/early	Introduction	Maturity	Low-volume, flexible product
Early/late	Introduction	Decline	Transition from low-volume flexible product to high-volume, low cost product
Late/late	Growth	Decline	High volume, low cost product

Figure 1-8. Life cycle entrance-exit strategies.

Early-Late. A company can enter the market when the product is first introduced and stay with it until the end of its life cycle. This strategy requires that manufacturing evolve from a low-volume, flexible production system into a high-volume, low-cost system, then back to a lower volume, low-cost system. Such a shift is a challenge because it means changing over to a whole new way of doing things, sometimes in a short time period if the product life cycle is short. Some companies minimize these shifts by dedicating one or more factories to an early-early strategy and others to a late-late strategy.

Late-Late. Some companies prefer to avoid product lines until the product has significant market appeal and will achieve high sales volumes by entering late and exiting late. Large companies, in particular, may accompany their entry with preemptive pricing, setting their prices considerably lower than those of their competitors to ensure the high-volume sales necessary for low unit costs. They can exploit their mass-marketing capabilities, establish distribution channels, and gain access to capital markets in order to finance the massive investment needed for high volumes and high efficiencies.

Designing Products to Match the Market

In designing a product, a firm must not only provide the specific features needed for the product to perform its basic function, but also must satisfy certain general objectives desired by the customer.

These general objectives can be divided into the following four classifications:

1. Cost

   Low cost. Being the lowest cost producer

   Low price. Offering the lowest price

2. Quality

   Performance. Offering products and services with unique, valuable features

   Product quality. Having better craftsmanship or consistency of the product or service

   Product and service reliability. Always working acceptably, enabling customers to count on the product or service performance

   After sale service. Making available extensive, continuing help with the product or service

3. Flexibility and variety

   Volume flexibility. Ability to change quickly and economically from low-volume production to high volume and vice versa

   Product and service flexibility. Ability to switch between different models or variants quickly to satisfy a customer or market

   Customization. Ability to quickly redesign and produce the product or service to meet the customer's needs

4. Time responsiveness

   Delivery reliability. Always meeting delivery promises, with a product or service that is never late

   Fast delivery time. Offering very short lead time to deliver products and services

   Speed of innovation. Being able to bring a range of new products and services to market quickly

Within a specific market, certain objectives may be more important to the customers than others. Each company must design the product to provide the objectives desired by its customers in a specific market.

Also, customers may assume that certain objectives will be available from all producers, basing their buying decisions on the remaining objectives. For example, in many of today's world markets the consumers expect that all producers will automatically satisfy all criteria in the first three categories (i.e., the producers will deliver high quality, customized products at a low price); thus competition centers on the last category, time. This expectation has given birth to what is called *Time-Based Competition.* Aggressive companies are altering their strategies from competing on cost, quality, and flexibility to competing on the basis of time (often measured in days, or even hours), with competitive cost, quality, and flexibility. We discuss this more in Chapter 13.

The general objectives desired by a particular market will not only affect the design of the product, but will also affect the design of the manufacturing capability that produces the product. These objectives may have an impact on the design of the manufacturing capability as follows:

1. Cost

   Special purpose equipment and facilities
   High utilization of capacity
   High productivity
   Low wage rates
   Low scrap

2. Quality

   Skilled educated workers
   Commitment to continuous improvement

Effective quality standards
Low inventory
Precision equipment

3. Flexibility and Variety

   Dependable, rapid suppliers
   Reserve capacity
   Multiskilled workers who can be shifted
   Versatile processing equipment
   Low setup time and cost

4. Time Responsiveness

   High rate of new product introduction
   Short lead times
   Minimum nonvalue added time
   Extremely effective planning and control system
   Concurrent engineering and manufacturing design

Strategic Process Design in Manufacturing

As discussed earlier, strategic manufacturing process design requires making highly interrelated decisions concerning the demand response strategy, the manufacturing process, and the planning and control system. The nature of the external environment and the market determine the nature of the product and the Demand Response Strategy. The nature of the product determines the manufacturing process. In turn, the demand response and manufacturing process determine the most appropriate planning and control system. We are most concerned with the planning and control system, because M&CRP is a planning and control system.

The classification of the response strategies, manufacturing processes, and planning and control systems used in the following pages is somewhat arbitrary and idealistic. In the real world, response strategies, manufacturing processes, and planning and control systems do not fit neatly into any one of our categories. Indeed, we may find a company that has two or more response strategies, processes, and/or control systems in the same plant. For example, a company may have a fabrication job shop with a small batch assembly line.

Demand Response Strategy. The Demand Response Strategy defines how a company will respond to consumer demand. For purposes of this discussion we will classify Demand Response Strategies into five categories:

1. Design-to-Order
2. Make-to-Order

3. Assemble-to-Order
4. Make-to-Stock
5. Make-to-Demand

Design-to-Order. In the Design-to-Order (or Engineer-to-Order) Demand Response Strategy, nothing is inventoried in the producer's system, not even the design. These products have not been made before, at least by this company. The customer usually asks for a quotation of cost and time from the producer. The quotation itself can be complex and costly. When the customer places the order, the producer first develops the design for the required product (which can involve considerable time and expense), receives customer approval of the design, and then orders the needed material. Upon receipt of the material, the producer fabricates the components, assembles the product, and ships it to the customer. In this strategy, the producer bears zero risk with respect to inventory investment. This demand response strategy is most appropriate for products that are new and/or totally unique. Ships, steel mills, military computers, bridges, and the prototypes and first production units of a new class of machine, can fall into this category. The operational focus is on specific customer orders rather than on parts.

Make-to-Order. In the Make-to-Order Demand Response Strategy, only the product designs and some standard raw materials are in inventory, that is, the products have been made before. The processing activities are tailored to each individual customer order. The order cycle begins when the customer specifies the product that he or she wants; the producer can assist the customer to prepare the specifications. The producer quotes a price and delivery time based on the customer's request. The quotation process itself is less costly and complex than Design-to-Order. The customer and producer frequently discuss alternatives to reduce cost, reduce time to ship, and/or meet the customer's actual needs more closely. If the customer accepts the quotation, the producer fabricates the components, assembles the product, and ships it to the customer. In this strategy, the producer bears a very limited risk with respect to inventory investment. As in Design-to-Order, the operational focus is on specific customer orders rather than on parts. Replacement machine parts, hand-crafted sailboats, and research computers can fall into this category.

Assemble-to-Order. In the Assemble-to-Order Demand Response Strategy, all subassemblies or modules are available in inventory. When a customer orders a product, the producer quickly assembles the modules and ships the final product. This demand response strategy is used by companies with modular products, in which several final products share common modules. In practice, the demand for the modules can be forecasted much more accurately than can the demand for the final product. Thus, these companies can respond to customer demand much more efficiently by forecasting and stocking the modules, and then assembling the final product only upon receipt of the customer order. In this strategy, the producer bears moderate risk with respect to inventory investment. The operational focus is primarily on modules and parts, and to some extent on customer orders. Automobiles, commercial computers, and sandwiches in restaurants fall into this category.

Make-to-Stock. In a Make-to-Stock Demand Response Strategy, the producer stocks the finished product in inventory for immediate shipment. Rather than starting with the customer, the cycle starts with the producer specifying the product, acquiring the raw materials, and producing it for stock. The customer orders the product if the price and specifications are acceptable. Assuming the product is indeed on hand, the producer ships immediately from stock. Operations focus entirely on replenishment of inventory; actual customer orders cannot be identified in the production process. At any particular time, the actual level of production may bear little correlation to the level of actual customer orders being received. The production system builds stock levels based on anticipated future orders, not current orders. This situation is especially true for seasonal products, such as lawn mowers, beach balls, and sleds. In this strategy, the producer bears total risk with respect to inventory investment. Many food products, toys, clothes, and telephones fall into this category.

Make-to-Demand. This is a totally flexible demand response that delivers the firm's product, with the quality and delivery time exactly as desired by the customer. This strategy is completely responsive to the customer's order, but can deliver the product with a speed approaching that of Make-to-Stock. It can use any combination of the other response strategies that are needed to meet customer demand. Depending on the competitive situation, designs, raw materials, components, assemblies, and/or finished products may be kept in inventory. This type of response has evolved in reaction to the recent emphasis on time-based competition and is discussed further in Chapter 13. This response is especially applicable to products in a declining stage of their life cycle, because those products require greater features and options coupled with low cost and rapid delivery.

Manufacturing Process (Flow). In this section we focus our attention on the three traditional manufacturing process designs: project, job shop, and line flow, with line flow being divided into three types: small batch line flow, large batch (repetitive) line flow, and continuous line flow. We conclude with a discussion of a relatively new process type, the Flexible Manufacturing System (FMS), and add a new process type that we have called Agile Manufacturing System (AMS).

Project (No Product Flow). In a Project, the materials, tools, and personnel are brought to the location where the product is being fabricated or the service is being provided. Strictly speaking, there is no product flow for a project, but there is still a sequence of operations.

The project form of operations is used when there is a special need for creativity and uniqueness. It is difficult to automate the manufacturing process on projects because they are only done once. Projects tend to have high costs and are difficult to plan and control, because they can be hard to define initially, and can be subject to a high degree of change and innovation.

Job Shop (Jumbled Flow). In a Job Shop or Jumbled Flow Process, products are manufactured in batches at intermittent intervals. Job shops organize equipment and labor into work centers by type (e.g., all lathes in one work center, all grinders in another work center). Products and jobs flow only to those work centers that they require. This results in a jumbled flow pattern, as shown in Figure 1-9.

Because they use general-purpose equipment and highly-skilled labor, job-shop operations are extremely flexible in responding to changes in product design or volume, but they are also rather inefficient. The jumbled flow pattern and product variety leads to severe problems in controlling inventories, schedules, and quality.

Line Flow. A line-flow processing system arranges the work stations in the sequence of operations that make the product, as shown in Figure 1-10. Line flow is sometimes called *product flow,* because the product follows the same sequential steps of production. All products require most of the same tasks, and all follow standard flow patterns. The automotive assembly line is

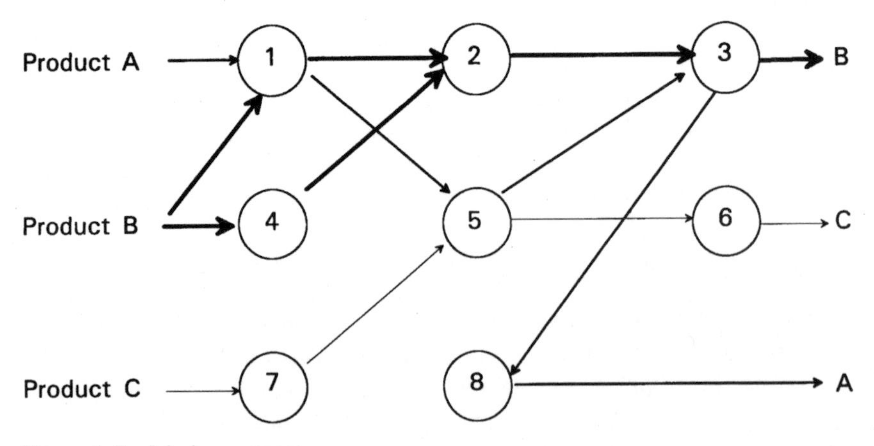

Figure 1-9. Job shop process.

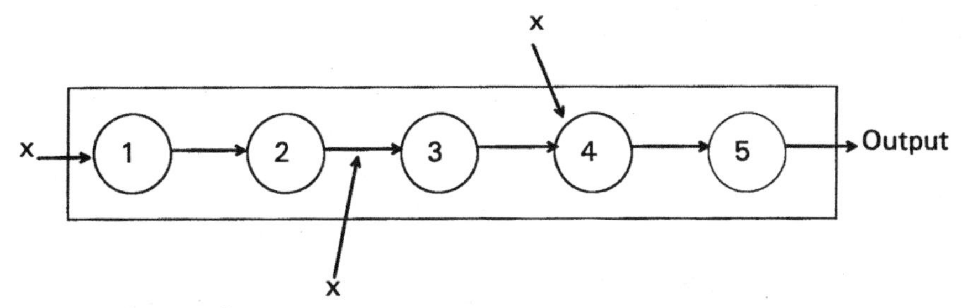

Figure 1-10. Line flow process.

a good example of a line flow process. There are three types of line flow: small batch, large batch (repetitive), and continuous.

- *Small-Batch (or Interrupted) Line Flow.* Small-batch line flow has all the characteristics of line flow, but it does not process the same product continuously. Rather it processes several products in small batches, with setups normally required between batches. Small-batch flow is used when the cost of a line process is justified, even though the items are not produced continuously. Relatively low-demand parts, assemblies, and nondiscrete items (e.g., pharmaceuticals) are often produced using interrupted or batch-flow production.
- *Large-Batch (or Repetitive) Line Flow.* Large-batch or repetitive line flow refers to the production of discrete products in large volumes. A large-batch flow line produces only a few products on the line, with long runs (large batches) of each product that require setups between batches. A repetitive flow line produces only one product in large volume, but the line does not operate continuously. Although the large-batch and repetitive lines are slightly different, they are sufficiently similar, for our purposes, that we can lump them together.
- *Continuous Line Flow.* Continuous line flow refers to the continuous production or processing of fluids, powders, basic metals, and other continuous products. Continuous line flow is used extensively in the process industries to process commodities such as sugar, petroleum, and other fluids, powders, and basic metals.

Flexible Manufacturing System (FMS). An FMS is totally automated, with a computer-integrated group of computer-controlled machines, or workstations, linked together by automated material handling for the completely automatic processing of various product parts, or the assembly of parts into different units. Some FMSs use an integrated engineering and manufacturing data base to automatically design products and processes, estimate materials and create suggested orders, track inventory, program machines, and perform all other activities of the manufacturing process. The purpose of the FMS is to respond accurately and rapidly to the needs of the customer. The FMS can respond rapidly to changes in product design, product volume, or product services. The flexibility and efficiency of FMSs is driving batch manufacturing toward cost-effective lot sizes of one. An FMS is capital-intensive.

Agile Manufacturing System (AMS). An AMS enables a company to achieve many of the benefits provided by FMSs, without using extensive automation. AMS is more a philosophy than a specific set of hardware. In one industry, an AMS will use JIT as the shop floor execution vehicle, because JIT is by far the most appropriate. In another industry, an AMS will use an automated system on the shop floor, because the technology is available and cost-effective. The hallmark of an AMS is its ability to support ruthless time-based competition, emphasizing quick response, flexibility, and efficiency. In a more general sense, an AMS is any manufacturing system that has the capability of being completely responsive to the demands of the customer.

Planning and Control System Strategies. Companies can select one, or more, of six Planning and Control System strategies as their information backbone, and, therefore, as the way in which they plan and execute their medium-term and short-term operations. These six strategies are:

1. Project Management
2. M&CRP-MRP II
3. Just-In-Time
4. Continuous Process Control
5. Flexible Control System
6. Agile Control System

Project Management. The Project Management (PM) Planning and Control System is primarily designed to manage projects. A project is a complex, often large-scale task that is unique or nonroutine for the performing organization that presents unique challenges to management. Network based scheduling techniques such as PERT (Program Evaluation and Review Technique), or CPM (Critical Path Method), are very effective for planning and controlling projects. They provide:

- a visual system to aid in planning,
- estimates of completion dates that can be easily revised as new information becomes available,

- a graphic picture of work assignments and their interrelationships, and a uniform vocabulary for communicating about them,
- identification of the critical path, the series of activities that determines the due date (if any task on the critical path slips, the due date for the entire project slips the same amount), and
- an effective means of comparing actual performance to the plan, so that the need for corrective action can be readily recognized.

In Chapter 11, we encourage a project team to use PC-based project management systems to manage an MRP II system implementation.

M&CRP-MRP II. M&CRP-MRP II recognize the source of demand for an item as being from a higher level assembly (dependent), or being from outside the company (independent). Dependent demand for components occurs only when a parent assembly is actually produced. Consequently, MRP can calculate requirements for dependent demand items based on the Master Production Schedule. MRP calculates the requirements for dependent components by multiplying the number of parent items to be produced by the number of components per unit, as specified in the bill of materials for the parent item. MRP calculates the timing of each requirement by offsetting by lead times, that is, backing up in time from the scheduled completion date for an assembly, by the amount of time required to make the item.

Production requires not only the availability of the necessary materials, but also the availability of adequate production capacity to perform the production operations. Thus, capacity requirements must be planned in coordination with material requirements. Capacity Requirements Planning (CRP) attempts to match the production plan and the production capacity of a company. A computer program for CRP uses the output of MRP to make detailed projections of the load on the production system, so that overloads, underloads, or bottlenecks can be identified in time to take corrective action.

MRP and CRP are discussed more fully in Chapters 4 through 6.

Just-in-Time (JIT). In the Just-in-Time, or Pull, planning and control system, the movement of work is controlled by the following operation; each workstation pulls the output from the preceding workstation, as needed. Only the final assembly line receives a schedule from the dispatching office. All other workstations and suppliers receive production orders from subsequent, or using, workstations. If a using workstation stops production for a time, the supplying workstations will also soon stop, because they will no longer receive production orders for more material.

Production orders can be communicated in many ways, including a shout, a wave, or by electronic means, but by far the most commonly used device is the *Kanban.* Kanban is a Japanese term that means signal, or visible record. When a consuming workstation needs material from a supply workstation, the consuming workstation sends the supply workstation a Kanban. No material can be moved or worked on without one of these cards.

Continuous Process Control. The Continuous Process Control system can best be described as a hierarchy, or layering, of functions. Starting from the bottom, the four major functional levels are:

1. *Process measurement and input-output control,* where the myriad of sensors and actuators sense temperatures and flow rates, and opening and closing valves and gates.
2. *Regulatory control,* and other direct process control, which is the control of process parameters, such as flow, temperature, and numerous other variables. Also, this level contains control logic used for equipment start and stop and, cycling through simple operations. The combination of regulatory control and sequence control is called *direct process control.*
3. *Process monitoring,* which presents all the process related data to the operators, giving them the ability to make changes as required.
4. *Process management,* which is at the highest level, that facilitates diagnostic work, alters plant operating conditions for maximum profit, and schedules production based on demand for the product.

Flexible Control System (FCS). This system controls a Flexible Manufacturing System. Because the FMS can effectively and efficiently manufacture a number of product types, from one-of-a-kind customized products to high-volume commodity products, the FCS must have this same flexibility. It also must be able to control all the resources required to make the product, such as tools, fixtures, NC tapes, material handling capability, and quality control equipment, in addition to the traditional resources of material, machines, and labor controlled by current versions of M&CRP.

Agile Control System (ACS). An Agile Control System is similar to a FCS except that it controls an Agile Manufacturing System rather than a FMS. The ACS is a hybrid of the best of JIT and MRP II. This system utilizes the order management, financial management, and communications capabilities of an MRP II system, including electronic links with customer and supplier, to minimize information transit time and errors. It uses JIT philosophy to identify and eliminate waste, and JIT techniques for shop floor control and scheduling.

Selecting a Manufacturing Process

The product and process life cycles can be viewed as two sides of a matrix, as shown in Figure 1-11. Across the top of the matrix the product life cycle

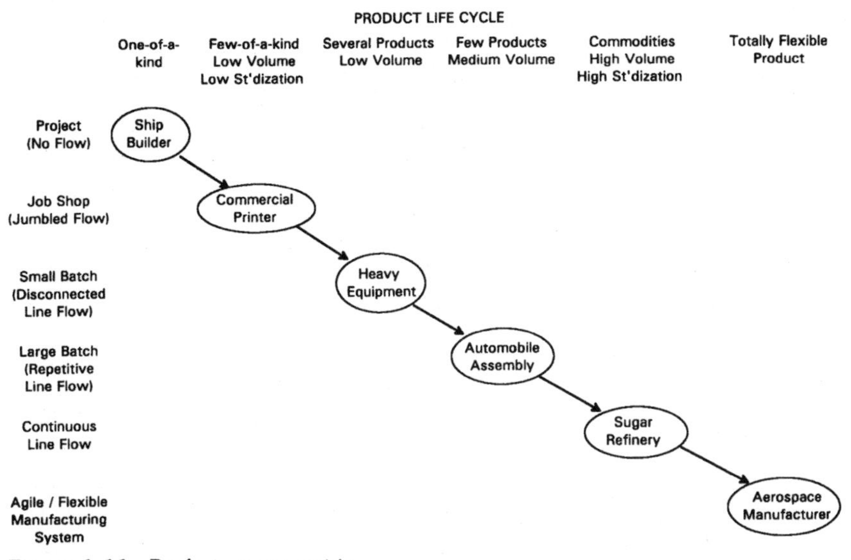

Figure 1-11. Product-process matrix.

varies from low-volume, one-of-a-kind products, through high-volume standard commodity products, ending with Totally Flexible Products, which includes any of the previous five types desired by the customer. As a product line matures, it can potentially move from the left side of the matrix to the right side, or it might remain at any column forever, such as building custom machines, or designing and printing custom brochures.

The rows of the matrix contain the type of process, ranging from a project (no product flow) through a continuous flow process, and ending with Agile-Flexible Manufacturing Systems that can perform any of the first five processes. The process can go through a life cycle just as the product does. The process moves from a fluid and flexible process (but not very efficient or standardized) at the top of the matrix, to an efficient and highly standardized (but much less fluid and flexible) process next to last, and then at the bottom of the matrix ends with a process that is both flexible and efficient. This matrix is called the Volume-Variety Matrix in the APICS Systems and Technologies curriculum.

As this matrix demonstrates, the type of manufacturing process a firm should select depends greatly on the nature of its products. To be most effective and efficient, a firm should select the manufacturing system that most closely matches its products, that is, lies on the diagonal of Figure 1-11. If the product and process are not matched (not on the diagonal), the firm's competitiveness will probably suffer. However, even being on the diagonal of the chart does not guarantee a competitive edge. For example, although a steel mill produces vast volumes of output efficiently, many steel mill customers consider them to be very inflexible with respect to scheduling to meet the customers' needs.

The optimum process strategy maximizes four objectives: cost, quality, flexibility, and time. When companies convert from being job-shop oriented to work-cell oriented, they substantially improve cost and quality, while dramatically reducing response time; they need to do so in such a way that they retain their flexibility.

Selecting Demand Response Strategy

Before we select the appropriate manufacturing process, we must consider Demand Response Strategy. Again, we can view the relationship between the manufacturing process and the demand response in a matrix format, as shown in Figure 1-12. Across the top of the matrix we have the five Demand Response Strategies previously described: Design-to-Order, Make-to-Order, Assemble-to-Order, Make-to-Stock, and Make-to-Demand. Down the left side of the matrix we have the five manufacturing process strategies from project to AMS-FMS.

A rectangle in the matrix indicates a match between the manufacturing process design and the Demand Response Strategy. A dark rectangle indicates a primary match, and a light rectangle indicates a secondary match. A discussion of each major area follows.

- The Project process primarily utilizes Design-to-Order, because many project efforts involve research and development or customized effort.

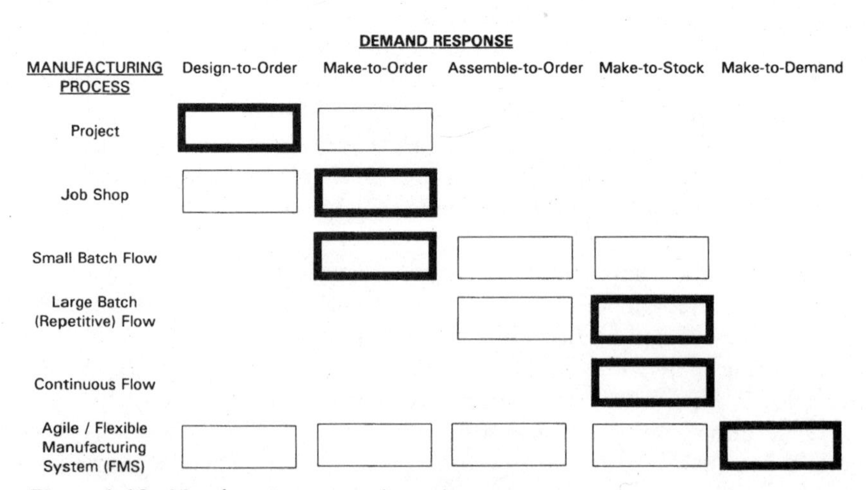

Figure 1-12. Manufacturing process-demand response.

However, projects for large products with available designs, such as locomotives and houses, may utilize Make-to-Order instead.

- Job shops tend to use either Design-to-Order for new products, or Make-to-Order for repeat products. Because most job shops tend to have a high percentage of repeat products, Make-to-Order is the primary mode.
- Small Batch Flow processes primarily use Make-to-Order, unless the product has a modular design, in which case they may use Assemble-to-Order.
- Repetitive flow processes primarily utilize Make-to-Stock, because they principally manufacture high volume standardized (commodity) products, which generally have short customer lead times. If the products are large, expensive, or modular, such as large computers and automobiles, Assemble-to-Order may be more cost-effective.
- Continuous Flow processes make highly standardized (commodity) products, and operate at a constant, or nearly constant, rate for highest efficiency. Consequently they tend to utilize Make-to-Stock.
- Agile and Flexible Manufacturing Systems will normally use the Make-to-Demand response strategy, because they have been specifically designed to respond to customers desiring a flexible, and time responsive output. However, because of their inherent flexibility, they can use any of the other four response strategies, as indicated in Figure 1-13.

Selecting Planning and Control System Strategies

The type of Planning and Control System design a firm should utilize depends greatly on the Manufacturing Process design and the Demand Response design, as shown in Figure 1-13. In this figure, a capital letter indicates major applicability and a small letter indicates minor applicability.

Project Management systems have been specifically designed to plan and control projects. They are also well suited to managing the startup of new, low-volume, products in any type of environment. M&CRP and MRP II systems manage job shops better than any other planning and control system. For a Small Batch Flow process, using either Make-to-Order, Assemble-to-Order, or Make-to-Stock, M&CRP systems have been the traditional choice.

The Just-in-Time (JIT) Pull system of planning and control is best suited for repetitive processes. An M&CRP system can be utilized for the planning aspects of repetitive processes, but it is not very effective for the control of these processes without major modifications. Here again, the Agile Manufacturing System can assist, because it combines JIT on the factory floor with the customer service, master scheduling, and forecasting strengths of an MRP II system.

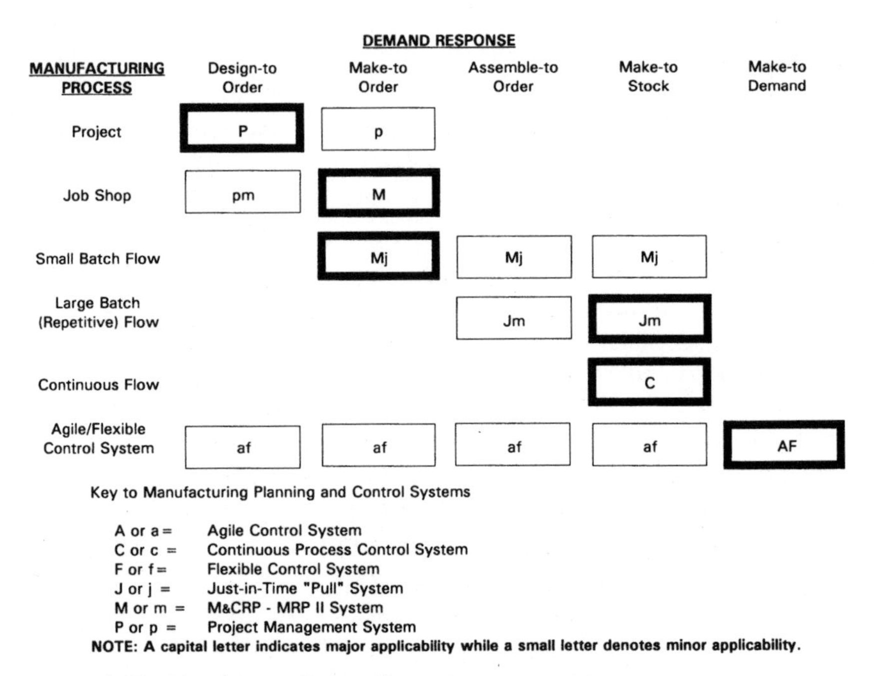

Figure 1-13. Manufacturing Process-Demand response matrix.

Continuous Flow processes tend to utilize the Continuous Process Control system, because these processes are too complex and dynamic to be accurately controlled by any other means. The Agile Control System and the Flexible Control System are most appropriate for Make-to-Demand, because these control systems, and their associated manufacturing systems, were especially designed for this response strategy. However, they can also be used for the other four response strategies, as indicated in Figure 1-13.

Manufacturing Configurations for Which M&CRP-MRP II Is Most Suitable

As indicated in Figure 1-13, the M&CRP-MRP II Planning and Control System is most applicable to the Job Shop Process using Make-to-Order, and to the Small Batch Flow Process, using either Make-to-Order, Assemble-to-Order, or Make-to-Stock. Throughout the book, we will refer to this as the "traditional" manufacturing configuration. Companies also can utilize M&CRP for other manufacturing processes and Demand Response Strategies by making certain additions and modifications to the basic system. These environments are discussed in more detail in Chapter 9.

Manufacturing Resources Planning (MRP II) System Overview

Because MRP II is the environment in which M&CRP operates, we conclude this chapter with an overview of MRP II and M&CRP. MRP II is an explicit and formal manufacturing information system that integrates marketing, finance, and operations. It encompasses all aspects of a manufacturing company, from business planning at the executive level, through detailed planning and control at the managerial and professional levels, through execution in the shop and purchasing, with feedback from each level to the levels above.

As indicated in Figure 1-14, the process begins with an aggregation of demand from all sources. Production, marketing, and financial personnel then work toward developing a production plan and master production schedule that best satisfies demand using available resources. The team must consider marketing and financial resources, as well as manufacturing resources, when developing the Production Plan and Master Production Schedule.

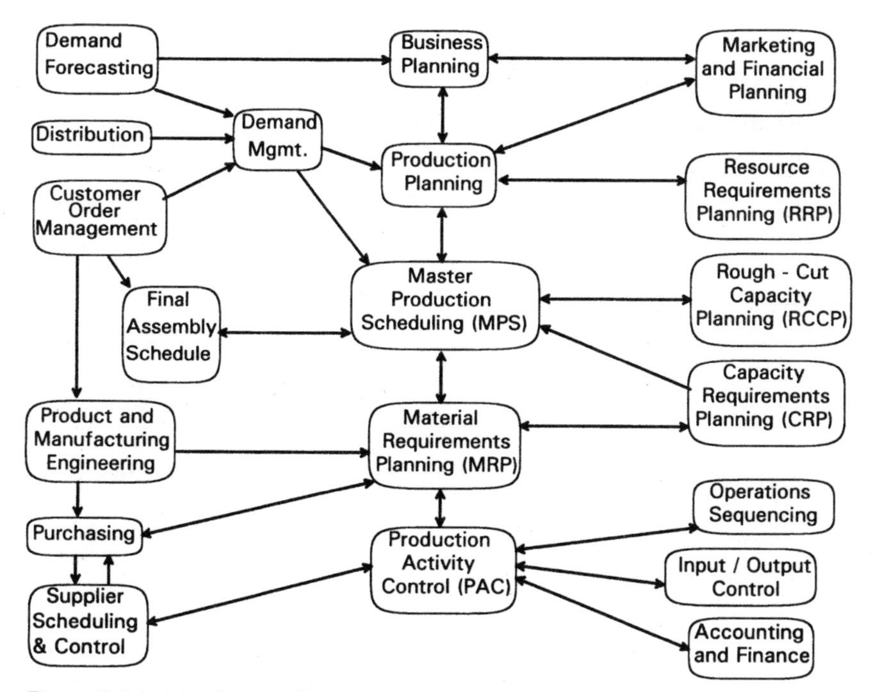

Figure 1-14. Manufacturing Resources Planning (MRP II) system.

At this point, the Material Requirements Planning (MRP) computer program calculates material requirements. Then Capacity Requirements Planning (CRP) compares the released and planned shop orders, by time period, to available capacity, by time period, to alert management to any major overloads and underloads. If the capacity plan is acceptable, the outputs from MRP become the basis for Shop Orders to the shop floor and Purchase Orders to outside suppliers.

This is a continuing process, with the master schedule being updated and revised frequently to achieve corporate goals with available resources. The computer is an important tool in MRP II, because MRP and CRP cannot be cost-effectively executed without a computer. The computer also provides the capability to simulate alternatives enabling managers to answer a variety of "what if" questions, thereby gaining a better appreciation of available options and their consequences.

M&CRP System Overview

Objectives of M&CRP

The objectives of M&CRP are to have the:

- Right part, at the
- Right place, at the
- Right time, in the
- Right quantity, with the
- Right capacity.

Implemented properly, M&CRP can assist a company to simultaneously minimize inventory and maximize customer service. M&CRP consists of two major capabilities:

1. MRP (Materials Requirements Planning)
2. CRP (Capacity Requirements Planning)

Figure 1-15 shows the M&CRP system divided into its two major subsystems: MRP and CRP, and its major interfaces. The SR and PO line (Scheduled Receipts and Planned Orders) is the output from the MRP subsystem that drives the CRP subsystem.

MRP Overview

MRP is the heart of an MRP II system. The MRP module plans material replenishment orders, both purchased and manufactured, which are re-

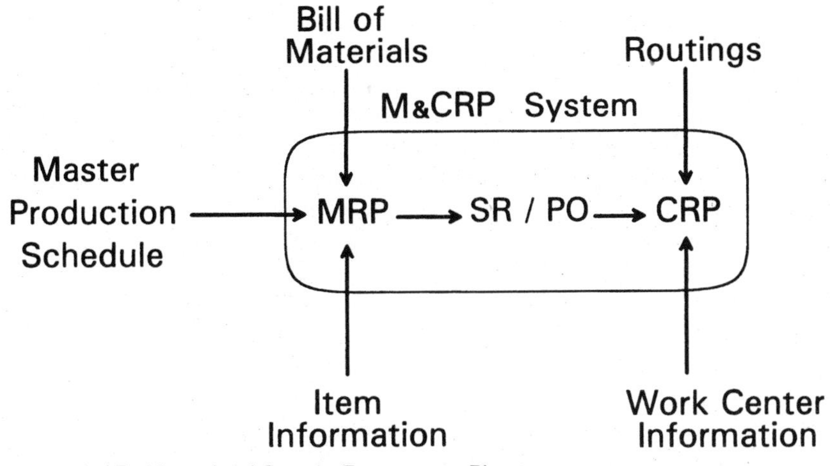

Figure 1-15. Material and Capacity Requirements Planning system.

quired to fill the customer orders and sales forecasts. These manufactured orders, in turn, form the basis for the operating of the manufacturing plant.

MRP projects future shortages of materials, and then plans to fill the projected shortages with exactly the right kind and quantity of material, at exactly the right time. It projects these material requirements, not by past history ("How many toaster trim pieces did we use last year?"), but by the actual requirements based on the latest build schedules, on-hand inventories, and Bills of Materials. A Bill of Materials is a statement, from Engineering, of the materials that are used to make a "parent," or manufactured item.

For example, the Bill of Materials for a ball-point pen could include a cap, a lower barrel, a refill, and a spring, as shown in Figure 1-16, below. If a pen manufacturer schedules an assembly order to make 1000 blue pens, starting on the 15th of next month, MRP uses the Bill of Materials to com-

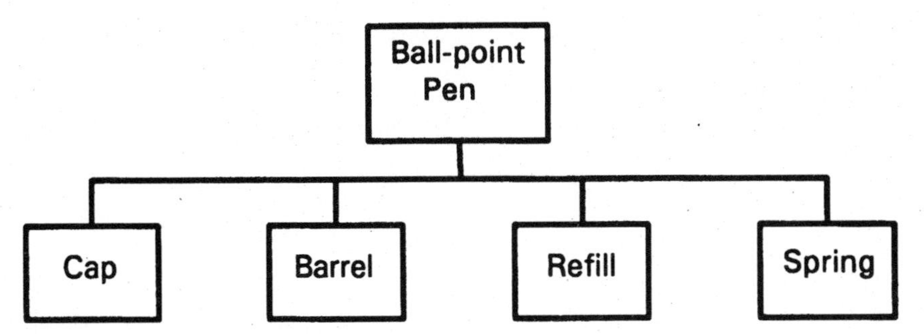

Figure 1-16. Bill of Materials for a ball-point pen.

pute that it will need 1000 caps, 1000 lower barrels, 1000 refills, and 1000 springs. For each of these components, MRP nets the requirement of 1000 against the projected on-hand quantity (computed by starting with the current stock-on-hand in the stockroom, adding incoming replenishment orders that will increase stock-on-hand, and subtracting any other projected issues that will decrease stock-on-hand). If the stock-on-hand will be sufficient on the 15th of next month to issue the required 1000 from inventory, MRP plans no additional orders for the component. If, however, the projected issue of 1000 from inventory will cause a shortage, MRP plans to procure the shortage amount. Continuing the example, if there will be 600 refills available on the 15th of next month (when the 1000 are needed to make the pens), MRP will create a suggested purchase order for 400 refills, due to be in stock on the 15th.

MRP uses the Bill of Materials to ensure that each work order is made from the proper components. Using the pens once more, let us assume that Design Engineering, in response to complaints from customers, has revised the pocket clip so it that will not snag on pockets. Design Engineering will inform MRP by putting a "change order" into the Bill of Materials, which states that the pen will start using the cap with the new clip on the 20th (and stop using the old clip after the 19th) of next month. MRP uses that information to determine which cap to order. This is illustrated in Figure 1-17. All orders for manufacturing pens on or before the 19th will use the old cap, with the old clip. All future orders will use the new cap. This gives Purchasing the visibility to start buying the new caps in time to support the production schedule, while using up the old caps that are in stock and on order, thereby minimizing the cost of the change.

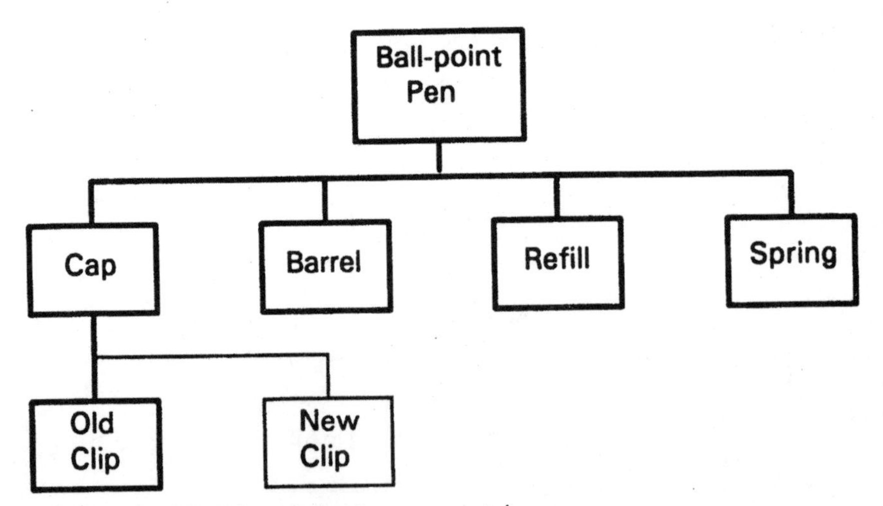

Figure 1-17. Bill of Materials showing component change.

As its output, MRP produces five critical plans:

Projected stock availability by part

Planned manufacturing orders

Planned purchase orders

Suggested manufacturing order actions and reschedules

Suggested purchase order actions and reschedules

These MRP plans tell *how to* make the business plan happen, at the real-world detailed level. If a company executes these plans as scheduled, it will meet its customer commitments and inventory level plans, thus fulfilling its business plan.

CRP Overview

Capacity Requirements Planning produces a report that shows the capacity required by work center and by time period, to actually execute the MRP plan. CRP produces exception reports that show which work centers are overloaded and which are underloaded for each time period. Production planners use this information to see, in advance, the projected capacity problems so they can develop methods to solve the problems. For example, if a work center is underloaded the week of June 15, but overloaded the following week, a production planner could pull some of the June 22 work to the June 15 week in order to level the load between the two weeks. The production planner notifies the rest of the company of the change by revising the dates on the affected orders. When MRP runs, it will plan to have the required material available to meet the new schedule.

CRP computes the projected manufacturing load by using the existing shop orders, at their current state of completion, plus all the future shop orders planned by MRP. CRP then places the load in each appropriate work center by reading the routing file for that order, and projecting when the order will be at each work center during its trip through the shop floor, based on its due date. While a Bill of Materials tells what parts go into a product, a routing tells *how* to make the product. The routing file, which is created and maintained by Manufacturing Engineering in most companies, specifies the amount of work required for a given quantity of parts.

Summary

We have described the salient features of a manufacturing company in order to provide a basic understanding of where and how M&CRP fits into manufacturing. In its most basic form, a manufacturing capability consists of four essential elements: inputs, outputs, transformation processes, and management, all operating in an environment.

Manufacturing must be designed in a hierarchical and concurrent manner, with the product and process design proceeding concurrently and interactively through the strategic, tactical, and operational design levels. The strategic process design requires decisions in three important areas: manufacturing process, demand response, and planning and control system.

In the area of manufacturing processes, the designer must choose between: project, job shop, small-batch line flow, large-batch (repetitive) line flow, and continuous line flow. In the demand response area, the designer has the following alternatives to choose from: Design-to-Order, Make-to-Order, Assemble to Order, Make-to-Stock, and Make-to-Demand. Make-to-Demand keeps inventories to a minimum, while simultaneously providing a high level of flexibility and timeliness.

In selecting the planning and control system, the designer chooses from the following alternatives: Project, M&CRP-MRP II, Just-in-Time, Continuous Process Control, and Agile-Flexible Control Systems. We suggested that the Agile Control System, which is a hybrid of MRP II and JIT, provides the greatest flexibility and timeliness, with the least investment cost. The M&CRP-MRP II planning and control system is most applicable to the Job Shop Process, using the Make-to-Order response and the small-batch flow line, using either the Make-to-Order, Assemble-to-Order, Make-to-Stock, or Make-to-Demand response.

We concluded the chapter with an overview of the MRP II and M&CRP systems.

Selected Bibliography

The following selected references are for those readers who desire to learn more about manufacturing organizations. The basic references are short introductory books that provide an easy introduction to manufacturing. The production-operation textbooks are much more comprehensive, and are designed for college-level courses. The books on manufacturing planning and control cover that aspect of manufacturing in considerable detail. The books on manufacturing strategy are much less detailed, but require a good understanding of manufacturing to read them. The books on special topics cover one or more specific aspects of manufacturing, and assume that the reader has a basic understanding of manufacturing.

Basic References

Amrine, H. T., J. A. Ritchey, and C. L. Moodie: *Manufacturing Organization and Management*, 5th ed., Prentice-Hall, Englewood Cliffs, NJ, 1987.

Brandenburg, R. G. (ed.): *What Every Manager Needs to Know About Manufacturing,* AmaCom, New York, 1983.
Fearoj, H. E. et al.: *Fundamentals of Production/Operations Management,* 3d ed., West, St. Paul, MN, 1986.
Fulmer, W. E.: *Managing Production: The Adventure,* Allyn and Bacon, Boston, 1984.

Production and Operations Management Textbooks

Adam, E. E., Jr. and R. J. Ebert: *Production and Operations Management,* 5th ed., Prentice Hall, Englewood Cliffs, NJ, 1992.
Buffa, E. S. and R. K. Sarin: *Modern Production/Operations Management,* 8th ed., Wiley, New York, 1987.
Chase, R. B., and N. J. Aquilano: *Production and Operations Management,* 5th ed., Irwin, Homewood, IL, 1989.
Dilworth, James B.: *Production and Operations Management,* 4th ed., McGraw-Hill, New York, 1989.
Evans, J. R., et al.: *Applied Production and Operations Management,* 3d ed., West, St. Paul, MN, 1990.
Fogarty, D. W., T. R. Hoffmann, and P. W. Stonebraker: *Production and Operations Management,* South-Western, Cincinnati, OH, 1989,
Gaither, N.: *Production and Operations Management,* 4th ed., Dryden Press, Chicago, 1990.
Heizer, Jay and B. Render: *Production and Operations Management: Strategies and Tactics,* 2d ed., Allyn and Bacon, Bmston, 1991.
Krajewski, L. J. and L. P. Ritzman: *Operations Management: Strategy and Analysis,* 2d ed., Addison-Wesley, Reading, MA, 1990.
Schonberger R. J., and Edward M. Knod: *Operations Management,* 4th ed, Business Publications Inc., Plano, TX, 1988.
Schroeder, R. G.: *Operations Management: Decision Making in the Operations Function,* 3d ed., McGraw-Hill, New York, 1989.
Stevenson, W. J.: *Production/Operations Management,* 3d ed., Irwin, Homewood, IL, 1990.

Introduction to Manufacturing Planning and Control

Blackstone, John H., Jr.: *Capacity Management,* South-Western, Cincinnati, OH, 1989.
Browne, Jimmie, J. Harhen, and J. Shivnan: *Production Management Systems: A CIM Perspective,* Addision-Wesley, Reading, MA, 1988, Chap 1.
Cox, James F., John H. Blackstone, and Michael S. Spencer: *APICS Dictionary,* 7th ed., APICS, 1992.
Deis, Paul, *Production and Inventory Management in the Technological Age,* Prentice-Hall, Englewood Cliffs, NJ, 1983.

Fogarty, Donald W., J. H. Blackstone, and T. R. Hoffman: *Production and Inventory Management,* South-Western, Cincinnati, OH, 1991.
Lunn, Terry with Susan A. Neff: *MRP: Integrating Material Requirements Planning and Modern Business,* Irwin, Homewood, IL, 1992.
Orlicky, Joseph: *Material Requirements Planning,* McGraw-Hill, New York, 1975.
Plossl, George W.: *Production and Inventory Control: Principles and Techniques,* 2d ed: Prentice-Hall, Englewood Cliffs, NJ, 1985.
Schultz, Terry: *Business Requirements Planning: The Journey to Excellence,* The Forum Ltd, Milwaukee, WI, 1984.
Smith, Spencer B.: *Computer Based Production and Inventory Control,* Prentice-Hall, Englewood Cliffs, NJ, 1989.
Vollmann, Thomas E., W. L. Berry, and D. C. Whybark: *Manufacturing Planning and Control Systems,* 3d ed: Irwin, Homewood, IL, 1992, Chap 1.
Wight, Oliver W.: *MRP II: Unlocking America's Productivity Potential,* CBI Publishing, Plano, TX, 1981.

Manufacturing Strategy

Blackburn, Joseph D. (ed): *Time Based Competition: The Next Battle Ground in American Manufacturing,* Business-One, Irwin, Homewood, IL, 1991.
Buffa, Elwood S.: *Meeting the Competitive Challenge: Manufacturing Strategy for U.S. Companies,* Wiley, New York, 1984.
Grant, Robert M., R. Krishnan, Abraham B. Shani, and Ron Baer: "Appropriate Manufacturing Technology: A Strategic Approach," *Sloan Management Review,* vol. 33, no. 1, Fall, 1991, pp. 43–54.
Gunn, Thomas G.: *Manufacturing for Competitive Advantage: Becoming a World Class Manufacturer,* Ballinger, Cambridge, MA, 1987.
Hayes, R. H. and S. C. Wheelwright: *Restoring Our Competitive Edge: Competing Through Manufacturing,* Wiley, New York, 1984.
Hayes, R. H., S. C. Wheelwright, and K. B. Clark: *Dynamic Manufacturing: Creating the Learning Organization,* Free Press, New York, 1988.
Hill, Terry: *Manufacturing Strategy: Texts and Cases,* Irwin, Homewood, IL, 1989.
McGrath, Michael E. and Richard W. Hoole: "Manufacturing's New Economies of Scale," *Harvard Business Review,* vol. 70, no. 3, May–June 1992, pp. 94–102.
Miller, Stanley S.: *Competitive Manufacturing: Using Production as a Management Tool,* Van Nostrand Reinhold, New York, 1988.
Moody, Patricia: *Strategic Manufacturing: Dynamic New Directions for the 1990s,* Dow Jones-Irwin, Homewood, IL, 1990.
Pannesi, Ronald T. and Helene J. O'Brien: *Systems and Technologies Certification Review Course Student Guide,* APICS, Falls Church, VA, 1992.
Skinner, Wickham: *Manufacturing: The Formidable Competitive Weapon,* Wiley, New York, 1985.
Stalk, George, Jr. and Thomas M. Hunt: *Competing Against Time,* The Free Press, New York, 1990.

Special Topics in Manufacturing

Black, J. T.: *The Design of the Factory with a Future,* APICS, 1991.

Gardner, J. A.: *Common Sense Manufacturing: Becoming a Top Value Competitor,* APICS, 1992.

Harmon, Roy L.: *Reinventing the Factory II: Managing the World Class Factory,* The Free Press, New York, 1992.

———and Leroy Peterson, *Reinventing the Factory: Productivity Breakthrough in Manufacturing Today,* The Free Press, New York, 1989.

Klein, Jan: *Revitalizing Manufacturing: Text and Cases,* Irwin, Homewood, IL, 1989.

2

Basics of Inventory Control, Priority Planning, and Capacity Planning

Introduction

All inventory management systems must answer the basic questions: *What to order*, *How much to order*, and *When to order*. This chapter provides the basic concepts and principles that answer those questions. We begin with a discussion of the concepts of manufacturing and distribution inventories, then introduce the related concepts of dependent and independent demand to help explain the difference between two of the major inventory control systems: order point and Material Requirements Planning. Also introduced are the basic concepts underlying Kanban, the most common inventory control system used with JIT.

We then develop the basic features of two order point inventory control systems: a fixed order *quantity* system and a fixed order *interval* system. This will provide the basis for comparing to, and better understanding, the MRP system.

The remainder of the chapter covers the basics of priority, capacity, and integrated priority and capacity planning. For every priority (output) planning activity there is, or should be, a comparable capacity (input) planning activity. The manufacturing system cannot produce the desired priorities unless the necessary capacity is available.

The process of priority and capacity planning proceeds in a hierarchical top-down fashion, with each pair of modules operating over progressively shorter planning horizons and in progressively greater degrees of detail, as shown in Figure 2-1.

We are primarily interested in the operation of the third level of the priority planning and capacity planning modules (commonly called Material and Capacity Requirements Planning), but we must understand the basics of the complete hierarchy to fully understand M&CRP. We will briefly discuss this hierarchy, and each layer in it, to better understand the relationship between priority planning and capacity planning.

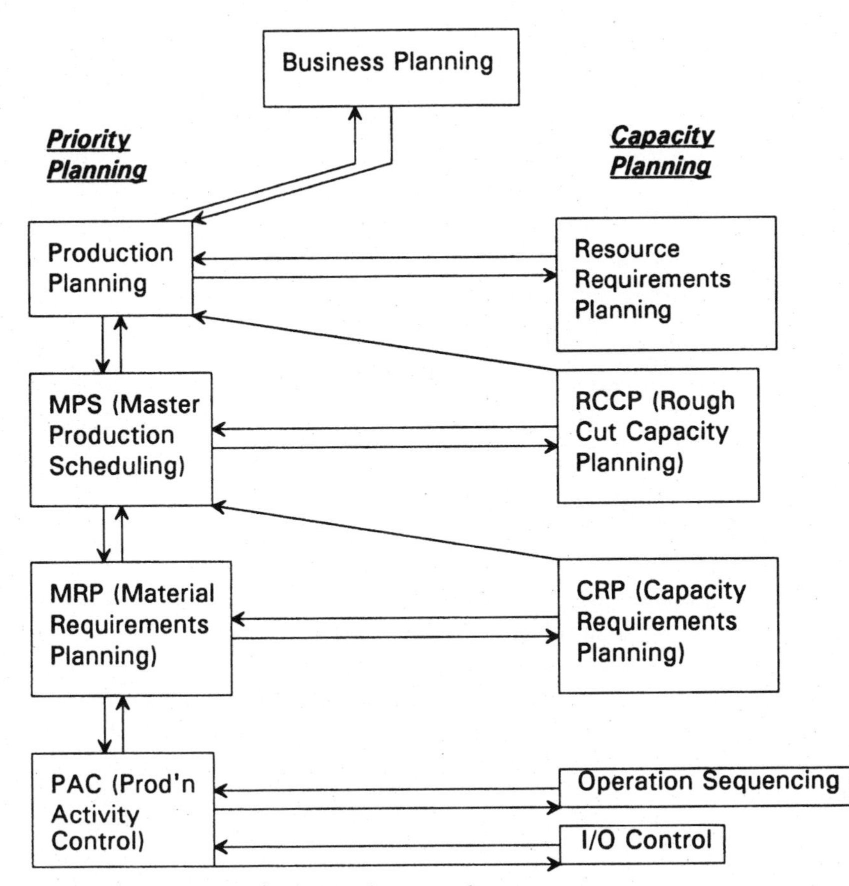

Figure 2-1. Hierarchy of priority and capacity planning.

Types of Inventory and Demand

Manufacturing versus Distribution Inventory

Failure to understand the differences between manufacturing and distribution inventories causes much of the controversy in deciding whether or not a certain inventory control procedure is applicable to manufacturing. Inventories within a manufacturing activity serve a different purpose than inventories of finished products or service parts (distribution inventories).

- *Manufacturing inventories* support the Master Production Schedule or the final assembly schedule. Most manufacturing operations require three basic types of manufacturing inventories: (1) Raw materials, (2) Components, and (3) Work-in-Process (WIP). *Manufacturing* considerations dictate the investment level. Manufacturing inventory levels can be calculated, based on the Master Production Schedule and the company's inventory procurement policies. The length of manufacturing lead times is the primary factor in determining the level of WIP inventories.
- *Distribution inventories* are primarily the finished products, service parts, and repair kits intended to satisfy customer needs. These inventories are the end result of the manufacturing process and are created from manufacturing inventories. The unpredictability of customer demand causes most of the uncertainty that plagues the distribution inventories manager. To ensure a certain reliability of product delivery, distributors generally define service levels and carry safety stocks. Many unit demands, originating from separate sources, combine to create the total demand for an item over a given period. *Marketing* considerations primarily govern the investment level.

In a distribution inventory environment, demand for each inventory item must be explicitly or implicitly *forecast.* Uncertainty exists at the item level. The successful inventory manager must determine what part(s) to order, what quantity to order, and when to reorder. The manager can answer the first and third questions by calculating a reorder point or interval, and the second by calculating an economic order quantity.

In a manufacturing inventory environment, uncertainty exists primarily at the master production level (will the customer demand materialize as forecast?). The inventory planner can easily calculate individual item demand from the master production schedule, eliminating the need to forecast individual item demand. Inventory, available prior to the time it is needed, or in a quantity greater than needed, is a waste of scarce resources. In a manufacturing environment, the planner does not need to (and should not) replenish inventory when it is low. Instead, the planner should order only what is required to cover production needs. Additionally, the planner can tie inventory delivery directly to the *time* of these needs to avoid having inventory on hand before the manufacturing process requires it.

Independent Demand versus Dependent Demand

The basic difference between distribution and manufacturing inventories is caused by the nature of the demand on these inventories. We classify the demand on a manufacturing inventory as *dependent,* because the demand for a manufacturing inventory item is derived from, or is dependent upon, the demand for a higher level item. *Dependent demand* items are typically subassemblies, component parts, or raw materials that will be used in the production of a higher level subassembly, or a final or finished product. In such cases, demand of subassemblies and component parts *depends* completely on the number of finished units that will be produced. For example, if a car has four wheels, the total number of wheels required for a production run is simply a function of the number of cars to be produced in the run. More specifically, if we schedule production of 100 cars, the production process requires $100 \times 4 = 400$ wheels. This illustrates two important principles:

1. The *production schedule* for the car *creates the demand* for wheels; we do not need to know whether or not there is a customer demand for the car. Although this point appears to be relatively minor, it has major implications on day-to-day operations.
2. Because we can calculate the exact quantity of wheels needed, and the exact date when they will be needed, *we have answered all three basic inventory management questions without requiring a forecast.* And because our precise calculations are much more accurate than any forecast, we do not need to forecast dependent (internal manufacturing) consumption of components, subassemblies, or raw materials.

On the other hand, *independent demand* items are finished goods, or other items of *distribution inventory.* Generally speaking, these items are sold, or at least shipped out of the factory, rather than used in making another product in the same factory. There is no feasible way to precisely determine how many of these items will be demanded during any given time period, because demand is typically somewhat random. Therefore, forecasting plays an important role in stocking decisions for independent demand items, whereas the master production plan determines stock requirements for dependent demand items.

Independent demand can range from smooth to very sporadic or "lumpy," depending on the industry and the customer-consumption patterns. Most independent demand items must be in stock virtually all the time, and usually require some Safety Stock.

Dependent demand can also range from smooth to very sporadic or lumpy, depending on how components are used in higher level assemblies. In many manufacturing companies, production planners tend to schedule large batches of production parts or assemblies to minimize the cost of setup time per unit, without appreciating the full impact of these large batches. Large batches create requirements for large quantities of components at specific points in time, with little or no usage at other times. However, dependent demand items need only to be stocked just prior to the time they will be needed in the production process. Moreover, the relative predictability of dependent demand (manufacturing) item usage means that there is little or no need for safety stock. Figure 2-2 contrasts the independent and dependent demand inventories.

Some items have both dependent and independent demand. These are often service parts. For example, an automobile manufacturer might sell shock absorbers directly to car dealers as replacement parts (independent demand), as well as use four shock absorbers for each car that they manufacture (dependent demand). In that case, the total demand for shock absorbers is the *sum* of the dependent demand, to be used in the manufacturing process, and the independent demand, to be shipped to customers.

Some manufacturers have realized that their *independent* demand, in reality, is the output from their customers' manufacturing plans. They have asked for copies of those plans, to reduce the uncertainty of that demand. Other manufacturers who distribute to several warehouses, and even to

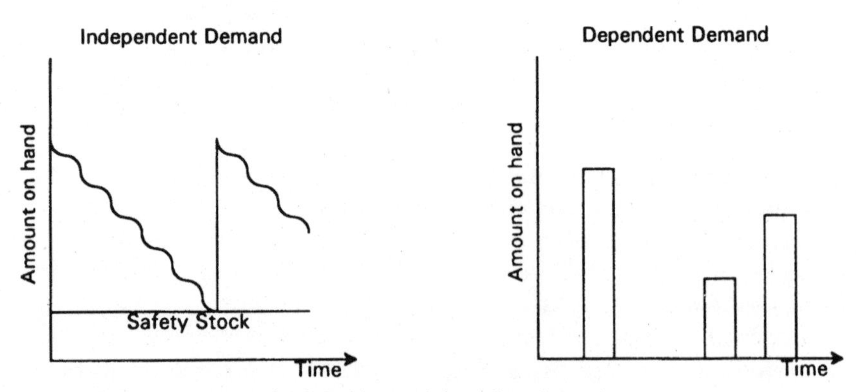

Figure 2-2. Comparison of independent and dependent demand.

some retail stores, have applied the time-phasing concepts of MRP to distribution inventories. This is called Distribution Resource Planning (DRP). Their intent is also to remove as much uncertainty as possible with respect to independent demand.

Three Inventory Systems

The *order point* inventory system attempts to ensure continuous physical availability of all items in inventory. An order point system, either manual or computer-based, monitors the depletion of the stock-on-hand of each inventory item, issuing a replenishment order whenever the supply drops below a predetermined quantity—the *reorder point.* Order point systems are thus reactive; they cannot predict with any accuracy when a part will actually fall below its reorder point, or when it will actually "stock-out."

By contrast, a *material requirements planning* (MRP) system plans to have on hand *only* the components required to support the Master Production Schedule, or Final Assembly Schedule. Unlike an order point system, MRP does not plan a replenishment just because the stock-on-hand of a component is low, or at zero. MRP translates a Master Production Schedule into time-phased requirements for each component, and then plans the coverage of such requirements for each inventory item needed to support the Master Production Schedule. An MRP system replans component requirements and coverage as a result of changes in either the Master Production Schedule, inventory status, or product structure. MRP systems are proactive; they calculate when a part will stock-out, and plan to have a replenishment order arrive just prior to that point.

Finally *Kanban* systems are again reactive; instead of relying on predictive calculations, as does MRP, Kanban systems rely on extremely fast replenishment response by the factory or supplier. In fact, the response is so fast that, in theory, the inventory never totally runs out.

Figure 2-3 graphically compares the performance of an order point inventory system with an MRP inventory control system and a Kanban system for a dependent demand component. For simplicity, we assume only one finished product with only one component. The demand rate for the finished product is fairly constant, and the top diagram shows the finished product's inventory level over time in a typical order point or MRP II environment. The second and third diagrams depict inventory levels for the component, using an order point system and an MRP system, respectively. Finally, the fourth and fifth diagrams show the inventory levels of the same finished product and component, using JIT and Kanban. The source of supply of the single component, purchased from an outside supplier or manufactured in-house, makes no difference in this example. A discussion of each of these systems follows.

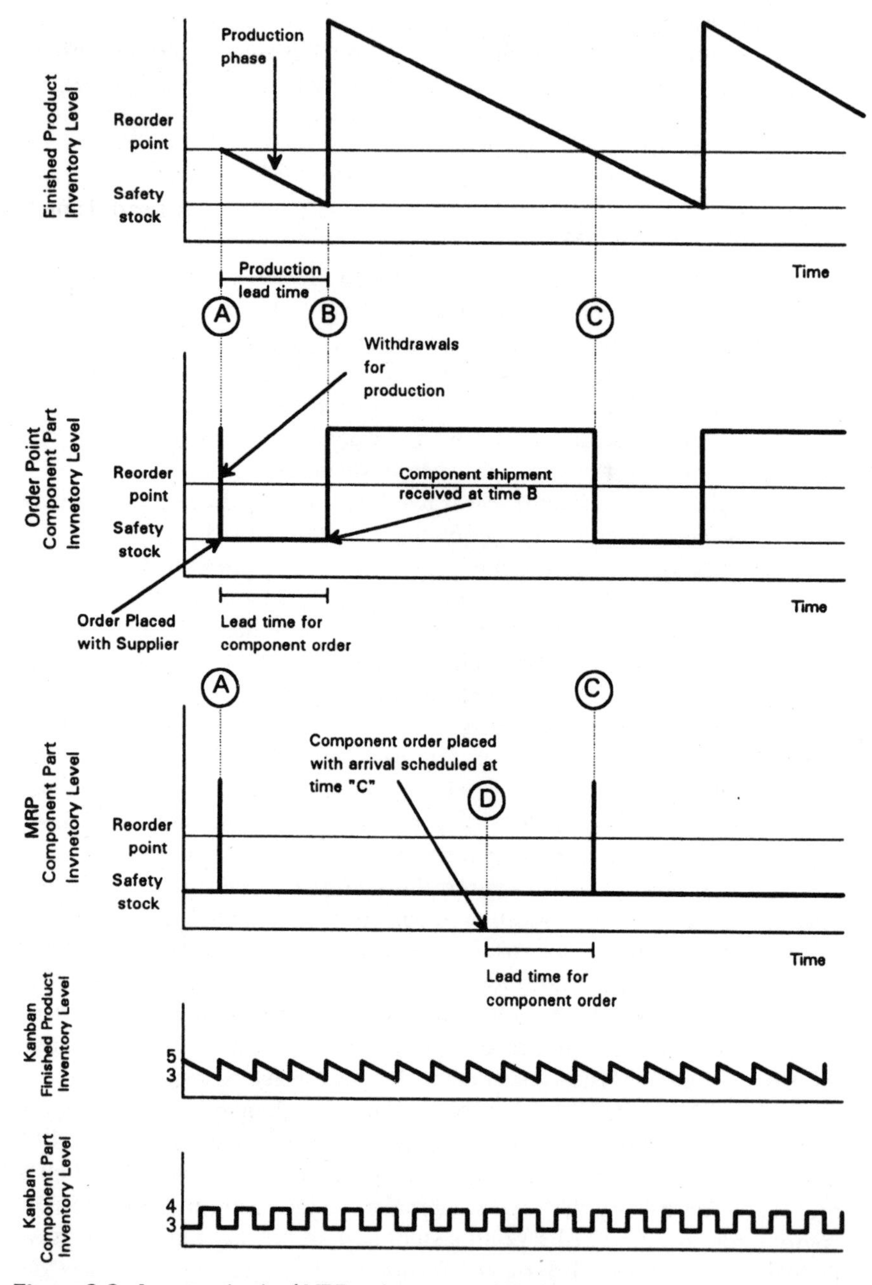

Figure 2-3. Inventory levels of MRP, order point, and Kanban.

Order point systems do not consider *vertical* (parent-component) *dependencies.* For the order point example, the second diagram shows that when the shop starts work on a shop order for the finished product (point *A* on the time axis), the stockroom attendant withdraws the component part. This causes the component inventory level to fall below its reorder point, so the order point system immediately places a replenishment order for the component, calculating the due date as "Today's Date plus Lead Time." The order arrives at point *B*. The component is not actually needed until point *C*, when components will be withdrawn for the next scheduled production of the finished part. Clearly the component inventory is not needed between points *B* and *C*; the investment is a waste of scarce resources.

The MRP system, as shown by the third diagram, dramatically reduces inventory levels when compared to Order Point. The MRP system places a replenishment order at point *D*, which is exactly one lead time prior to point *C*, so that the components will arrive at exactly point *C*. This intentionally avoids the added cost of holding the component inventory from point *B* to *C*. In this respect, MRP plans the inventory replenishment to arrive "Just-in-Time," minimizing stock on hand.

Additionally, order point systems do not consider *horizontal dependencies* (dependencies between components of the same parent). For example, if assembly *X* uses components *Y* and *Z*, but component *Y* will arrive three weeks late from a supplier, an order point system has no way to utilize this information when deciding when to order component *Z*. MRP can provide this information to humans for their action. Again, the order point system creates unnecessary inventory because it assumes that there are no dependencies between parts.

Finally the Kanban example, in the bottom two diagrams, shows that a well-implemented Kanban system has even less inventory than most MRP systems. This is because Kanban in normally implemented as part of a Just-in-Time effort, and JIT regards inventory as waste (and therefore directly focuses on substantially reducing inventories). However, Kanban, like JIT, requires several environmental prerequisites, such as extremely high and predictable quality, very short response times (often measured in hours, or even minutes), and relatively stable production patterns.

JIT deliberately levels the production load for the parent, scheduling the same number each day (and perhaps even each hour). Because the load for the parent is level, the requirements for the component are also level. JIT does not plan specific replacement lots or orders, instead, it relies on the ability of each supplier to keep the input area of the downstream work center filled with the proper quantity of parts. In this example, only four containers of the component part will fit in the input queue for the parent part. In JIT, the only time a supplier can make a component, or deliver a raw material, is when there is room to store it in the input queue of the downstream work center. When the downstream work center uses one container of the component to make the parent, the work center that supplies the component will make one more container of component, and put it in the downstream work center's input area. In our example above, the maximum inventory of the component is only four containers; the minimum is zero. The probable stocking level is close to four containers, if the supplying work center is responsive.

An MRP system looks into the future and calculates when items are needed; it is a proactive, or "push," system. An order point system calls for action when inventory breaks through the order point; it is a reactive, or "pull," system. MRP facilitates formal and explicit planning. Order point systems for dependent demand items make life more exciting by increasing the number of surprises, and decreasing the time managers have to cope with capacity management problems. (Most managers prefer less exciting lives.) JIT is a reactive or pull system, depending on the stability of the schedule for the end items, and the ability of the suppliers to react very, very quickly to keep the input queues of the downstream work centers filled.

Types of Inventory, Demand and Applicable Inventory Systems

Figure 2-4 summarizes the important differences between the inventories, demand, and inventory systems for manufacturing and distribution.

Characteristic	Manufacturing	Distribution
Items in Inventory	Inputs to Manufacturing System: Raw Materials, Components, and Work-in-Process	Outputs of Manufacturing System: End Items, Service Parts, and Repair Kits
Type of Demand	Dependent Demand	Independent Demand
Source of Demand	Master Production Schedule	Forecasted and Actual Customer Orders
Nature of Demand	Dependent on demand for items in MPS. Demand can be calculated.	Demand independent of demand for any other item. Demand must be forecast
Type of Inventory System	Material Requirements Planning System (MRP) or Just in Time (JIT)	Statistical Order Point System or Distribution Resources Planning (DRP)

Figure 2-4. Comparison of manufacturing and distribution inventories, demand and inventory systems.

Order Point Inventory Control Basics

In this section we describe the basics of order point inventory systems so that you might better understand the MRP system by comparing these two approaches. We will not discuss the more sophisticated features of order point systems, because those are not necessary to understand the MRP system. There are two basic types of order point inventory systems.

1. The fixed order *quantity* system first determines the order quantity by using the traditional Economic Order Quantity (EOQ) model, which minimizes the costs of holding and ordering inventory. When to order occurs when the quantity on hand drops below the Reorder Point (ROP). The Reorder Point is the sum of the safety stock, held to reduce the probability of stock-out during lead time, plus the expected demand during order lead time. Reorder point models are developed for only two situations: constant demand rate and variable demand rate. Both of these assume constant replenishment lead times.
2. The fixed order *interval* system first determines when to order by establishing a fixed review period *R*. The length of the review period can be selected to minimize the sum of the ordering and holding costs, or to satisfy other criteria the company considers relevant. Once the review period is selected, the maximum inventory *M* needed to cover demand during the review period plus lead time, can be calculated. The order quantity is the difference between the maximum inventory and the inventory on hand at the end of the review period.

For both order point inventory control systems, the order due date is determined by adding the standard order lead time to the date the order is placed. We will concentrate on the fixed-order quantity system, because it is this system that MRP has essentially replaced in the manufacturing environment.

Kanban systems are basically a type of simple, manual, order point system. In a Kanban system, the order quantity is the amount that fits into one container or skid. The order point is reached whenever a skid or container is used in a downstream work center, and the Kanban card is returned to the replenishment station.

Objectives of Inventory Planning and Control

Inventory planning and control has two main objectives:

1. *Maximize the level of customer service* by having the right goods, in sufficient quantities, in the right place, at the right time, and

2. *Minimize the cost* of providing the desired level of customer service, including inventory holding costs, setup costs, ordering costs, and shortage costs.

Traditional manufacturing environments consider these two objectives to directly oppose each other. Attaining high levels of customer service by stocking additional inventory causes high costs, and attaining lower costs by cutting inventory levels usually causes low levels of customer service. Consequently, most traditional inventory decisions, especially those involving order point, are trade-offs involving a compromise between cost and the level of customer service. In practice, management may select a desired level of customer service, in which case the goal of inventory control is to attain that level at the lowest possible cost. Conversely, management may set cost levels, in which case the goal is to attain the highest possible level of customer service under those cost conditions. In MRP and JIT environments, companies can come much closer to maximizing *both* of these objectives.

The decision maker's challenge is to achieve a balance in stocking levels, avoiding both overstocking and understocking. The two fundamental decisions relate to the *timing* and *size* of orders (when to order and how much to order). The majority of material in this section is devoted to assisting managers in making these two decisions.

Inventory Costs

The costs of holding inventories can be categorized into three major types: holding, ordering, and shortage costs.

1. *Holding or carrying costs* relate to physically holding items in storage. They include interest, insurance, taxes, depreciation, obsolescence, deterioration, spoilage, pilferage, breakage, and warehousing costs (heat, light, rent, and security). Holding costs also include opportunity costs associated with having funds tied up in inventory that could be used elsewhere. The significance of the various components of holding cost depends on the type of item involved. For example, one major fast-food hamburger chain has a policy that forces the store to discard any prepared hamburgers that are over one-half-hour old. Pharmaceuticals and some chemicals have well-defined shelf lives. Fashion clothing has a "life" of only one season. Conversely, an automobile can await purchase for several months (at least until the end of the model year). A building can await purchase indefinitely.

Holding costs are stated in either of two ways: cost as a percentage of unit price, or as a dollar amount per unit. In any case, typical annual holding costs range from 20 to 40 percent of the value of an item according to Generally Accepted Accounting Practices, and higher according to Activity-

Based Costing. Thus, to hold a $1 item for a year could cost from 20 to 40 cents, or more.

2. *Ordering costs* are associated with ordering and receiving inventory. These costs include determining how much is needed, preparing purchase orders, inspecting goods upon arrival for quality and quantity, and moving goods to storage. Ordering costs are generally expressed as a fixed-dollar amount per order, regardless of order size.

When a firm produces its own inventory instead of ordering it from a supplier, it can directly see and control the additional ordering costs of the manufacturing setup (preparing equipment for the job by adjusting the machine, changing cutting tools, delivering the components and raw materials to the starting work center, and so on). The ordering costs listed earlier still apply to the company ordering the items. If a firm orders from a supplier, the supplier incurs these manufacturing costs and includes them in the purchase price.

3. *Shortage costs* result when demand exceeds the supply of inventory available. The costs can include the opportunity cost of not making a sale, loss of customer goodwill, lateness charges, and similar costs. These costs can be extremely high; one very large retailer penalizes suppliers $25,000 for the first late delivery, $50,000 for the second, and $100,000 for the third! If a supplier misses a scheduled delivery window (measured in hours, rather than days) three times in a twelve-month period for a major electronics manufacturer, the supplier loses that manufacturer as a customer.

To overcome this problem, a company can incur additional charges in the form of premium freight, overtime, expediting charges, special handling and the like. Furthermore, if the shortage occurs in an item carried for internal use (to supply an assembly line), it can cause lost production or downtime. Such costs can easily run into hundreds of dollars a minute, or more. Because the shortage of *any* part, even a *C* part, can shut down an entire manufacturing line, the *ABC* method of inventory control can severely understate the potential stock-out cost of *C* parts in manufacturing companies. To compensate, some companies invest in substantial quantities of safety stock for *C* items; this is not too costly, because these are *C* items. Shortage costs are usually difficult to measure, so they are often subjectively estimated. Shortage costs are excluded from classic order point formulas that we present later in this chapter.

Fixed Order Quantity System

Fixed order quantity systems are the most widely used of the various order point systems. This section focuses on how such systems answer the fundamental questions of inventory management: what to order, how much to order, and when to order. For ease of understanding, we present the answer to the second question first.

How Much to Order

In the fixed order quantity inventory system, shown in Figure 2-5, an order of a fixed quantity (the reorder quantity) is placed whenever the amount on hand falls to, or below, a specified level (the reorder point). This system assumes that the inventory level is under continuous review. In this system, the size of the order is fixed, but the time interval between orders depends on actual demand.

In the fixed order quantity system, the question of how much to order is frequently determined by using an *Economic Order Quantity* (*EOQ*) model. EOQ models identify the optimal order quantity by trading off costs that are assumed to be outside the control of the company. In manufacturing companies, both MRP and JIT have successfully challenged this underlying assumption. There are several versions of the EOQ model; we present only the most basic one.

The basic EOQ model computes the order size that minimizes the sum of two factors: the annual cost of *holding* inventory and the annual cost of *ordering* inventory. The unit purchase price of items in inventory is not generally included in the total cost, because the EOQ model assumes that the order size does not affect unit cost. The basic model involves the following assumptions:

- Holding costs can be accurately computed, and are linear
- Ordering costs can be accurately computed, and are linear
- There is only one product involved

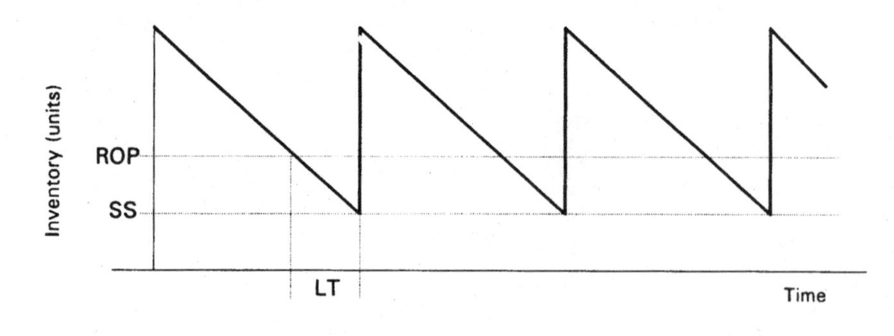

Figure 2-5. Fixed order quantity inventory system.

- Annual usage (demand) requirements are known
- The usage rate is reasonably constant
- Lead time does not vary
- Each order is received in a single instantaneous delivery
- There are no quantity discounts

Some of these assumptions may not hold true in a real-world manufacturing company. To the extent that they are untrue, this order point system becomes less applicable.

The diagram of an inventory cycle in Figure 2-5 illustrates the implications of some of these assumptions. The inventory cycle begins with receiving an order of *Q* units. The stockroom clerk withdraws units at a constant rate in response to very smooth demand. When the quantity on hand is just sufficient to satisfy demand during order lead time (the time between submitting an order and receiving it), the purchasing department submits an order for *Q* units to the supplier. Because the EOQ model assumes that both the usage rate and the lead time do not vary, the model assumes that the order will arrive at the precise instant that the inventory on hand reaches the Safety Stock level. Thus, the model establishes the order due date to avoid having excess stock-on-hand and to avoid stock-outs.

The optimal order quantity reflects a trade-off between holding costs and ordering costs. As order size varies, one type of cost will increase, while the other one decreases. Relatively small order quantities cause a low average inventory, which will result in low holding (carrying) costs. However, small order quantities will necessitate frequent orders, which the EOQ model assumes will increase ordering costs. Conversely, ordering large quantities at infrequent intervals, will reduce ordering costs, but result in higher average inventory levels and, therefore, increased holding costs. These two extremes are illustrated in Figure 2-6.

Thus, the ideal solution will typically be an order size that causes neither a few very large orders, nor many small orders, but one that lies somewhere between those two extremes. The exact amount to order will depend on the relative magnitudes of holding (carrying) and ordering costs. A discussion of each of these factors follows.

Annual holding (carrying) cost equals the average amount of inventory on hand multiplied by the cost to carry one unit for a year, even though any given unit would probably not be held for a year. Mathematically, the cost of holding 12 units for one month equals the cost of holding one unit for a year. Based on the assumption of smooth demand (withdrawal from stock in small, even amounts), the EOQ model assumes that the average inventory is simply one-half of the order quantity *Q*. The amount on hand decreases steadily from *Q* units to 0, for an average of $Q/2$ units. Using the

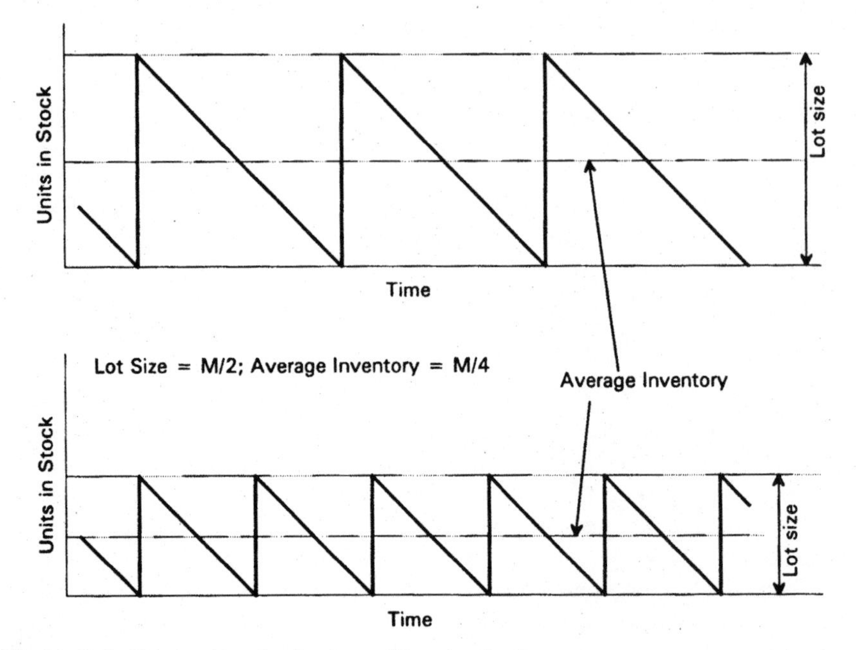

Figure 2-6. Relationship of order size and inventory level.

symbol H to represent the average annual holding cost per unit, the total annual holding cost is

$$\text{Annual holding cost} = \frac{QH}{2} \tag{2-1}$$

Holding costs are thus a linear function of Q. Holding costs increase, or decrease, in direct proportion to changes in the order quantity Q, as illustrated in Figure 2-7*a*.

On the other hand, total ordering costs for a year will decrease as order size increases, because for a given annual demand, a larger order size requires fewer replenishment orders. For instance, if annual demand is 1200 units, and the order size Q is 100 units, there will be 12 orders during a year. But if $Q = 200$ units, only 6 orders will be needed. And if $Q = 300$ units, only 4 orders will be needed. In general, the number of orders per year will be D/Q, where D = annual demand and Q = order size. Unlike holding costs, ordering costs are relatively insensitive to order size. Regardless of the amount of an order, certain functions must be performed, such as preparing the purchase or manufacturing order, placing the order with the ven-

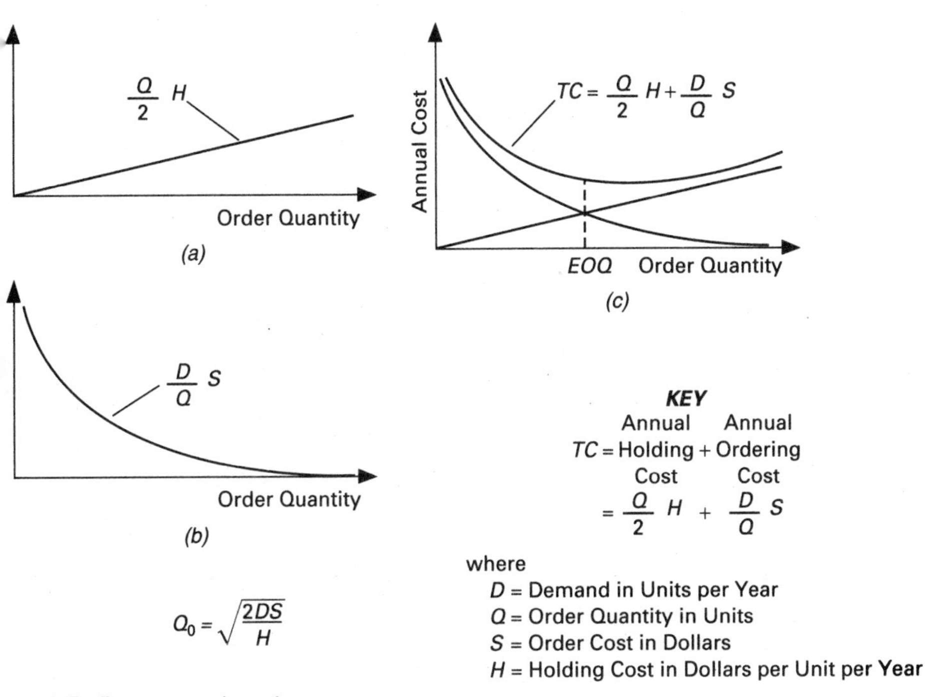

ure 2-7. Determining the order quantity.

dor or in the shop, expediting to insure timely arrival, inspecting the order as it arrives, storing the items, and informing the computer that x has arrived. Purchase orders also require that accounting receives an invoice, matches it to the PO and receiving report, and pays the vendor. Inspection of the shipment to verify quality and quantity characteristics is not strongly influenced by order size, because large shipments are sampled rather than completely inspected. We can, therefore, consider ordering cost as fixed cost, with little loss of accuracy. *Annual ordering cost* is a function of the number of orders per year and the ordering cost per order:

$$\text{Annual ordering cost} = \frac{DS}{Q} \tag{2-2}$$

where S = ordering cost. (To help remember, consider that S originally meant *Setup* cost, for manufacturing orders.)

Because the number of orders per year, D/Q, decreases as Q increases, annual ordering cost is inversely related to order size, as illustrated in Figure 2-7*b*.

The total annual cost (TC), associated with holding and ordering inventory when Q are ordered each time, can be calculated as follows:

$$\text{TC} = (\text{Annual holding cost}) + (\text{Annual ordering cost}) \tag{2-3}$$

$$= \frac{QH}{2} + \frac{DS}{Q}$$

where D = demand, in units per year
Q = order quantity, in units per order
S = order cost, in dollars per order
H = holding cost, in dollars per year

Note that D and H must be in the same time units, for example, years. Figure 2-7*c* reveals that the total cost curve is U-shaped and that *it reaches its minimum at the quantity where holding costs equal ordering costs.* Thus, given the annual demand, the ordering cost per order, and the annual holding cost per unit, we can compute the optimal EOQ. (The EOQ formula is most easily derived by using calculus.)

$$\text{EOQ} = \sqrt{\frac{2DS}{H}} \tag{2-4}$$

The minimum total cost is:

$$\text{TC}_{\text{min}} = \frac{EOQH}{2} + \frac{DS}{\text{EOQ}} \tag{2-5}$$

Holding costs are frequently stated as a cost (c) times an interest rate (i), rather than as a dollar amount per year. The "interest" cost is much more than the cost of money to the company. It includes factors such as storage, shrinkage and obsolescence and can range from 20 to 40 percent.

As an example, a large bakery buys flour in 100-pound bags. The bakery uses an average of 1344 bags a year. Preparing an order and receiving a shipment of flour costs $3 per order. Flour costs $35 per bag; the interest cost is 40 percent per year. With that data, we can answer the following questions.

1. What is the Economic Order Quantity?
2. What is the average number of bags on hand?

3. How many orders per year will there be?
4. What is the total cost of ordering and holding flour?

In this example, the data is: $D = 1344$ bags/yr; $S = \$3$/order; $c = \$35$/bag; $i = 40\%$.

1. EOQ $= \sqrt{\frac{2DS}{ci}} = \sqrt{\frac{(2)(1344)(3)}{(35)(.4)}} = 24$ bags

2. Average Inventory $= \frac{Q}{2} = \frac{24}{2} = 12$ bags

3. Orders/yr $= \frac{D}{Q} = \frac{1344}{24} = 56$ orders/yr

4. TC $= \frac{QH}{2} + \frac{DS}{Q} = \frac{24(14)}{2} + \frac{1344(3)}{24} = \$336$

When to Order and What to Order

EOQ models answer the question of how much to order, but they do not address the questions of what and when to order. The latter is the function of models that determine the *Reorder Point (ROP)* in terms of a *quantity*. The Reorder Point occurs when the quantity on hand drops to a pre-specified amount, which is the sum of the quantity expected to be used during lead time, plus an extra cushion, or Safety Stock, to reduce the probability of a stock-out during lead time. The four factors which determine the reorder point quantity are:

1. Rate of demand (usually based on a forecast)
2. Length of lead time
3. Extent of demand and lead time variability
4. Degree of stock-out risk acceptable to management

Reorder-point models can be developed for the following four situations:

Constant demand rate, constant lead time

Variable demand rate, constant lead time

Constant demand rate, variable lead time

Variable demand rate, variable lead time

For the purposes of comparing order point inventory systems with MRP inventory systems, we will limit our development of models to the first two situations (constant lead time), and will use the following symbols for these models:

Demand	*Lead Time*
d = Constant daily demand	LT = Constant lead time
d_{avg} = Average daily demand	LT_{avg} = Average lead time
σ_d = Standard deviation of daily forecast demand	σ_{LT} = Standard deviation of lead time demand

The models assume that any variability in either demand rate or lead time can be adequately described by a normal distribution (bell-shaped curve), shown in Figure 2-8. However, this is not a strict requirement; the models provide reasonable reorder points even in cases where the curve is substantially different from bell-shaped.

Constant Demand and Constant Lead Time. The risk of stock-out comes from three primary sources:

1. Unanticipated demand during the order lead time,
2. Increased order lead time, and
3. Unacceptable quality of products when received.

We assume that suppliers and our own plants produce and ship products that meet our quality standards. Unfortunately, in many manufacturing companies, this assumption is not very realistic. We briefly discuss the issue

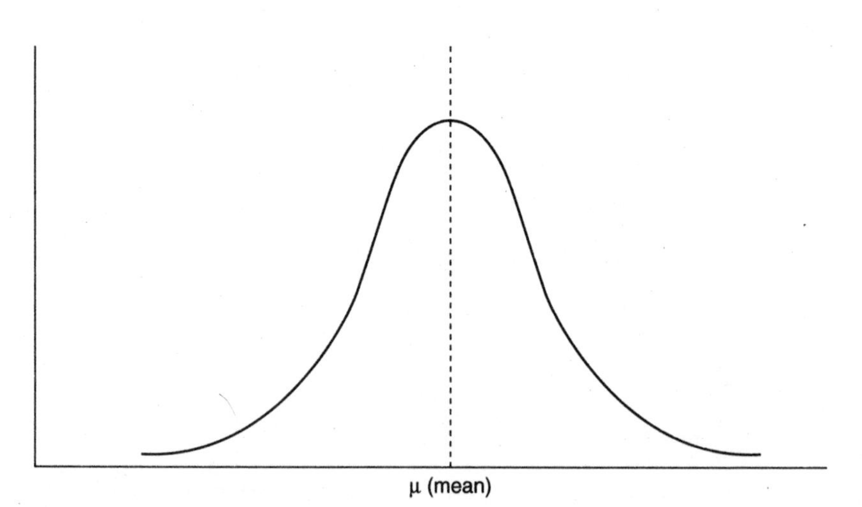

Figure 2-8. Normal distribution.

of quality in Chapter 9; both JIT and TQM (Total Quality Management) directly address it.

The risk of stock-out is minimized if we know, with certainty, both the demand (usage) rate and lead time. In such a case, the ROP is equal to the product of usage rate and lead time; we need no extra cushion of Safety Stock.

$$\text{ROP} = d \times \text{LT} \tag{2-6}$$

For example, if a company uses 20 units of a particular component each day, and the total replenishment lead time is 3 days, the reorder point for this component would be 20 × 3 = 60 units.

Variable Demand and Constant Lead Time. Most of the time, we do not know the exact demand rate. The amount requested during the lead time might actually be greater, or less, than the amount that was used to calculate the reorder point. If the demand is greater than expected, we will run out of stock before the replacement order arrives, unless we carry additional stock as a cushion. (Conversely, if demand is less than expected, we will have excess stock.) *Safety stock,* SS, is the additional amount added to the reorder point to provide a cushion in case of greater demand than expected. After we add safety stock to our reorder point formula, the formula is:

$$\text{ROP} = (d_{\text{avg}} \times \text{LT}) + \text{SS} \tag{2-7}$$

If we could replenish inventory on a moment's notice, there would be no reason to be concerned about demand uncertainty. Whenever inventory reached zero, we would restock. In fact, this is the JIT approach; JIT deliberately reduces lead times and minimum order quantities, with the final intent to be able to have a replacement order of one unit available almost immediately. The risk of stock-out increases as the lead time increases, and as the variability of demand increases. When stock-out costs are high and demand is very unpredictable, the financial risk can be quite sizable. In mathematical terms, the variability of the demand is probabilistic, which means that it can be predicted by using probability theory.

Safety Stock. The two methods of determining the amount of Safety Stock needed are the rule-of-thumb method and the service-level method.

- *Rule-of-Thumb.* In the rule-of-thumb method, the inventory managers, or higher level manager arbitrarily set ratios to guide the establishment of reorder points and Safety Stocks. They may decide that the on-hand quantity be twice (or 1.5 times, 1.2 times, or some number times) the projected use during lead time. This method recognizes that demand may be higher than expected, or that lead time may be longer than expected. This method uses intuition rather than mathematical formulas to calculate probabilities of stock-out, cost of inventory, and cost of stock-out.

- *Service Level.* The concept of service level is a better method of determining the safety stock. The service level is a measure of how effectively a company supplies demanded goods from its stock-on-hand (i.e., how well it prevents stock-outs and lost sales). Specifically, *service level* is the probability of covering the demand that occurs during the inventory replenishment lead time. This is also the probability that we will *not* run out of stock during the inventory cycle.

If demand during lead time is symmetrical (evenly distributed) and unimodal (with one peak), it can usually be represented by the normal distribution that we introduced in Figure 2-8. Figure 2-9 depicts a normal distribution of demand during lead time. In Figure 2-9, the bell-shaped curve is on its side, with the highest point (the highest probability) at the Safety Stock level. As computed by the Reorder Point formula, the average or expected demand would just use up the nonsafety-stock inventory during the lead time. If no Safety Stock were carried, we would expect the com-

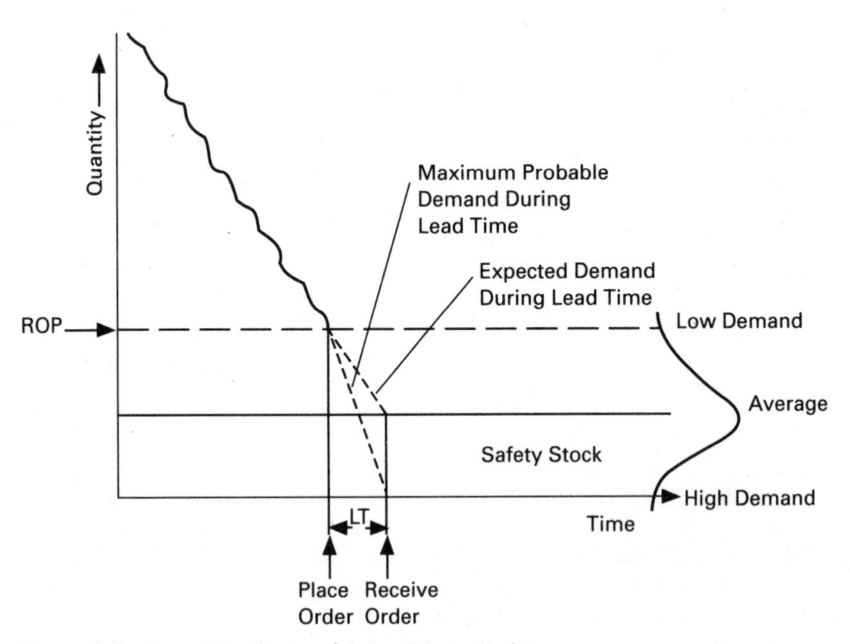

Figure 2-9. Normal distribution of demand during lead time.

pany to run out of stock during approximately 50 percent of the order cycles.

The standard deviation (σ) and Mean Absolute Deviation (MAD) are useful measures of dispersion (the amount of variation) of actual individual demand values from the mean value (the expected or forecast value; in this case, the demand over lead time). σ and MAD can be calculated as follows when (N) samples of actual demand (X) have been measured, and the mean (average) demand (μ) is known:

$$\sigma = \frac{\sqrt{\Sigma(X_n - \mu)^2}}{N} \tag{2-8a}$$

$$\text{MAD} = \frac{\Sigma \mid X_n - \mu \mid}{N} \tag{2-8b}$$

$\sigma \approx 1.25$ MAD (One standard deviation is approximately 1.25 MAD) (2-8c)

where N = number of samples taken
X_n = value of each individual demand n
μ = mean (average) of demand
σ = standard deviation
MAD = Mean Absolute Deviation

We can express the service level as a percentage of the area inside the normal curve, and then calculate the Safety Stock required to provide a specified service level in terms of standard deviations from the mean. Conversely, if the number of units of Safety Stock is specified, we can calculate how much protection against stock-out (that is, service level) the Safety Stock provides. An abbreviated table of safety factors (SF) for normally distributed demand is given in Figure 2-10. Mathematically speaking, the SFs are the number of standard deviations (or MAD equivalents) required to describe the percentage of area under the normal curve specified by the service level. In other words, the SFs state how many standard deviations of inventory we must hold as Safety Stock to achieve a specified customer service level. Safety Factors for additional values of service level can be calculated by entering a normal distribution table with the service level.

For normally distributed demand, the reorder point can be calculated by using the SFs in Figure 2-10 as follows:

$$\text{ROP} = (d_{\text{avg}} \times \text{LT}) + \text{SS} \tag{2-9a}$$

$$\text{SF}\sigma_{\text{LT}} = \text{SF}(\sigma_d)\sqrt{(\text{LT})} \tag{2-9b}$$

Service level (% Cycles w/o stock-out)	Safety factor using:	
	Standard deviation	Mean Absolute Deviation
50	0	0
75	0.67	0.84
80	0.84	1.05
85	1.04	1.3
90	1.28	1.6
94	1.56	1.95
95	1.65	2.06
96	1.75	2.19
97	1.88	2.35
98	2.05	2.56
99	2.33	2.91
99.5	2.57	3.2
99.99	4	5

Figure 2-10. Safety factors for a normal distribution.

As the replenishment lead time increases, the risk of stock-out also increases, but by the square root of the lead time, rather than linearly. This formula supports our contention that long lead times are costly, and should be reduced where possible. The unit of time for the standard deviation and the lead time must be the same. For example, if our measurement of demand is daily, the standard deviation will also be daily, and the lead time must be expressed in days.

As an example, a pharmacy dispenses a generic drug at an average rate of 50 tablets per day. The pharmacy records show that daily demand is approximately normal and has a standard deviation of 5 tablets per day. Lead time is four days, and the pharmacist wants a stock-out risk that does not exceed 1 percent. As discussed earlier, higher service levels require higher stocking levels (reflected in higher safety stocks), and thus higher cost. The pharmacist wants to know:

1. How much Safety Stock should be carried?
2. What is the Reorder Point?
3. What service level (SL) would a Reorder Point of 216 tablets provide?

The data provided in this example can be organized as follows: $d = 50$ tablets/day; $\sigma_d = 5$ tablets/day; LT = 4 days; Stock-out Risk $\leq$ 1 percent.

1. To find SS, we must use Eq. (2-9*b*). We know all the variables except SF. Therefore we must find SF.

$$SL = 100\% - \text{risk} = 100\% - 1\% = 99\% = .99$$

$$SL = .99 \rightarrow SF = 2.33 \quad \text{(from Figure 2–10)}$$

$$SS = SF(\sigma_d)\sqrt{(LT)} = 2.33 \times 5\sqrt{4} = 23.3 \text{ tablets}$$

2. Using Eq. (2-9a),

$$ROP = (d \times LT) + SS$$

$$= (50 \times 4) + 23.3 = 223.3 \text{ tablets}$$

3. Using Eq. (2-9b),

$$SF = \frac{SS}{\sigma_d\sqrt{LT}} = \frac{\text{ROP} - \text{Expected Demand}}{\sigma_d\sqrt{LT}}$$

$$= \frac{216 - 200}{5\sqrt{4}} = 1.6$$

From Figure 2-10, SF = 1.6 → SL = .945, or 94.5 percent of the time the pharmacist would have enough stock when a customer needed that drug. Conversely, the pharmacist would have a stock-out rate of 5.5 percent.

Fixed Order Interval System

Fixed order interval systems are also widely used. We discuss below how a fixed order interval system answers the fundamental questions of inventory management: what to order, how much to order, and when to order. For ease of understanding, we present the answer to the second question first.

How Much to Order

In the fixed order interval system shown in Figure 2-11, the time interval between orders is fixed, such as weekly, semimonthly, monthly, bimonthly, or quarterly, but the size of the order varies according to the usage since the last review. The planner reviews the inventory level only at the end of the interval, and orders the difference between the current inventory and the target maximum.

A fixed order interval inventory system involves determining the amount of an item in stock at a specified, fixed time interval, and placing an order for a quantity which, when added to the quantity on hand, will equal a predetermined maximum level. The *maximum inventory level, M,* is the sum of

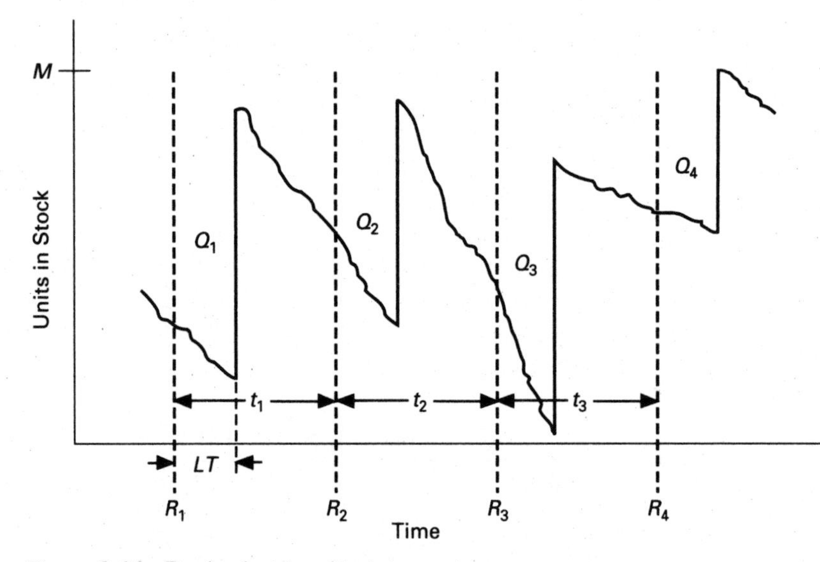

Figure 2-11. Fixed order interval inventory system.

the anticipated demand during the review period, the anticipated demand during lead time, and the Safety Stock, or:

$$M = d(R + \text{LT}) + \text{SS} \tag{2-10}$$

where M = maximum inventory level
d = demand rate
R = review period duration
LT = lead time duration
SS = safety stock

And, with I = inventory on hand and Q = order quantity, the order quantity equals the maximum level minus the inventory on hand, or:

$$Q = M - I, \text{or} \, Q = d(R + \text{LT}) + \text{SS} - I \tag{2-11}$$

If the demand follows a normal distribution rather than being constant, we can determine the required Safety Stock in the same manner as we did for the EOQ model, except the period of uncertainty is the lead time plus the review period rather than just the lead time, or:

$$\text{SS} = \text{SF} \times \sigma_{(R+\text{LT})}, \text{ or SS} = \text{SF} \times \sigma_d \sqrt{R + \text{LT}} \tag{2-12}$$

As indicated by comparing Equations (2-9) and (2-12), the fixed order *interval* system must carry more Safety Stock than the fixed order *quantity* system in order to provide the same service level. This is because in the fixed order quantity system, the Safety Stock has to protect against fluctuations in demand only over the lead time, whereas with the fixed order interval system, it has to protect against fluctuations in demand over the lead time *plus* the review period.

As an example, a lab orders a number of chemicals from the same supplier every 30 days. Lead time is 5 days. Eleven 25 ml jars are on hand. Daily use of the chemical is approximately normal, with a mean of 15.2 ml per day and a standard deviation of 1.6 ml per day. The desired service level is 95 percent. The lab technician wants to determine:

1. What amount of Safety Stock should be carried?
2. What maximum inventory M should be carried?
3. How much should be ordered?

1. $SS = SF \times \sigma_d \sqrt{R + LT} = 1.65 \times 1.6\sqrt{30 + 5} = 15.62$ ml

2. $M = d(R + LT) + SS = 15.2(30 + 5) + 15.62 = 547.2$ ml

3. $Q = M - I = 547.62 - 275 = 272.62$ ml

where $R = 30$ days; LT = 5 days; $I = 275$ ml; $d = 15.2$ ml/day; $\sigma_d = 1.6$ ml/day; SL = 95%; SF = 1.65 (from Figure 2-10).

When to Order and What to Order

If we calculate R, the review period, to minimize the sum of the ordering and holding costs, this means that the Q of Equation (2-11) is the same as the EOQ of Equation (2-4), and we can develop an approximate expression for R as follows:

$$R = \left[\frac{EOQ}{d}\right] - LT = \left[\frac{\sqrt{(2DS)/H}}{d}\right] - LT \qquad (2\text{-}13)$$

Note that when an item appears for reordering, our system has also defined *what* to order. We order the parts that belong in this ordering cycle, whether we use a computer report or a manual system.

However, in using this system, there are many other variables besides the ordering and holding costs for a single item that should be considered in selecting R:

- If many different items are ordered from the same source, review time may be a composite that minimizes the ordering and holding costs for all items, rather than for any one item.

- If both ordering costs and holding costs are very low, R can be selected to optimize some other objective(s), such as: minimizing transaction costs, minimizing shipping costs, or maximizing convenience.
- If the necessary information is available, R can be selected to minimize some combination of the above or to minimize total inventory costs.

Despite the increased cost of Safety Stock, the fixed interval inventory system can be advantageous in a number of situations:

- If the cost of continuous review is quite expensive and the cost of the item is relatively inexpensive.
- If there are many small issues to and from inventory, posting transactions to and from inventory are very expensive, and the cost of the item is relatively inexpensive.
- Groups of items are purchased from a common supplier, and the total costs per item are greatly reduced by combining the items into one periodic order.
- Grouping items to generate a full carload (or truckload) shipment, or to utilize production capabilities available only at certain intervals.

Even when an MRP inventory system is installed, a fixed interval inventory system may be used for controlling portions of the inventory, if any of the above conditions exist. For example, nuts and bolts used on the assembly line may be replenished at periodic intervals rather than being issued for each product. If the cost of the item is low, the benefits of decreased transaction cost may outweigh the increased holding cost.

Why Order Point Systems Fail for Planning Dependent Demand

We have discussed the basic mechanics and potential application of two major order point systems. We now present the reason why they are incapable of supporting dependent demand items, which defines manufacturing inventories.

When a shop order for an assembly or subassembly with multiple components is released to the shop floor, the stockroom clerk must procure the proper quantity of each one of the needed components and raw materials. Unfortunately for companies who are using order point systems to manage dependent demand items, the service level of the entire order is not the same as the service level for each item. Instead, it is the *product* of the service level of each direct (first-level) component. In other words, it is the serv-

	Service Levels		
Number of Components	90%	95%	98%
5	59%	77%	90%
10	35%	60%	82%
15	21%	46%	74%
20	12%	36%	67%
50	1%	8%	36%

Figure 2-12. Probability of having all components in stock in an order point system.

ice level of the first direct component, *multiplied* by the service level of the second direct component, *multiplied* by the service level of the third component, and so on, until there are no more components. The probabilities of having all direct components in stock when you want to assemble their parent are summarized in Figure 2-12.

As an example, a production planner has just released a shop order for an assembly that has 5 direct components. Each component has a service level of 95 percent. What is the probability that the stock clerk will find all 5 components on the shelf, in the quantities required?

The probability of finding all 5 components on the shelf is $.95 \times .95 \times .95 \times .95 \times .95 \times .95 = .77$, or 77 percent. For 10 components, the probability is 60 percent. If we have chosen a 90 percent service level (because management has decided that we need to reduce inventories), the probabilities are 59 percent for 5 components, and 35 percent for 10 components! Even with a 98 percent service level (with a substantially higher investment in inventory), we only have an 82 percent chance of having 10 components on the shelf at once. Even more frustrating, the component that is short is "always" a different one. Using order point systems to manage dependent demand inventories inevitably causes constant shortages, backorders, and expediting.

Summary of Order Point Inventory Control Systems

We have examined the basics of two order point methods of managing inventory for independent demand items (finished goods, service parts, etc.): the fixed order *quantity* system and the fixed order *interval* system.

When inventory replenishment is based on the EOQ and ROP models, *fixed quantities* of items are ordered at *varying time intervals*. In the *fixed order interval* model, orders of *varying quantities* are placed at *fixed time intervals*—just the opposite. This is summarized below.

	Quantity	Interval
Order quantity	Fixed	Variable
Order interval	Variable	Fixed

For the fixed order quantity system, we developed the basic EOQ model, which minimizes the sum of the ordering and holding cost to determine *how much* to order, and the basic ROP model to compute *when* to order. We pointed out the trade-offs between higher customer service and higher Safety Stock. For the fixed order interval system, we developed the formula to determine the maximum amount of inventory M to be carried, and the amount to be ordered Q, at the end of each fixed interval, where R is the difference between the existing inventory I and the maximum inventory M. We indicated that R could be selected to minimize the sum of the ordering and holding costs or to satisfy other appropriate objectives. These are summarized in the following table:

	Fixed quantity	Fixed interval
How much to order	EOQ	$M - I$
When to order	ROP	R

There are a number of differences between these two approaches to inventory control. Both are sensitive to demand fluctuations just prior to reordering, but in somewhat different ways. In the fixed order quantity model, a higher than normal demand causes a *shorter time* between orders, whereas in the fixed-interval model, the result would be a *larger order* size. Another difference is that the fixed order quantity model requires close monitoring of inventory levels in order to know *when* the amount on hand has reached the reorder point, whereas the fixed order interval model requires only a periodic review of inventory level (i.e., a single inspection just prior to placing an order), to determine *how much* is needed.

The fixed order quantity system is best for those *independent* demand situations where the objective is to minimize the ordering and holding costs. We saw in Figure 2-3 that MRP is considerably better at reducing ordering and holding costs (especially holding costs) for *dependent* demand items than a fixed order quantity system. We demonstrated with Figure 2-12 that order point systems are not a good choice for insuring availability of dependent stock items. Consequently, the fixed order quantity order point system should rarely, if ever, be used for dependent demand items.

However, there can be situations of *dependent* demand where minimizing ordering and holding costs are not the paramount objectives. In these situations the paramount objectives may instead include reducing transaction costs, reducing continuous review costs, and/or combining many orders for a single supplier to substantially reduce unit costs. Here a fixed interval inventory system may be the best choice in lieu of using MRP for dependent demand items. Or, to accomplish the same results, a planner can choose a fixed order interval lot size for an MRP-planned part. We cover this in more detail in Chapter 6.

Kanban

Kanban Basics

While Kanban inventory systems are similar in approach to order point systems, Kanban systems are intentionally more simple; they replace complex formulas for determining order points or lot sizes with visual controls. JIT's goal is "stockless production," in which each work center is working on only one piece, with no other inventory around! This pull system has two prerequisites:

1. The movement of all upstream materials must be tightly synchronized to the actual shipment or withdrawal of end items, and
2. The total amount of inventory in the system must be tightly constrained.

Visualize the flow of inventory throughout an entire manufacturing company as one continuous pipeline, with very limited storage capacity in the middle of the pipeline. Before a supplier can add any parts at the beginning of the pipeline, a customer has to remove a part from the end of the pipeline. Another example is the childhood game of musical chairs: in JIT, each part has to have a "chair," or authorized space; otherwise it cannot be made or acquired.

Kanban cards are a common way, but by no means the *only* way, of implementing a JIT pull system. If all work centers and storage spaces were arranged contiguously, workers could operate on the visual signals alone and would not need Kanban cards. However, this is not true in many cases. Each Kanban card represents the authority to produce one storage unit of one

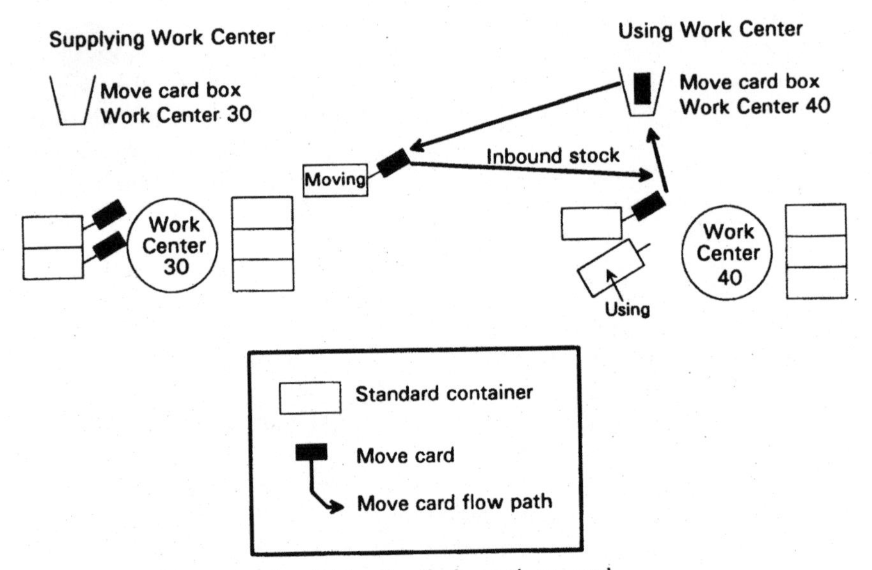

Figure 2-13. Movement of materials using Kanban with one card.

item. A Kanban card can represent one piece, one shipping container, or one skid or pallet of an item. Kanban systems use the following rules:

- A Kanban card must always be attached to each container that contains parts,
- Parts must always be stored and moved in standard containers, which contain the identical number of parts,
- "Customer" work centers must come get parts from supplying work centers or signal their requirements by a "move" card, and
- A work center can produce a container of parts only when they have an unattached production card; they attach the production card to the full container as soon as they finish.

The movement of materials through a JIT-oriented factory is shown in Figure 2-13.[1]

When to Order and What to Order

When a final assembly work center manufactures a finished item, the assembly process consumes various subassemblies and purchased parts from their storage spaces at that work center. Each time a subassembly or pur-

[1] R. W. Hall, *Attaining Manufacturing Excellence*, Dow-Jones Irwin, Homewood, IL, 1987.

chased part container is pulled from its storage space, the worker removes a Kanban production card from the container which holds the part and places it in an outgoing card bin. This card is returned to the supplying work center, or vendor, as the signal to replace the parts in the container that was just used. This card tells the supplying work center which part to make, directly answering the question, "What to Order?"

How Much to Order

By definition, the order quantity is defined as one standard container for each Kanban card. Inventory levels are controlled very simply by adjusting the number of production Kanban cards. Adding cards adds inventory in the pipeline; removing cards reduces the inventory. The number of cards can be decreased as the supplying work centers and vendors reduce their lead times. For example, if the final assembly area uses 2 containers of a purchased part per day and the supplier has a 3-day lead time, the company will probably have 8 to 10 Kanban cards—2 for the day's use, 6 for the supplier's lead time, and 2 as safety stock. When the supplier can reduce its delivered lead time to 1 day, the manufacturer can reduce the number of Kanban cards to 4 or 5—2 for the day's use, 2 for the supplier's lead time, and possibly 1 as safety stock.

Kanban Summary

Kanban is the most simple and straightforward way to control inventories. Kanban requires no formal planning or record keeping, and generally does not involve a computer system at all. It is more applicable for manufacturing rather than distribution. Like order point systems, it is reactive; MRP is proactive. Kanban works best in an environment where the product line, mix, and volume are relatively stable. Kanban can have difficulty signaling the timing of engineering changes.

Priority Planning and Capacity Planning

Manufacturing planning includes planning both the outputs and the inputs of the manufacturing operation.

- *Priority planning* determines the products or priorities of the manufacturing operation to meet customer demand, such as: what products are needed, how many are needed, and when they are needed.
- *Capacity planning* determines the resources or capacity needed by the manufacturing operation to produce the desired output, such as: machine hours, labor hours, and financial resources.

Priority Planning

Priority planning encompasses a large number of decisions that can affect several organizational levels. Figure 2-14 shows that these decisions can be grouped into four broad levels: strategic planning, tactical planning, operational planning, and execution and control. These four levels of decisions differ markedly in terms of level of management responsibility and interaction, scope of the decision, level of detail of required information, length of planning horizon needed to assess the consequences of each decision, and degree of uncertainty and risks inherent in each decision. Each of these levels requires different procedures and data to support decision making.

These decision levels form a natural hierarchy. Higher level decisions have longer lead times, longer planning horizons, and utilize aggregated data, such as total manpower requirements and total demand. The lower level decisions have shorter lead times, shorter planning horizons, and require detailed information on individual products, machines, and workers. When upper level decisions are made, they provide guidance and constraints for lower level decisions. In turn, decisions at a lower level provide feedback to improve the quality of future upper level decisions.

Capacity Planning

Capacity planning determines the level of capacity (resources or inputs) needed to achieve scheduled production, compares this with available capacity, and adjusts capacity levels or production schedules. Capacity planning includes, at a minimum, the resources of manufacturing labor,

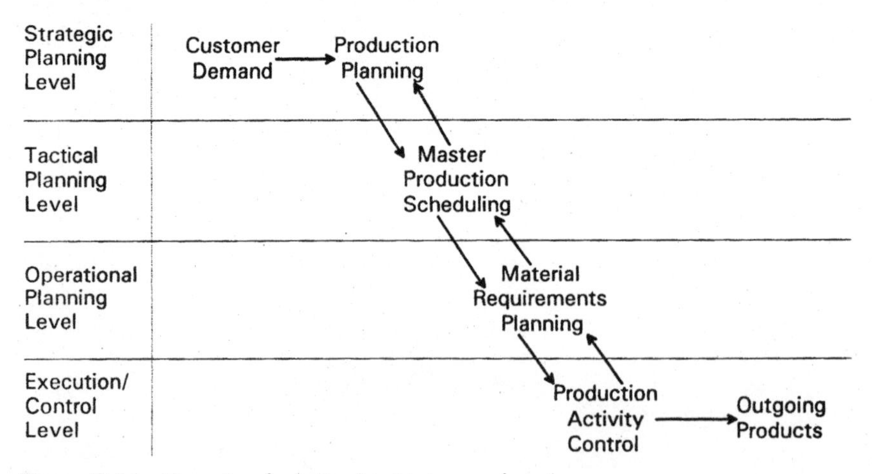

Figure 2-14. Hierarchy of priority planning in manufacturing.

machine hours, facilities, warehouse space, tools, fixtures, and engineering. Some manufacturers also include less traditional categories, such as cash and energy. In MRP II systems, capacity planning does *not* include materials, because they are handled by the priority planning functions (e.g., MPS, MRP).

Successful manufacturing planning and control requires effective capacity planning, which means having the required capacity at the right time. Insufficient capacity causes shortages, failure to meet production targets, overdue shipments to customers, frustrated production managers, and a loss of confidence in the formal system. On the other hand, excess capacity will result in low utilization of resources, high costs, and reduced profit margins.

In materials-driven companies, planners decide on sales rates of product families first (priority planning), then compare the capacity requirements of their plans to the actual capacity that could be made available. In capacity-driven companies that have enormous investments in fixed facilities, such as refineries, some chemical manufacturing plants, paper mills, and basic metals, the process is reversed. These companies review their available capacity first, then ask what products they can sell to fill the capacity. These facilities tend to run 24 hours per day, 7 days per week; these manufacturers are focused on keeping the machinery running, because the cost of stopping is too high.

To plan capacity in a product-oriented company, a planner or scheduler starts by preparing a production schedule. Next, the planner determines the level of capacity required to produce this schedule and compares the required level with the available capacity. If they are equal, or close enough, the process stops. If not, the planner prepares a new capacity plan that adjusts available capacity to match required capacity. If the new plan is satisfactory, the process stops. If the best capacity plan is not satisfactory, then the scheduler adjusts the production schedule and compares the revised production schedule to the capacity plan. The planner continues this process until the production schedule and capacity plan are both acceptable and compatible.

Integrated Priority Planning and Capacity Planning

The manufacturing system cannot produce the desired priorities (outputs) unless it has the necessary capacity (inputs). Because each Priority Planning activity has a complementary Capacity Planning activity, there is a hierarchy of Capacity Plans that corresponds to the hierarchy of Priority Plans presented in Figure 2-14. Similar to the priority-planning modules, the capacity-planning modules operate over progressively shorter planning horizons and produce reports in progressively greater degrees of detail. As indicated in Figure 2-1, there are four levels in the integrated priority and capacity planning hierarchy:

1. Production Planning and Resource Requirements Planning
2. Master Production Scheduling and Rough-Cut Capacity Planning
3. Material Requirements Planning and Capacity Requirements Planning
4. Production Activity Control and Input/Output Control

Although we are primarily interested in the third level, or material and capacity requirements planning (M&CRP) we must understand the basics of the complete hierarchy to fully understand M&CRP. We will briefly discuss this hierarchy, and each layer in it, to better understand the relationship between priority planning and capacity planning.

Production Planning and Resource Requirements Planning. At the strategic level, *Production Planning* is the process of setting the overall level of manufacturing output to best satisfy the planned level of sales and inventory. The production plan defines the manufacturing rate, usually stated as a monthly rate for a period of two years or more, for each product group and projects the resulting inventory. The Production Plan must be consistent with the Business Plan, which is the input to the process. Production Planning is a top-management responsibility that requires consensus across the functional departments. It establishes the framework for master production scheduling and manufacturing execution.

Resource Requirements Planning (RRP) evaluates the Production Plan to determine if the long range resources, such as land, facilities, machinery, and work force are available, or can be made available. At the RRP level, products are often aggregated into groups, or families, of similar items, and a "typical" item in the family is used to calculate the load for the entire family. If the necessary resources are available, the production plan can be executed as is. If they are not available, the production plan must be changed, or steps taken to acquire more resources. When the required and planned available resources are equal, the Production Plan can then be passed down to the next level, the Master Production Schedule, for execution.

Master Production Scheduling (MPS) and Rough-Cut Capacity Planning (RCCP). At the tactical level, the *Master Production Schedule* breaks down the Production Plan to show for each time period (usually a week), the quantity of end items to be produced over the tactical planning horizon (usually one year). Whereas the production plan shows a production rate for a group of products, the MPS schedules quantities of specific end items in specific time periods.

Rough-Cut Capacity Planning determines whether the planned resources are sufficient to carry out the Master Production Schedule. The RCCP uses definitions of unit product loads called product-load profiles, bills of capacity, or bills of labor. Multiplying the loads per unit by the quantities of products scheduled per period, gives the total load per period for each work center. For ease of computation, RCCP usually includes the few "critical" work centers. RCCP is more detailed than RRP because RCCP calculates the load for all the scheduled items, and in the actual time periods, where they are scheduled. When the Rough-Cut Capacity Planning process indicates that the Master Production Schedule is feasible, the MPS is passed down to the Material Requirements Planning process to determine the needed raw materials, components, and subassemblies. In a capacity-oriented company, if RCCP indicates problems with the MPS, the planners change the MPS (either by rescheduling customer orders, or by telling Sales and Marketing to sell the unsold capacity).

Material Requirements Planning and Capacity Requirements Planning. At the operational level, *Material Requirements Planning* develops planned orders for the raw material, components, and subassemblies needed to satisfy the Master Production Schedule. MRP also recommends rescheduling open orders when due dates and need dates are not in agreement. MRP uses Bills of Material and inventory data, in addition to the Master Production Schedule, as inputs. MRP usually covers a year, but may use monthly, rather than weekly, time buckets in the latter part of the year.

Capacity Requirements Planning compares the required capacity to the projected available capacity for both the open manufacturing orders and the planned manufacturing orders generated by the MRP system. CRP uses routing files and work-center information to calculate the scheduled load at work centers, assuming infinite capacity. If the projected capacity differs from that required by the projected load, planners can recommend corrective actions to top management, including reducing or rescheduling orders, hiring or laying off workers, reassigning workers, subcontracting, or alternate routings. When CRP indicates that the load of existing released orders plus the MRP schedule of planned orders is feasible from a capacity viewpoint, the planned orders are released to Production Activity Control for execution.

Production Activity Control, Input/Output Control, and Operations Sequencing. At the execution and control level, *Production Activity Control (PAC)* develops detailed short-range schedules using component due dates from MRP and detailed routings. The PAC schedules are usually in terms of days (sometimes hours), and tend to cover from one to three months. PAC involves planning, releasing, and controlling shop orders.

Input/Output Control monitors the quantity of work arriving at a work center and the amount leaving it. Production planners compare actual arrivals and completions to planned amounts, taking corrective actions such as overtime, transferring workers among work centers, alternate routings to transfer load to other work centers, or splitting and/or overlapping operations.

Operations sequencing is a simulation technique for short-term planning and the priority dispatching of jobs to be run at each work center, based on current capacity, priority, routings, and other information.

Production Activity Control represents the execution and control of manufacturing plans that have been developed in the previous planning levels. This is the level where the work is actually accomplished. This level also generates valuable feedback, which is used by the higher levels to improve their planning.

Summary

Manufacturing inventories serve only to satisfy the needs of the master production schedule or final assembly schedule. The demand on a manufacturing inventory is *dependent,* because the demand for an item is dependent upon the demand for a higher level item. On the other hand, the demand for items in the distribution inventory (finished goods, spare parts, or other items that are shipped out or sold) is *independent,* because this demand is not dependent on the demand for another product made by the company.

The three basic types of inventory planning systems are MRP, Order Point, and Kanban.

1. An MRP system translates a master production schedule into time-phased net requirements for each component, and then plans and re-plans coverage for these requirements to implement the schedule. Of the three systems, MRP is the only one that is proactive; it can be aware of a future change to a bill of materials and plan components accordingly to minimize overstock of the old item and shortages on the new item.
2. There are two order point inventory control systems: the fixed order *quantity* system and the fixed order *interval* system. Both order point systems are reactive; they predict future usage based on past usage.
   - The fixed order *quantity* system utilizes the EOQ model to determine how much to order. The EOQ model calculates an order size that minimizes the sum of the annual costs of holding and ordering inven-

tory. The fixed order quantity system calculates a reorder point to determine when to order. The reorder point is the sum of the quantity expected to be used during lead time, plus an extra cushion, or Safety Stock, to reduce the probability of a stock-out during lead time.

- In the fixed order *interval* system, the time interval between orders is fixed (e.g., weekly, monthly, or quarterly), but the size of the order varies according to the usage since the last order. This system orders a quantity that will return the inventory on hand to a predetermined maximum level. The Safety Stock can be determined in the same manner as for the EOQ model, except the period of uncertainty is the lead time plus the review period, rather than just the lead time.

Comparing the two order point systems, the fixed order *quantity* system should probably never be used for dependent demand items, but the fixed order *interval* system may be useful if special conditions are present. Order point systems virtually guarantee shortages and stock-outs of dependent demand items.

3. Kanban systems are common signaling systems in JIT environments. A Kanban card is the visual signal that authorizes production of a standard container of a specific part. JIT systems are reactive pull systems that dramatically minimize inventories and rely on quick response at each level of manufacturing and supply.

Priority planning and capacity planning form a hierarchy of short-range plans, from the strategic to the detailed. Priority planning and capacity planning must be integrated. For every priority planning activity there is, or should be, a comparable capacity planning activity.

Selected Bibliography

APICS *Inventory Management Reprints,* the American Production and Inventory Control Society, Falls Church, VA, 1991.

Blackstone, John H., Jr.: *Capacity Management,* South-Western, Cincinnati, OH, 1989, Chaps. 2 and 3.

Brown, R. G.: *Decision Rules for Inventories Management,* Holt, Reinhart & Winston, Orlando, FL, 1967.

———*Advanced Service Parts Control,* Material Management System, Inc., Norwich, VT, 1982.

Deis, Paul: *Production and Inventory Management in the Technological Age,* Prentice-Hall, Englewood Cliffs, NJ, 1983, Chap. 6.

Fogarty, Donald W., T. R. Hoffmann, and P. W. Stonebraker: *Production and Operations Management,* South-Western, Cincinnati, OH, 1989, Chap. 2.

———J. H. Blackstone, Jr., and T. R. Hoffman: *Production and Inventory Management,* 2d ed., South-Western, Cincinnati, OH, 1991.

———:*Inventory Management: An Introduction* (Training Aid), APICS, 1983.

Gilbert, James P. and Richard J. Schonberger: "Inventory-based Production Control Systems: A Historical Analysis," *MCRP Reprints,* APICS, 1991, pp. 17–21.

Greene, J. H. (ed.): *Production and Inventory Control Handbook,* 2d ed., McGraw-Hill, New York, 1987.

Hall, R. W.: *Attaining Manufacturing Excellence,* Dow Jones-Irwin, Homewood, IL, 1987.

———: *Zero Inventories,* Dow Jones-Irwin, Homewood. IL, 1983.

Lunn, Terry with Susan A. Neff: *MRP: Integrating Material Requirements Planning and Modern Business,* Irwin, Homewood, IL, 1992, Chap. 2.

Mather, H.: *How to Really Manage Inventories,* McGraw-Hill, New York, 1984.

———and George W. Plossl: "Priority Fixation Versus Throughput Planning," *MCRP Reprints,* APICS, 1991, pp. 27–51.

Orlicky, Joseph: *Material Requirements Planning,* McGraw-Hill, New York, 1975, Chap. 1.

Plossl, G. W.: *Production and Inventory Control: Principles and Techniques,* 2d ed., Prentice-Hall, Englewood Cliffs, NJ, 1985.

Schultz, Terry: *Business Requirements Planning: The Journey to Excellence,* The Forum Ltd, Milwaukee, WI, 1984, Chaps. 4 and 7.

Silver, Edward A. and Rein Peterson: *Decision Systems for Inventory Management and Production Planning,* 2d ed., Wiley, New York, 1985.

Smith, Spencer B.: *Computer Based Production and Inventory Control,* Prentice-Hall, Englewood Cliffs, NJ, 1989, Chap. 5 and 6.

Tersine, Richard J.: *Principles of Inventory and Materials Management,* 3d ed., Elsevier, New York, 1988.

Vollman, T. E., W. L. Berry, and D. C. Whybark: *Manufacturing Planning and Control Systems,* 3d ed., Dow Jones-Irwin, Homewood, IL, 1992.

Wight, Oliver W.: *MRP II: Unlocking America's Productivity Potential,* CBI Publishing, Plano, TX, 1981.

3

The Major Inputs to M&CRP

BOM and MPS

"Garbage in; garbage out." ANONYMOUS

Introduction

How can we understand what M&CRP can and cannot do, if we do not first understand what goes into it? We need to know not only the nature and characteristics of these inputs, but also how they are generated, who is responsible for generating them, and how they may be changed if they produce unsatisfactory results. This chapter, therefore, describes the major inputs to the M&CRP system, including:

- The Product Definition process, which is part of the operational phase of Product-Process design. From the perspective of M&CRP, the Bill of Materials (BOM) is the most important output from Product Definition.
- The Master Production Schedule (MPS) process and the nature of the MPS output. The Master Production Scheduling process depends greatly on the nature of the Product Structure as defined by the BOM.

While describing the Product Definition process, we discuss the four data elements in the Item Master that are critical to this process. We differentiate between the three major types of Bills of Material (Engineering, Manufacturing and Planning), and we describe Low-Level Coding in a Bill of

Materials. We discuss Phantom Bills of Material and Planning Bills of Material, including Modular Bills of Material, Inverted Bills of Material, and Super Bills of Material, and present a brief description of managing engineering changes.

The fundamentals of Master Production Scheduling are discussed and the time-phased format is introduced. We discuss both single-level and two-level master schedules, and describe the two most commonly used time fences.

This chapter concludes by integrating the major factors that determine which items to Master Schedule, including:

- Product structure
- Type of competition
- Type of demand response (see Chapter 1)
- Number of products or modules in the product line

Although you will not become an expert on these topics, we provide sufficient background information to help you to fully understand how and why M&CRP operates.

Overview of the Major M&CRP Inputs

The major inputs to M&CRP are shown in Figure 3-1. Note that the M&CRP system has been divided into two subsystems: MRP and CRP to better indicate the entry point for the various inputs.

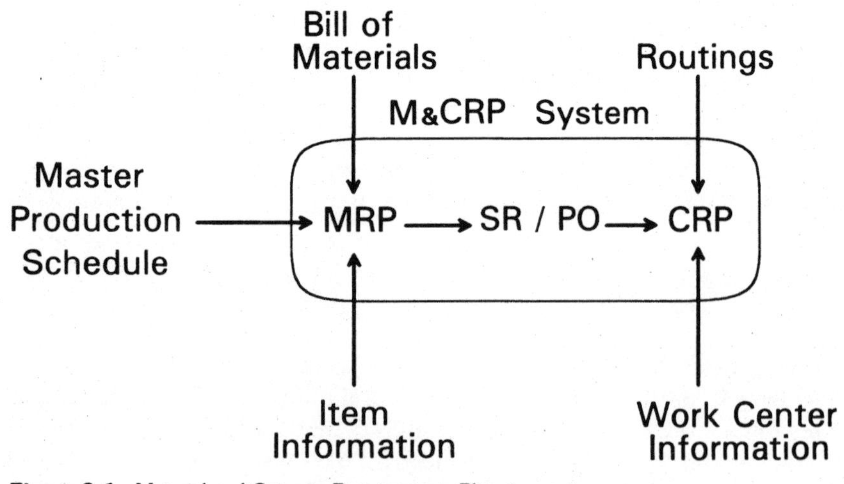

Figure 3-1. Material and Capacity Requirements Planning system.

- Bill of Materials (BOM) provides product structure information on the item.
- Master Production Schedule (MPS) provides the schedule of end items to be produced, which is the "driver" for M&CRP.
- Item Information provides data on each item, such as identification and description, planning information, cost data, and inventory records, from the Item Master File. In many systems, Inventory Records are maintained as a separate file to separate on-hand balance data from permanent data in the Item Master File.
- SR-POR (Scheduled Receipts and Planned Order Releases) is the output from the MRP subsystem that drives the CRP subsystem and Purchasing.
- Work Center Information contains data on each Work Center in the factory, which is needed to perform Capacity Requirements Planning.
- Routings describes the path that each item takes through the factory.

Sources and Responsibilities for M&CRP Inputs

The major inputs to the M&CRP System come primarily from two sources. The MPS comes from previous stages of the Manufacturing Planning and Control System. The BOM, Routing, and Work Center Information result from the Product and Process procedure introduced in Chapter 1 and shown in Figure 3-2. The Item Information comes from both sources.

For the M&CRP system to properly plan and control the product and process, the design process must provide accurate data about each. As shown in Figure 3-2, the product design developed in Tactical or Intermediate Product Design strongly influences the operation of the MPS, including:

- Which items are scheduled by the MPS
- Whether the MPS will use a manufacturing or planning Bill of Materials
- Whether or not the MPS will use a modular bill
- How the products will be grouped for ease of forecasting and manufacturing

We will see how these decisions affect the Master Production Scheduling process, and the resultant input to M&CRP, later in this chapter.

A number of engineering groups are responsible for product and process design. The R and D (Research and Development) Engineer and the Design Engineer are primarily involved in product design, whereas the Manufacturing Engineer and the Industrial Engineer are primarily in-

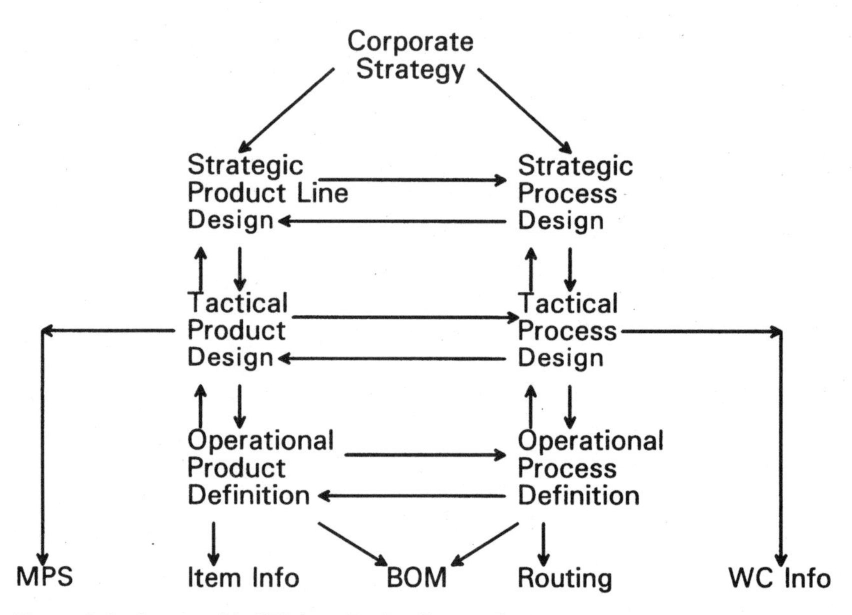

Figure 3-2. Inputs to M&CRP from Product-Process design.

volved in process design. As discussed in Chapter 1, the more progressive companies are using concurrent design (see Figure 3-3) to integrate these engineering design functions and to achieve faster product development with greater manufacturability of the developed products.

The R and D Engineer starts the product definition process, working closely with Marketing to define the customer's needs and then develop a concept and a functional specification that expresses these needs. The Design Engineer develops a detailed engineering design of the product. In addition to drawings, specifications, and other technical information, the detailed design includes an *Engineering Bill of Materials* for the product. Also, most of the item's identifying and descriptive information in the Item Master File (such as item number, item description, drawing number, group technology code, make or buy code, and unit of measure) will come from the engineering design.

The Manufacturing Engineer completes the product definition by developing the assembly chart and the *Manufacturing Bill of Materials* for the product and its components. To develop these documents, the Manufacturing Engineer must have a comprehensive knowledge of manufacturing processes and the capabilities of the manufacturing machines and operations that will be used, as well as the product. Companies are now including purchasing and materials professionals early in the design process, to speed the overall process, and to improve the quality of the design decisions.

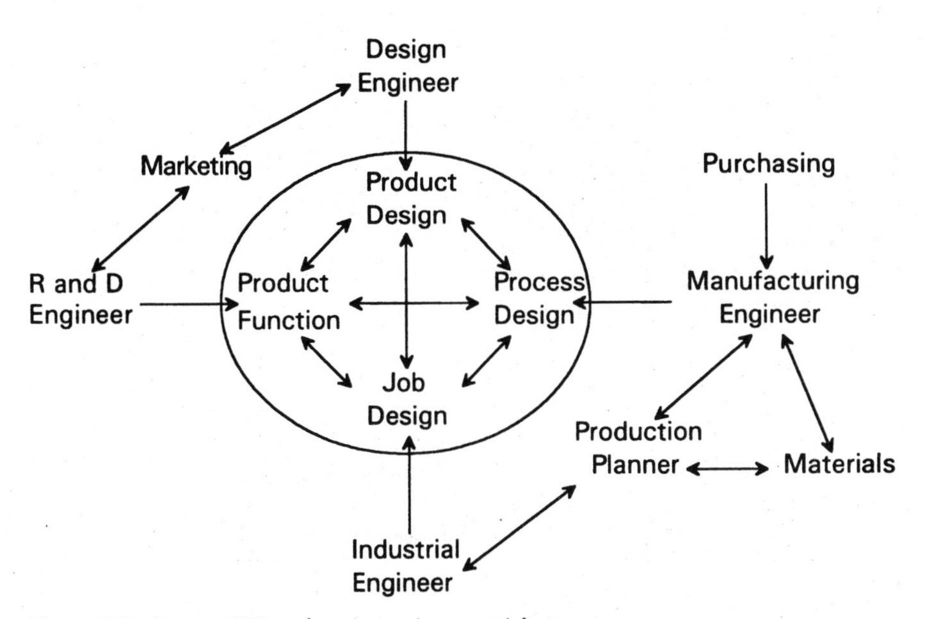

Figure 3-3. Responsibilities of product and process definition.

The Manufacturing Engineer and the Industrial Engineer provide the tactical design of the *manufacturing process.* The Manufacturing Engineer designs and/or selects the machines, tools, and fixtures that will be used to manufacture the firm's products. The Industrial Engineer designs and/or selects the material handling equipment, and designs the factory layout so that the equipment will be arranged in the optimum manner.

As the factory layout is being designed, the Production Planner works with the Manufacturing Engineer and the Industrial Engineer to identify the optimal *work centers.* The Manufacturing Engineer can then develop the process plans, or *routings,* for each item. Based on job analyses, the Industrial Engineer develops operation sheets and training materials for these jobs.

Product Definition

As noted in the previous section, the inputs from Product Definition to M&CRP come via two files: the Item Master File and the Bill of Materials file (also called the Product Structure file). We use these same categories in the following discussion.

Item Master File

The *item master file* contains one single record for each item in a company's manufacturing inventory: products, assemblies, components, materials, and supplies. The record contains four categories of information: (1) identification and description, (2) planning information, such as what order quantity to use, (3) cost data for use in accounting, and (4) inventory data. (In some M&CRP systems inventory data records reside separately in *inventory records*.) Each item may have over 100 pieces of data. We describe below the four data elements that are significant inputs to the M&CRP system. We cover others in Chapter 8.

Item Number (or part number). For any inventory-management system, including M&CRP, each inventory item must be uniquely identified. All items that have the same item number must be totally interchangeable. Each item must have one, and only one, item number. This includes raw materials and subassemblies. The two major issues in developing an item-numbering system for M&CRP are:

1. Whether alphabetic characters should be used in the item number. Using alphabetic characters (or a combination of alphabetic and numeric characters) has these disadvantages:

Alphabetic characters are more prone to error; the combination of alpha and numeric characters is substantially more prone to error

Alphabetic characters slow data entry speeds

Seven numeric digits can provide 10 million item numbers, and provide the foundation for check digits

However, the advantages of using alphabetic characters, or a combination of alpha and numeric characters, include:

Alphabetic characters tend to be easier for humans to remember and utilize

Bar coding, or other electronic means of data entry, eliminates most of the disadvantages of using alphabetic characters

Practitioners have debated the issue for decades. All other things being equal, we suggest that companies restrict their part numbers to straight numeric. However, the issues of speed and accuracy of data entry, which favor straight numeric item numbers, are becoming less relevant due to the increased use of electronic data entry technologies, such as bar coding and EDI.

2. Whether or not the item numbers should be "significant." A significant item number contains one or more characters that give some information about the item. For computer-based information systems, significant numbers have several disadvantages:

The numbers become physically large because of positional notation requirements and/or reserved blocks of numbers.

Many numbers are reserved for future use that may never be used.

Conversely, as product lines change, a block of numbers can become too small, and must "overflow" into a different area, which starts to seriously compromise the original significance of the design.

Most important, whenever any significant area changes, the item number has to change. Changing item numbers is costly and disruptive.

On the other hand, significant item numbering schemes enable data to be recognized and sorted more easily by humans. Some significance in part numbers can be quite beneficial. For example, humans can relate more quickly to "LM20-001" than to "13742," both of which refer to a 20-inch lawn mower.

The debate over alphabetic versus numeric item numbers has included significant versus nonsignificant numbering as well. Perhaps the most intelligent compromise for many companies is a "semi-significant" part number, in which the first few characters define the major product line, followed by nonsignificant numbers that create a unique identification. Even this, however, is prone to future renumbering, because an item that started its life as a component for product family *X* can be used by *Y*. When *X* is discontinued in a few years, we have a component whose initial description is now obsolete, but whose fit, form, and function have not changed. Therefore, we suggest that purely nonsignificant numbers can be the best long-term choice.

Whichever item numbering system is used, we strongly recommend that you store each data element separately in its own field, so that it can support easy computer access for inquiry and analysis.

Because of the cost, we do *not* encourage companies to renumber all their parts as a prerequisite to bringing up M&CRP. Even more important, renumbering parts can slow the entire implementation process, significantly delaying the substantial benefits.

Whether or not subassemblies should be assigned item numbers depends entirely upon how the product is assembled, and not on the design. Every item that will be stocked, however briefly, for M&CRP purposes *must* have an item number. However, if the subassembly is assembled and immediately goes to another work station or into the final assembly at the same work station, and is never stocked, no item number is required. See Phantom Bill of Materials, later in this chapter, for a more detailed discussion.

User Code. The user code indicates which users will review the item in their reports. The user codes enable reports to be tailored to the needs of the various departments, while the data is taken from common files. Items are included or excluded depending upon the destination of the report. For example, three MRP planners could each receive a unique report containing only those items for which he or she is responsible. A second user code could separate the same parts on a different report, by buyer.

Scheduling Code. The scheduling code tells which system has the responsibility for scheduling the item: M&CRP, MPS, or FAS (Final Assembly Schedule).

Low-Level Code. The low-level code is the lowest level at which an item occurs in any bill of materials. It will be discussed in more detail in the section on manufacturing bills of material, later in this chapter.

Types of Bills of Material

M&CRP recognizes three types of bills of material: engineering, manufacturing, and planning. The Engineering Bill of Materials is the first bill and is used primarily by Product Engineering and Purchasing. The Engineering Bill of Materials lists all the items required to build a particular product, and identifies the engineering specifications for each item. It is normally accompanied by an assembly drawing, which is an exploded view of how all the items fit together in the final product. The Engineering Bill of Materials does not indicate how the product will be manufactured, and does not identify the subassemblies produced in the manufacturing process. However, because it defines all the purchased items, Purchasing can start procuring parts, especially those with long lead times. The Engineering Bill of Materials was also discussed in Chapter 1.

From the Engineering Bill of Materials and the Assembly Drawing, Manufacturing Engineers and Production Planners develop the Manufacturing Bill of Materials. The process of developing a Manufacturing BOM from the Engineering BOM, and other information, is often called "structuring the Bill of Materials." The Manufacturing Bill of Materials defines for M&CRP precisely which components are combined into which subassemblies and assemblies.

Planning Bills of Material are used for forecasting and planning. Planning Bills of Material do not represent actual products that will be made. Rather they represent composite, or pseudo (false) products, that are created to facilitate forecasting, customer order entry, and/or Master Production Scheduling.

M&CRP primarily utilizes Manufacturing and Planning Bills of Material, which we describe in the following paragraphs.

Manufacturing Bills of Material

To explain the basic concepts of Manufacturing BOMs, we will utilize a simple product, a ball-point pen. The Cross Section of the Assembled Product and the Assembly Drawing for the ball-point pen are shown in Figure 3-4. The *Assembly Drawing* is an exploded view of the product that shows how all the items fit into the Assembled Product. The Engineering Bill of Materials (often called the Parts List) for the ball-point pen is shown in Figure 3-5. It lists the items needed to make the product, and provides the technical specifications required to make or buy the parts, but it does not show the necessary sequencing to manufacture the product.

An Assembly Chart showing how the ball-point pen can be assembled is shown in Figure 3-6. Manufacturing Engineers and/or Industrial Engineers use the Assembly Chart to document the most efficient method to assemble a product and to train personnel in the assembly procedure. The Assembly Chart portrays essentially the same information as the Manufacturing Bill of Materials, but in a somewhat different format.

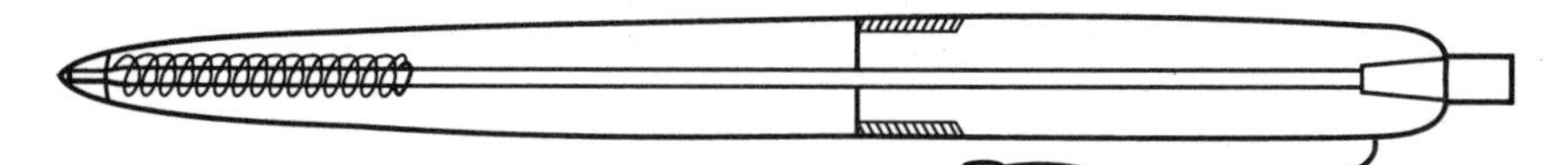

Figure 3-4*a*. Cross section of assembled ball-point pen.

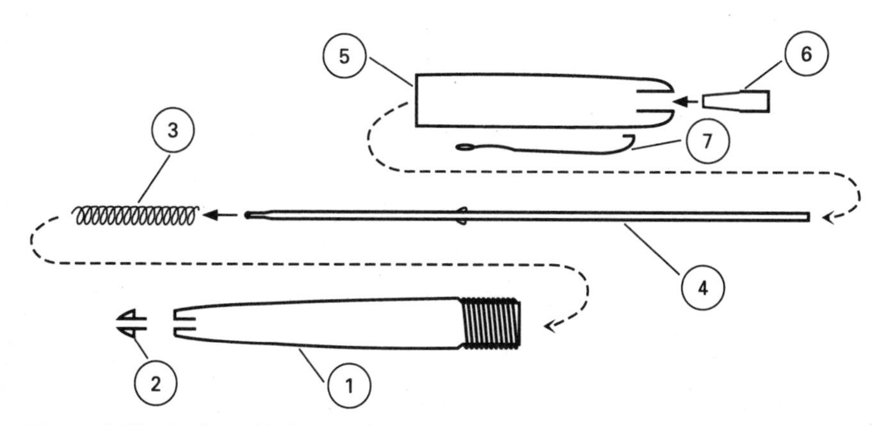

Figure 3-4*b*. An Assembly Drawing for a ball-point pen.

Item no.	Description	Quantity	Make or buy	Drawing file no.
1	Barrel	1	Make	26,079
2	Tip	1	Buy	26,080
3	Spring	1	Buy	20,091
4	Refill	1	Buy	20,026
5	Cap	1	Make	26,048
6	Plunger	1	Buy	26,032
7	Clip	1	Buy	26,054

Figure 3-5. Engineering Bill of Material (parts list) for ball-point pen (item No. 100).

Manufacturing Bill of Materials (in Chart Format). The Manufacturing BOM in chart format (often called *product structure* or *tree structure*) shown in Figure 3-7, shows essentially the same information as the Assembly Chart in Figure 3-6, but in a slightly different format. They both differ from the Engineering Bill of Materials, shown in Figure 3-5, in that they show the sequence in which the items are assembled to make the final product, and they show intermediate subassemblies.

The BOM in Figure 3-7 covers four levels:

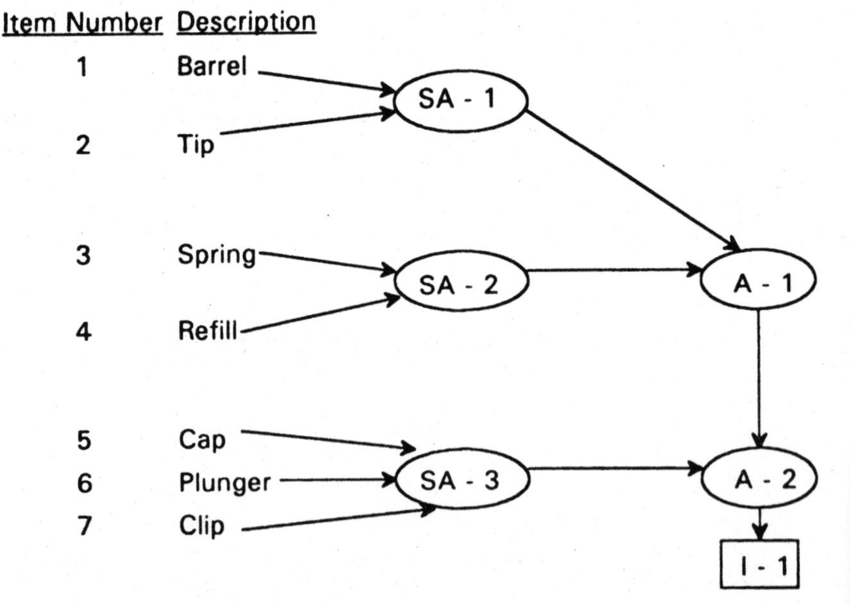

Figure 3-6. Assembly Chart for ball-point pen.

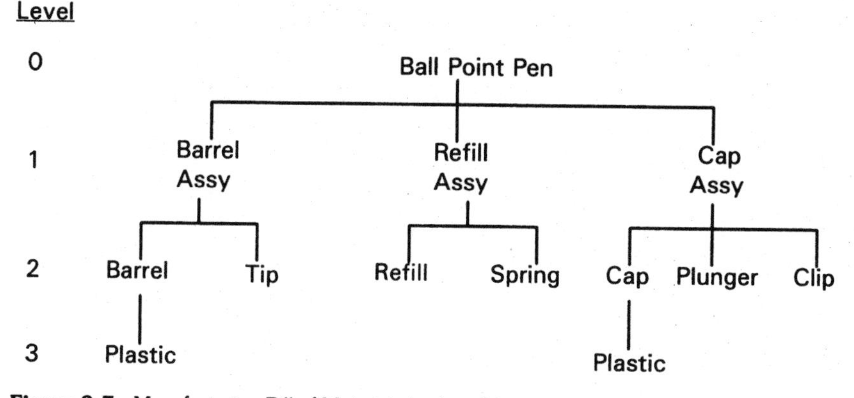

Figure 3-7. Manufacturing Bill of Materials in chart format.

- Level 0 includes the end product, the ball-point pen,
- Level 1 includes the assemblies that directly make up the ball-point pen,
- Level 2 includes the components that make up the three assemblies, and
- Level 3 includes the raw materials that are used to manufacture the components. Raw materials are listed only for the barrel and cap, because the other components are purchased as indicated in the Engineering BOM, Figure 3-5. The raw materials are also purchased because they have no further parts under them.

Each level in the BOM represents a manufacturing operation. At level 0, the barrel, refill, and cap assemblies are assembled into the final product. At level 1, the seven components are assembled into three assemblies, and at level 2, the raw materials are molded into the components. (At level 3, the purchased parts are acquired from a supplier.)

Indented Bill of Materials. The chart format for the Manufacturing Bill of Materials greatly assists manual preparation or evaluation, because it provides an easily understood picture of how the product is assembled. However, it is not suitable for computer use. For computer use, the *Indented Bill of Materials* is most practical. Figure 3-8 shows a multilevel Bill of Materials for a ball-point pen, in Indented Format, and provides exactly the same information as Figure 3-7; only the format is different.

Any multilevel bill can be broken down into a number of single-level bills. For example, six single-level bills, shown in Figure 3-9, provide the same information as in the multilevel bill of Figure 3-8.

Computers store Bills of Material in the single-level format so that the structural information concerning an item will be stored only in one loca-

Item no.	Description	Quantity
101	Ball-Point Pen	
11	Barrel Assy	1
1	Barrel	1
8	Plastic	1
2	Tip	1
12	Refill Assy	1
3	Spring	1
4	Refill	1
13	Cap Assy	1
5	Cap	1
8	Plastic	1
6	Plunger	1
7	Clip	1

Figure 3-8. Multilevel Bill of Materials for ball-point pen in indented format.

tion. No matter where the subassembly or component is used, the same Bill of Materials will be retrieved by the computer. Storing the same information in multilevel form would result in needless duplication and the increased probability of obtaining differing information for the same part. Computers create multilevel screens and reports by simply chaining together the appropriate sequence of single-level Bills of Material.

A *Summarized Bill of Materials* lists each item in a product only once, and provides the total quantity of each item used in that product. Summarized bills do not show assembly sequence relationships. For example, if the barrel and cap of the ball-point pen required the same type and quantity of plastic, the plastic (item No. 8) would only be listed once in our summarized bill, with a quantity of two, as shown in Figure 3-10. The computer forms a Summarized Bill by condensing all single-level Bills of Material required to make an item.

Low-Level Coding in Bills of Material. The Low-Level Code refers to where an item fits in the product structure. The final product is at level 0. The components or subassemblies used directly in making the final products are at level 1. Components used in making level 1 items are at level 2, and so forth. (See Figures 3-7, 3-8, and 3-9 for examples.) We discuss low-level codes in detail in Chapter 4, because MRP requires low-level codes for its processing logic. Cost roll-up programs start at the lowest level in a Bill of Materials and "roll" costs upward; the low-level code facilitates this process. Low-level codes are maintained by the Bill of Material maintenance processing program as it maintains Bill of Material relationships.

Item no.	Description	Quantity
101	Ball-Point Pen	
11	Barrel Assy	1
12	Refill Assy	1
13	Cap Assy	1
11	Barrel Assy	
1	Barrel	1
2	Tip	1
12	Refill Assy	
3	Spring	1
4	Refill	1
13	Cap Assy	
5	Cap	1
6	Plunger	1
7	Clip	1
1	Barrel	
8	Plastic	1
5	Cap	
8	Plastic	1

Figure 3-9. Six single-level Bills of Material for ball-point pen, in indented format.

Phantom Bills of Material. Phantom assemblies are assembled components that are used immediately in making a larger assembly, and are not stocked. If, however, there is the possibility that occasional inventories of a phantom assembly may exist owing to production overruns, returns, or other reasons, an item number should be assigned, and a phantom Bill of Materials developed, to enable the M&CRP system to control these phantom items. Another reason for using phantoms is to allow the grouping of items for purposes other than manufacturing (for example, for Design Engineering convenience).

The Bill of Materials for the parent includes the phantom as a child, indicating that it is a phantom. When the parent is being produced, the order-release program, in some M&CRP systems, checks to determine if any units of the phantom are in stock. If so, the computer prints a pick list so they will be withdrawn and used. If not, the order-release system skips the phantom and directly requisitions the phantom's components. The process of skipping the phantom and calling out its components is called *blowing through.* Other M&CRP systems prohibit a phantom from being stocked, so the order-release process does not check for any phantoms on-hand in the stockroom.

Item no.	Description	Quantity
1	Barrel	1
2	Tip	1
3	Spring	1
4	Refill	1
5	Cap	1
6	Plunger	1
7	Clip	1
8	Plastic	2
11	Barrel Assy	1
12	Refill Assy	1
13	Cap Assy	1

Figure 3-10. Summarized Bill of Materials for ball-point pen.

Where-Used Reports are the inverse of normal Bills of Material. Normal bills of material go from parent to child, whereas the Where-Used Reports go from child to parent. In a where-used report, the part being reported on is printed as level 0; its immediate parents are at level 1; its grandparents are at level 2, and so on. Figure 3-11 shows a Single-Level Where-Used Report for the same ball-point pen used for the normal BOMs. This report lists all the parents in which the child (plastic) is used directly, together with the quantity of the child to produce each parent. This report can be used in engineering change analysis when the engineer wants to determine what assemblies will be affected by a change in a component.

A *Multilevel Where-Used Report* shows the assembly in which the item is used directly, the higher level assembly in which that assembly is used directly, and so forth, up to the product or highest level end item, as shown in Figure 3-12. Planners, schedulers, and engineers use this report for the same purposes as the single-level report for more complex products.

The *Summarized Where-Used Report* for item No. 8 (Plastic) is shown in Figure 3-13. This report lists all parents in which the item is used directly, or indirectly. Cost accountants use this report primarily in determining the effect of component cost changes.

Item no.	Description	Quantity
8	Plastic	
1	Barrel	1
1	Cap	1

Figure 3-11. Single-level Where-Used report for item No. 8.

Item no.			Description	Quantity
8			Plastic	
	1		Barrel	1
		11	Barrel Assy	1
		101	Ball-Point Pen	1
	5		Cap	1
		13	Cap Assy	1
		101	Ball-Point Pen	1

Figure 3-12. Multilevel Where-Used Report for item No. 8.

Planning Bills of Material

The Manufacturing Bills of Material discussed in the previous section were developed from the viewpoint of manufacturing and are used in making and scheduling the end product. Many companies find it desirable to develop a different type of Bill of Materials, called a *planning bill,* to facilitate sales forecasting, order entry, and particularly Master Production Scheduling.

We will investigate planning bills primarily from the viewpoint of Master Production Scheduling. In developing planning bills, we are interested in developing a Bill of Material for some item, other than the actual end product, that will assist us in master scheduling. From this viewpoint we can divide planning bills into two types, each with subtypes:

1. Planning bills with scheduled item that is a component of an end product. The scheduled items are physically smaller than the end product.

   Modular Bill of Material
   Inverted Bill of Material

2. Planning bills with scheduled item that has end products as components (Super Bills). The scheduled items are physically larger than the end product.

   Super Bill of Material
   Super Family Bill of Material
   Super Modular Bill of Material

Item no.	Description	Quantity
1	Barrel	1
5	Cap	1
11	Barrel Assy	1
13	Cap Assy	1
101	Ball-Point Pen	2

Figure 3-13. Summarized Where-Used Report for item No. 8.

Planning Bills with Scheduled Items that are Components of an End Product

Modular Bills. A modular bill groups subassemblies and parts based on whether they are unique to a specific product option or common to all configurations of the product. Each group is called a module. The module, rather than the end product, is scheduled in the MPS. To understand why modular bills can be useful, let's consider an automobile that has the options listed in Figure 3-14.

The total number of automobiles with different combinations of options that could be sold is $3 \times 3 \times 2 \times 4 = 72$. If the autos are made-to-stock and the MPS includes all 72 possible configurations, then planners would have to forecast each of the 72 configurations; for example, four-door body, V-8 engine, automatic transmission, and accessory package #3.

By using modular bills, and scheduling the modules in the MPS, planners would only need to forecast each module or option. In our example this would require only 12 forecasts $(3 + 3 + 2 + 4)$. To use modular bills, the Master Scheduler schedules each *module* in the MPS, and MRP schedules the subassemblies and components to build each module. The Final Assembly Schedule, using an end product bill of materials, dictates which modules to actually assemble to specific customer orders. Firms having modular products and modular planning bills will normally use the Assemble-to-Order Demand Response Strategy discussed in Chapter 1.

The advantages of the modular planning bill increase rapidly with the number of options for a product. Not only are there fewer items to schedule in the MPS, but the forecasts for the modules are much more accurate than the forecasts for specific configurations.

Inverted Bills of Material. A single component, or raw material, such as petroleum, iron, wood pulp, or chocolate, can be converted into many unique products. Figure 3-15 shows an inverted Bill of Material for petroleum.

The most logical place to forecast and Master Production Schedule these

BODY	ENGINE	TRANSMISSION	ACCESSORIES
2-door	V-6	Manual	No accessories
4-door	V-8	Automatic	Acc'y package #1
Convertible	Diesel		Acc'y package #2
			Acc'y package #3

Figure 3-14. Modular options for an automobile.

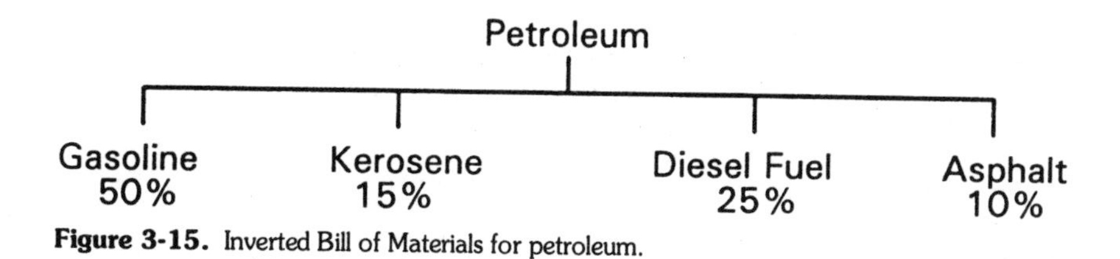

Figure 3-15. Inverted Bill of Materials for petroleum.

products is at the raw material level rather than at the end product. Forecasts at the aggregate raw material level are more accurate than forecasts at the individual end-product level. Inverted bills are based on the assumption that the usage percentages are relatively constant and predictable. Planners can include contingency inventory by inflating the percentages, so that they total more than 100 percent, or by increasing the Master Production Schedule for the aggregated product, petroleum. Planning, using inverted bills, is common to the process industries and is discussed further in Chapter 9.

Planning Bills with Scheduled Items that Have End Product as Components

Super Bills of Material. Planning Bills in which the items to be scheduled are larger than the end product are called *Super Bills.* Specifically, a Super Bill is a single-level Bill of Material in which the parent is a pseudo (not real) assembly, and the children are *real* end products. The quantity of each child is a fraction of the total forecast for the parent. For this reason, the Super Bill is sometimes called a *ratio bill* or a *percentage bill.* The fraction for each child is normally based on past sales information, although it could also reflect projected sales trends.

Super Bills require a two-level MPS for scheduling. Under a two-level MPS system, the Master Scheduler enters total demand for the super (and pseudo) assembly into the top level of the MPS. The MPS system then explodes the top level, using the fractions in the Super Bill, calculating production schedules for the actual end products in the lower level of the MPS. The schedule of end products in the lower level will drive M&CRP just as a normal MPS output does. We describe this process in more detail in the section on master production scheduling later in this chapter.

As an example, suppose that a manufacturer of lawn and garden equipment makes push lawn mowers, self-propelled lawn mowers, riding lawn mowers, rototillers, and garden tractors. They could use the Super Bill pictured in Figure 3-16 to obtain forecasts and manufacturing requirements for each end item. To do so, the forecast of total customers would be multiplied by the appropriate sales fraction for each end item.

All Super Bills have the same basic format, a pseudo assembly parent with

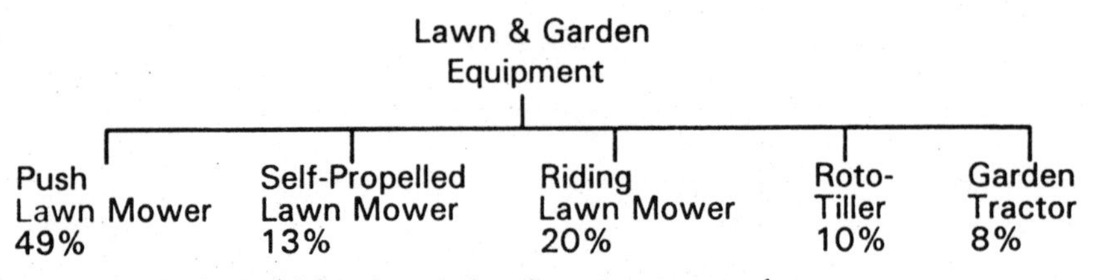

Figure 3-16. Super Bill for a lawn and garden equipment manufacturer.

buildable end products as children. The pseudo assembly is used to drive the upper level of a two-level MPS. The second level is a schedule of end products, which is a percentage of the top level schedule. The only difference between types of Super Bills is in how the pseudo assembly is developed. If this is not specified, or is different than the two methods we will describe next, the bill is simply called a Super Bill.

Super Family Bills of Material. To improve the accuracy of demand forecasting, many companies form families of products with similar demand patterns. The aggregate (family) forecast is normally much more accurate than the forecasts for any one product. To use the aggregate forecast they must develop a *Super Family Bill of Material,* consisting of the family (pseudo) assembly as the parent and the individual family end products as the children. The quantity per parent for the end items is the historical, or forecasted, percentage of sales for each end item. For example, a manufacturer of hand tools might have families for hammers, screwdrivers, chisels, drills, wrenches, and pliers. The Super Bill for the drill family is presented in Figure 3-17.

Frequently, companies with very broad product lines will divide the products into groups, or families, in order to simplify planning and control. Some companies make 5,000 to 10,000 end products that can be grouped logically into from 10 to 30 families. The appearance and use of the Super Family Bill is the same as the regular Super Bill; the only difference is in how the parent pseudo product is determined.

D101
Drills

D11 1/16" Drill 12%
D12 1/8" Drill 16%
D13 3/16" Drill 18%
D14 1/4" Drill 23%
D15 3/8" Drill 14%
D16 1/2" Drill 10%
D17 3/4" Drill 7%

Figure 3-17. Super Bill for drill family.

Super Modular Bill of Material. A Super Modular Bill of Material is a combination of a Super Bill and a Modular Bill. In it the parent is an unbuildable group of modules that is used for planning purposes only. The children are the modules that can appear in a final product. Using the automobile from the previous example, and the indented format, a Super Modular Bill for the automobile is shown in Figure 3-18. Because this automobile has an engine that is 50 percent V-6, 40 percent V-8, and 10 percent diesel engine, it is obviously not buildable or operable. However, it can be used to forecast and build the various modules needed to build real automobiles. In most situations, practitioners prefer a Super Modular Bill to a Modular Bill because the pseudo assembly of modules can usually be forecasted more accurately than can each individual module.

Managing Engineering Changes

Engineering changes, if not properly planned and controlled, can result in high costs owing to scrapping obsolete product, excessive tooling cost, production inefficiency, and late shipments. Nevertheless, maintaining market position usually requires ongoing product improvement, and the pace of

	Percent
Automobile	100
Body	100
2-door	40
4-door	50
Convertible	10
Engine	100
V-6	50
V-8	40
Diesel	10
Transmission	100
Manual	30
Automatic	70
Accessories	80
Accessory Package 1	40
Accessory Package 2	30
Accessory Package 3	10
Common Parts	100

Figure 3-18. Super Modular Bill for an automobile.

change continues to increase. In the computer industry, for example, a product that used to have a life of three years now lasts less than nine months. The challenge is how to implement engineering changes efficiently with the least disruption throughout the company.

Requests for engineering changes can originate anywhere within the company, or from customers, for a number of reasons:

Improved performance

Improved safety

Reduced cost

Compliance with government regulations

Most companies have an engineering change committee that reviews the change, approves or disapproves the change, and decides when approved changes will go into effect. In regard to implementation, there are two basic types of change: (1) *Mandatory or Immediate,* and (2) *Discretionary or Optional.* The *Mandatory* type of change involves a serious defect in the product, such that it will not perform or is unsafe. In this case, management will immediately halt production and implement the change, without regard to cost, delay in production, disruption of other operations, or any other factor.

Most changes fall into the *Discretionary* category. For Discretionary changes, management has considerable leeway in how and when to implement the change. The two methods of implementing change rely on (1) a fixed date on which the change will be effective, called an *Effectivity Date,* or (2) a *serial number, lot number, work order number,* or the like. We discuss these in more detail later.

Regardless of the method used to implement a change management must consider the following factors in determining the timing of a change:

Availability of new items: raw materials, tooling, equipment

Need for recalling or retrofitting products in use

Time for minimizing cost, including obsolete inventory, tooling, equipment alterations

Time for depletion of existing product and component inventories

Need to retain supply of current parts for customer service

Time of changes to other products

Time of model changeover or start of a new contract

Impact on our plant or supplier's plant capacity

Importance of change to competitive position

Need for documentation changes

Time for approval by regulatory body

The engineering change coordinator enters the change(s) into the Bill of Materials. One major point: the change affects *only* the individual parent-component relationships. For example, if item *X* is being phased out of parent *A* on October 15, entering this change into the BOM only affects that particular usage of item *X*. It can still be used in many other places. This action does *not* render item *X* obsolete. In fact, item *X* can be phased back into the BOM for parent *A* at a later time.

Some minor engineering changes require modifications only in the Item Master File, or Routing File, and not in the BOM. When the "fit, form, or function" remain the same, the new part is interchangeable with the old and retains the same item number. For example, the change might only affect a tolerance or the finish of a surface, or it might be a change to correct errors on the blueprint.

If the new part is not interchangeable with the old, then the part number and the BOM is changed. Further, the parent assembly of the new part must be checked to determine if it is interchangeable with the old assembly. It not, its number must also be changed. The analysis continues upward, level by level, until the assemblies are found to be interchangeable, or the final product is reached.

There are two approaches to revising the BOM.

1. The engineering change coordinator prepares a new BOM that incorporates the changes, but does not put the change into the M&CRP system in advance. On the effectivity date or serial or lot, the old BOM is deleted and the new BOM is entered. This approach has two major drawbacks:
   - It fails to communicate the upcoming change through MRP to the materials planning department; they will be likely to keep ordering parts that may soon be obsolete, and
   - It loses any ability to track the engineering change in the BOM itself, which may be critical for field service at a later date.
2. As soon as the decision is made, the engineering change coordinator enters the effectivity date on which the old part will be made noneffective and the new part added. The MRP II system will execute the change automatically. Once the effectivity date or serial or lot number arrives, the new design goes into effect. This approach is illustrated in Figure 3-19, where we decided to use a plastic plunger, instead of a metal plunger, in our ball-point pen to reduce costs. For temporary variances, it is possible to use stop as well as start dates. This second method quickly

Item no.				Description	Quantity	Start	Stop
101				Ball-Point Pen			
	11			Barrel Assy	1		
		1		Barrel	1		
			8	Plastic	1		
		2		Tip	1		
	12			Refill Assy	1		
		3		Spring	1		
		4		Refill	1		
	13			Cap Assy	1		
		5		Cap	1		
			8	Plastic	1		
		6		Plunger, metal	1		02/15
		66		Plunger, plastic	1	02/16	
		7		Clip	1		

Figure 3-19. Effectivity date entered into Bill of Materials.

communicates scheduled changes to all concerned and facilitates the coordination needed to successfully make the change.

Of the two methods used to control the timing of the change, effectivity dates are by far the most common. They are also easier to use. The second method of implementation involves using something other than a date to control the effectivity of the change. Serial number, lot number, and work-order number have been used; serial number is the most popular of this group. Serial number effectivity is considerably more complex than date effectivity. First, the specific subassembly where the change occurs must be identified, according to the end product serial number into which it will eventually be assembled. The net result of serial number effectivity is that the identity of the serial number of the end product must be carried all the way down the entire product structure.

The Master Production Schedule must contain both product identifying model and serial numbers. Products below the cutoff serial number in the MPS will use the existing bills, and products at and above the cutoff serial will use the changed bill. When the change occurs at the middle and lower levels of the bills, tight discipline in the shop is needed to ensure using the proper components. Because such discipline is difficult to obtain and it can be expensive, serial numbers should be assigned only after final assembly begins or, better still, after production has been completed.

The Master Production Schedule

The Master Production Schedule is a statement of what end items (shippable products, including finished goods, replacement parts and spares) a company plans to produce, by quantity and time period. The MPS disaggregates and implements the production plan. Whereas the production plan is stated in aggregate terms, the MPS is stated in specific configurations with item numbers that are contained in the Item Master and BOM files. The MPS performs four primary functions:

1. Provide the principal input to the M&CRP system
2. Schedule production and purchase orders for MPS items
3. Provide the foundation for determining capacity and resource requirements
4. Provide the basis for making delivery promises

The MPS has five major inputs, as shown in Figure 3-20.

1. The *production plan* provides a set of constraints on the MPS. The MPS must sum to the production, inventory, and resource levels in the production plan.
2. The MPS must take into account all types of *demand data* for the items being scheduled including: sales forecasts, customer orders, distribution warehouse requirements, interplant requirements, service demand forecasts, and Safety Stocks.

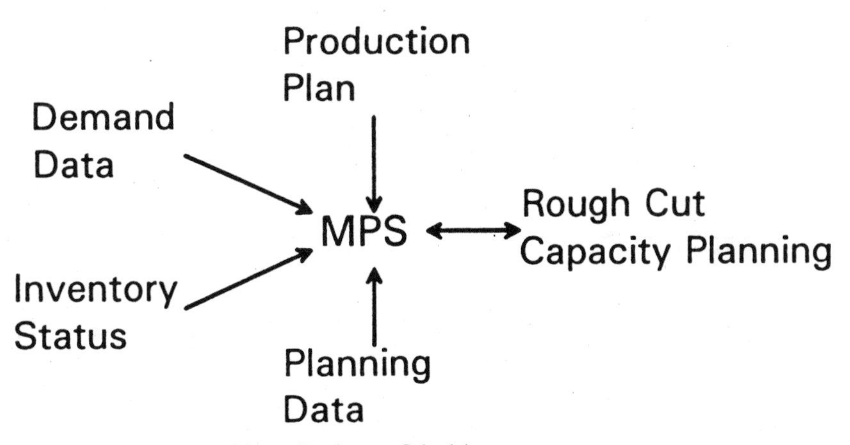

Figure 3-20. Inputs to Master Production Schedule.

3. The MPS must know how much is available to accurately determine how much to order. This requires the *inventory status* information: on-hand inventory, allocated stock, released production and purchase orders, and firm planned orders.
4. The Item Master File provides *Planning Data* on each item to guide the MPS planning process, such as: lot-sizing rule to be used, shrinkage factor, safety stock, and lead time.
5. *Rough Cut Capacity Planning* determines the *capacity requirements* to implement the Master Production Schedule, verifying the schedule's feasibility or causing the Master Scheduler to revise the schedule.

Fundamentals of Single-Level Master Production Scheduling

Figure 3-21 shows a time-phased record of a Master Production Schedule for a mountain bicycle (item B21). This chart projects what each of these fields will look like at various dates in the future. By so doing, it enables the Master Scheduler to understand the consequences of his or her decisions before the decisions become firm and affect the company.

We elaborate on each of the rows below:

- The chart shows the data for each field at the end of a *time period.* For MPS, the time periods are usually weeks, although you can use days if you wish. To conform to prevalent industry practice, we use weeks in this chapter. For greater precision and utility, we suggest you consider using days for the first several periods, then weeks, then months.
- *Item* is the item number of the part that we are master scheduling. In this example, our item is B21, a 21-speed mountain bicycle.
- *SS* (*Safety Stock*) is the amount of stock that you want to keep in reserve, rather than using routinely. The Safety Stock is for the item being master scheduled. In this example, we have decided on a Safety Stock level of 10 units.

		Week									
Item: B21 SS 10	OH	1	2	3	4	5	6	7	8	9	10
Forecast		10	10	10	10	10	20	20	20	20	20
Available	10	10	10	10	10	10	10	10	10	10	10
MPS		10	10	10	10	10	20	20	20	20	20

Figure 3-21. A Chase demand MPS response to seasonal sales.

- *Forecast* is the quantity that Sales and Marketing intend to sell during that time period. In this three-line chart, the forecast is the only representation of actual customer demand. In this example, Sales and Marketing have forecast a demand of 10 units per week for the first five weeks and 20 units per week for the last five weeks.
- The *Available* row indicates the units available in inventory at the *end* of that period. The first column shows the current on-hand, which is the starting Available Inventory. Our current on-hand in our example is 10. The MPS system calculates Available quantities as outlined below.
- The *MPS* row indicates the number of units scheduled for production that week. The MPS planner enters and maintains the MPS quantities; they are not calculated by the MPS system. These quantities represent the commitment by the MPS planner (also known as the Master Scheduler or MPS scheduler) to build these units in the specified periods. This MPS row, by item number, is the primary input to MRP.

The relationship between the various entries in the record are indicated in Equation (3-1). The MPS scheduler may schedule any MPS quantity desired, although the Available Inventory should remain above the Safety Stock.

$$\begin{pmatrix}\text{Available}\\ \text{Inventory}\\ \text{at end of}\\ \text{current}\\ \text{period}\end{pmatrix} = \begin{pmatrix}\text{Available}\\ \text{Inventory}\\ \text{at end of}\\ \text{previous}\\ \text{period}\end{pmatrix} - \begin{pmatrix}\text{Forecast}\\ \text{demand}\end{pmatrix} + \begin{pmatrix}\text{MPS}\\ \text{quantity}\end{pmatrix} \tag{3-1}$$

To illustrate how MPS computes the Available quantity, let us refer again to Figure 3-21. We start with 10 on hand. In period 1, we subtract the period 1 forecast (10 units), add the MPS quantity (10 units), and finish the week with 10 units. Periods 2 through 5 are identical. In period 6, our forecast and our MPS quantity each increase to 20, so we have 20 coming in and 20 going out each week. This still leaves a balance of 10 on hand at the end of each week. Why 10? Because that is the Safety Stock; we want to keep those in reserve in case we have more actual customer orders than we have forecast for a given period.

The Master Scheduler can schedule any quantity of any item to be produced, whether there is demand or not. For example, if a company is experiencing a downturn, management could decide to keep building product (knowing that inventory will increase). Conversely, the Master Scheduler could build fewer units than Sales and Marketing has forecast, or even fewer units than customers have actually ordered (risking a dramatically short career with the company). The Master Production Schedule is the

Item: B21 SS 10	OH	1	2	3	4	5	6	7	8	9	10
Forecast		10	10	10	10	10	20	20	20	20	20
Available	10	15	20	25	30	35	30	25	20	15	10
MPS		15	15	15	15	15	15	15	15	15	15

Figure 3-22. A level production MPS response to seasonal sales.

steering wheel of the manufacturing and materials organization. Where the MPS goes, the rest of the manufacturing and materials organization follow.

In Figure 3-21, the MPS scheduler has chosen to schedule a quantity that would exactly satisfy the demand forecast, that is, a "chase demand" strategy. However, he or she may choose to follow a "level production" strategy, as shown in Figure 3-22, by selecting a production rate that is constant over the period and will not reduce inventories below the Safety Stock level. These two examples represent the two extremes in strategy; there are many combinations in between. *The master production schedule is a build schedule*—not a forecast—and is under the complete control of the MPS planner.

Figure 3-23 presents the same situation as Figure 3-21, but incorporates a lot size of 30 units. When the Available Inventory falls below the Safety Stock level of 10, an MPS quantity of 30 is scheduled. This results in a lumpy MPS, which in turn results in lumpy demand in the M&CRP system.

In Figure 3-24, we expand the MPS format to include:

- *Customer Orders,* which represents the company's backlog of customer orders for shipment each week.
- *Available to Promise* indicates the *number of units from a particular production batch that have not been allocated to a customer order and are thus available to promise to a new customer.* As shown in Figure 3-24, the Available to Promise for the first period is calculated by adding the MPS quantity to the on-hand inventory, subtracting Safety Stock, and then subtracting the total

Item: B21 SS 10; OQ 30	OH	1	2	3	4	5	6	7	8	9	10
Forecast		10	10	10	10	10	20	20	20	20	20
Available	10	30	20	10	30	20	30	10	20	30	10
MPS		30			30		30		30	30	

Figure 3-23. A chase demand MPS response with lot size = 30.

Item: B21 Week

SS 10 OQ 30	OH	1	2	3	4	5	6	7	8	9	10
Forecast		10	10	10	10	10	20	20	20	20	20
Orders		10	7	8							
Available	10	30	20	10	30	20	30	10	20	30	10
Avail to Promise		5			30		30		30	30	
MPS		30			30		30		30	30	

ATP = (On Hand + MPS - SS) - Total Customer Orders until Next MPS Receipt

ATP (Week 1) = (10 + 30 - 10)- (10 + 7 + 8) = 30 - 25 = 5

Figure 3-24. MPS with customer orders and Available to Promise.

of the actual orders that are scheduled before the next MPS scheduled receipt. Only the first period can use the on-hand. All subsequent periods use only the incoming MPS quantity; they do *not* use carryover from a previous MPS batch or on-hand inventory.

Designing the Single-Level MPS System

Selecting the Planning Horizon. MPS plans must cover a period at least equal to the total time required to accomplish the plan. For the MPS to provide the basis for procuring raw materials, the MPS horizon must extend through the *cumulative lead time,* as indicated in Figure 3-25. This is the total elapsed time from first ordering the raw materials until completing the final product. Actually, the horizon should extend for some period beyond the cumulative lead time, typically three to six months, so that Purchasing can make informed decisions on the size of raw materials orders, particularly where quantity discounts, or expected price increases, are involved. If management uses the MPS in simulation mode for long-term rough-cut capacity planning of personnel, machines, warehouse space, and so forth, the MPS horizon should be considerably longer in order to cover the longer procurement times of these long-range resources.

Selecting Time Fences. Changes in the MPS become more difficult, disruptive, and costly as they are made closer to product completion. To stabilize schedules and to make sure that changes are properly weighed before

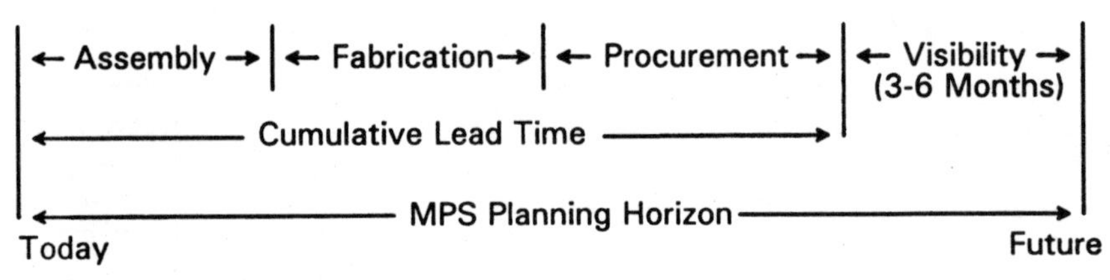

Figure 3-25. MPS planning horizon.

they are approved, master schedules can be divided into several zones, with different procedures governing schedule changes in each time zone. *Time Fences* separate the time zones.

The *Demand Time Fence* (*DTF*) and the *Planning Time Fence* (*PTF*), as shown in Figure 3-26, are the most common time fences. The DTF is set at the final assembly time and the PTF is set at the cumulative lead time.

- Inside the DTF, only emergency changes should be made, and only with the approval of the Manufacturing Manager.
- Between the DTF and the PTF, the MPS planner can change the product mix, subject to the availability of material and capacity. The planner cannot change the production rate without ensuring that material and other resources can be adjusted to accommodate the new production rate.
- Outside the PTF, the MPS planner can freely change the production rate to meet anticipated changes in demand. Beyond the PTF, the MPS has two functions:

  Provide an input to rough-cut capacity planning, and thus provide a basis for making decisions on procuring long-range resources that require long lead times

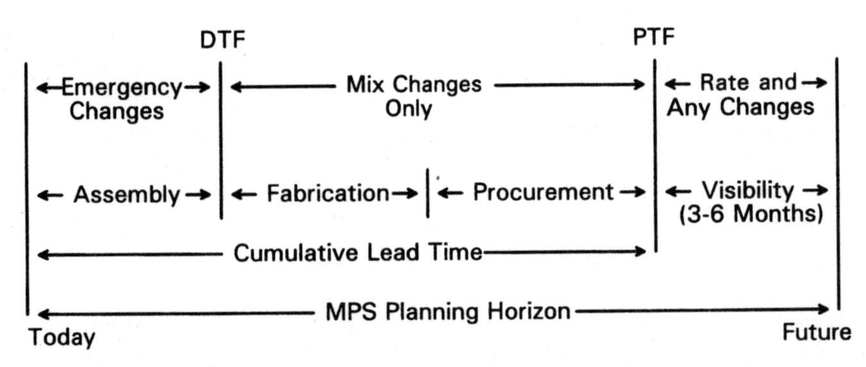

Figure 3-26. MPS Time Fences.

Provide greater visibility over long-lead-time components and raw materials, thus giving Purchasing the ability to work more closely with suppliers.

Planning and managing the MPS is a critical ongoing decision process, involving marketing and manufacturing in trade-offs in the allocation of resources to products, orders, and customers. If an organization has, *and follows,* a clear, written statement of policy governing changes to the MPS by zone, they can avoid considerable friction and controversy.

The Two-Level MPS System

A company can schedule an end product in the MPS by using a Manufacturing Bill and the single-level MPS system, as illustrated in Figures 3-21 through 3-24. However, scheduling a *pseudo product* from one of *the Super Planning Bills,* discussed in Figures 3-16 through 3-18, requires using *a two-level MPS.* To illustrate the operation of a two-level MPS, we will use: (1) the transmission portion of the Super Modular Planning Bill for an automobile in Figure 3-18, and (2) the demand data in Figure 3-23, both of which are reproduced at the top of Figure 3-27.

Figure 3-27 shows MPS requirements for the automobile in the first record, and calculates the requirements for the manual and automatic transmissions in the second and third records. The MPS quantities for the automobile, multiplied by the usage fractions (30 percent for manual and 70 percent for automatic) become the forecast requirements for the manual and automatic transmissions. The MPS quantities for the manual and automatic transmissions will drive M&CRP. Customer orders and Available to Promise can also be incorporated into the two-level MPS, as they were for the single-level.

As the example above illustrates, a two-level MPS is actually quite simple. It merely harnesses the power of the computer to perform calculations that can simplify the effort for the Master Production Scheduler. The top level never actually gets built; its only purpose in life is to explode into the bottom level, which can and does get built.

Selecting MPS Items

Selecting items to be scheduled by the MPS deserves special attention. This selection is important, not only because it affects how the MPS operates, but because it also affects how the M&CRP system and the overall manufacturing planning and control system operates. The basic criteria governing the selection are:

Item Description	Usage Fraction
Automobile	1.0
Manual Transmission	.3
Automatic Transmission	.7

AUTOMOBILE — Week

SS 10; OQ 30	OH	1	2	3	4	5	6	7	8	9	10
Forecast		10	10	10	10	10	20	20	20	20	20
Available	10	30	20	10	30	20	30	10	20	30	10
MPS		30			30		30		30	30	

x.3 x.7

MANUAL TRANSMISSION — Week

SS 10; OQ 20	OH	1	2	3	4	5	6	7	8	9	10
Forecast		9			9		9		9	9	
Available	10	21	21	21	12	12	23	23	14	25	
MPS		20					20			20	

AUTOMATIC TRANSMISSION — Week

SS 10; OQ 40	OH	1	2	3	4	5	6	7	8	9	10
Forecast		21			21		21		21	21	
Available	10	29	29	29	48	48	27	27	46	25	25
MPS		40			40				40		

Figure 3-27. Two-level MPS for automobile transmissions.

1. The items scheduled should be end products, unless there are clear advantages for items smaller than end products (modular or inverted planning bills), or larger than end products (super family, super modular, or other super planning bills). Scheduling end products in the MPS causes it to be the same as the Final Assembly Schedule.
2. The number of items should be kept small. Management cannot make effective decisions relative to the MPS if the number of MPS items is very large.
3. It should be possible to forecast demand for the items (unless they are Made-to-Order). The item scheduled must be closely related to the item sold.
4. Each manufactured item must have a BOM, so that the MPS can explode through the BOM to determine the need for components and materials. (A BOM is not necessary for purchased parts which are master

scheduled. For example, an electronics assembly house might master schedule the various sizes of finished metal cabinets that it buys from an outside supplier because of capacity constraints at the supplier.)

5. Collectively, the items selected should account for most of the productive capacity that will be required. Otherwise, Rough-Cut Capacity Planning will not be reliable.
6. The MPS items should facilitate translating customer orders into shippable product configuration. With simple standard products, catalog numbers can actually be specific MPS items. With more complicated products, the customer's choice of product options must be translated into a number of MPS items that will be assembled under the FAS to fill the order.

Major Factors Determining MPS Items. The major factors influencing the selection of MPS items are shown in Figure 3-29. They are:

Product structure

Type of competition

Type of demand response

Number of products or modules in the product line

The product structure refers to the shape of the product's Bill of Materials in the chart format. Figure 3-28 shows typical product structures. Most products have the *standard, tree,* or *pyramid* structure, which has more subassemblies than products, and more components than subassemblies. However, some products, such as automobiles and computers, have fewer subassemblies or modules than final products. These products have a *modular* or *hourglass* product structure. Finally, some products not only have

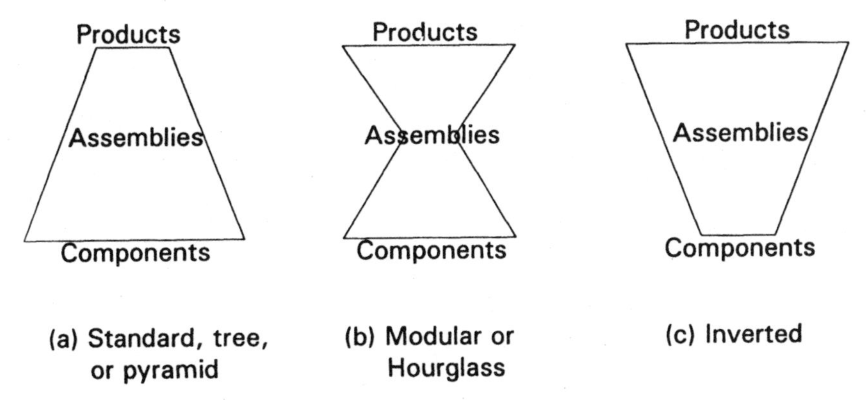

Figure 3-28. Typical product structures.

fewer subassemblies than products, but also have fewer components and materials than subassemblies. The products have an *inverted* product structure, which is typical of process industries, such as petroleum, paper, and glass.

Type of Competition. Time is the key competitive factor that influences the type of Demand Response and, consequently, what MPS items should be scheduled. For our purposes, we need only answer the question, "Is time an important factor in our competition?"

Type of Demand Response. This refers to the various demand response strategies available (MTS—Make-to-Stock, ATO—Assemble-to-Order, MTO—Make-to-Order, DTO—Design-to-Order, and MTD—Make-to-Demand), which we defined in Chapter 1.

Number of Products or Modules. The number of products or modules in the product line determines the size and complexity of the product line. A large and complex product line usually calls for some kind of Super Planning Bill.

MPS Items. For a single-level MPS, the MPS will schedule an end product using a manufacturing BOM, and the M&CRP will perform the first explosion. For a two-level MPS, the MPS will schedule the pseudo product using a Super BOM on the first level, and will also explode the Super BOM to obtain the schedule of end products or modules on the second level.

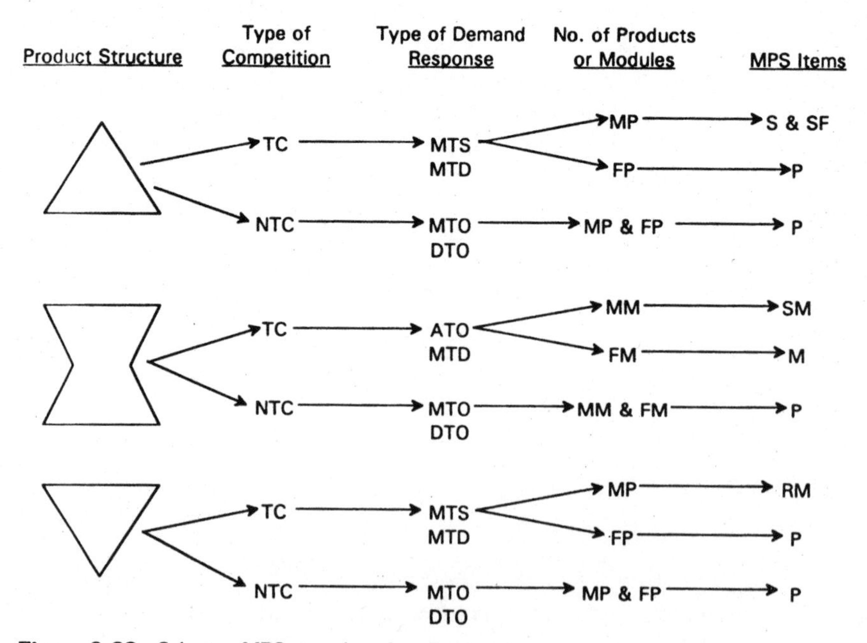

Figure 3-29. Selecting MPS items based on Product Structure and other factors.

Key to Symbols in Figure 3-29

Product Structure

The standard, tree, or pyramid structure, which is characteristic of most products

The modular or hourglass structure: products with many options or modules

The inverted structure: many products made from a few components or raw materials

Time Competition

TC Firm is engaged in Time Competition; meeting customer demand in an expeditious manner is important

NTC Firm is not engaged in Time Competition; other factors are more important than time.

Demand Response (Described in Chapter 1)

MTD Make-to-Demand

MTS Make-to-Stock

ATO Assemble-to-Order

MTO Make-to-Order

DTO Design-to-Order

Number of Products or Modules

MP Many products

FP Few products

MM Many modules

FM Few modules

MPS Items

P End *P*roduct from a Manufacturing BOM

M *M*odules or options from modular products

RM *R*aw *M*aterial from Inverted Planning BOM

S Pseudo Product from a *S*uper Planning BOM

SF Pseudo Product from a *S*uper *F*amily Planning BOM

SM Pseudo Product from a *S*uper *M*odular Planning BOM

As indicated by Figure 3-29, if the product structure is standard, if the firm operates in time competition, if it utilizes an MTS demand response, and if it has many products, the firm should use their MPS to schedule the pseudo product from either a Super BOM or a Super Family BOM. We recommend the Super Family BOM, if feasible. For all other situations, the MPS should schedule end products using manufacturing bills, making the MPS and FAS one and the same.

If the product structure is modular, the picture is more complex. If there is no time competition, the company can consider introducing it to their industry. The leader in time competition can capture extensive market share, while increasing profitability. Lacking any time competition, the company should select a MTO demand response and schedule the end product of a manufacturing bill in both the MPS and the FAS. However, if there is time competition, they should select the ATO demand response and schedule the modules in the MPS, if there are only a few modules. If there are many modules, they should create a Super Modular Planning BOM and schedule its pseudo product.

If the firm has an inverted product structure, has time competition, and has many products, the MPS should schedule the lowest item (usually raw materials). Under all other conditions, it should use the end product of the manufacturing bill.

Summary

In this chapter we have described the principal options for the two major inputs:

- Bills of Material, and
- Master Production Schedule.

For Bills of Material, we outlined the differences between an Engineering Bill (used for purchasing long lead time components), a Manufacturing Bill (used for MRP to time-phase all components and subassemblies), and Planning Bills (used for Master Scheduling, order entry, and sales forecasting). Bills are created through the interaction of both engineers and nonengineers. MRP needs to have low-level codes to operate properly, so the Bill of Materials maintenance program automatically maintains the low-level code each time a parent-component relationship is changed.

Additionally, we discussed the two major types of Planning Bills of Material, those that schedule items smaller than (below) the end product, such as Modular Bills and Inverted Bills, and those that schedule items larger than (above) the end product, called Super Bills. We encourage you to use whichever of these Planning Bills will help you manage and control the smallest number of items, and will support your efforts to provide extremely responsive service to your customers.

For item information, we suggested that an item number be numeric, although alphabetic is an acceptable alternative as long as you address the data entry and error factors. We further preferred a nonsignificant item number, while recognizing the advantages that a semi-significant item number can provide.

The Master Production Schedule and the Product Structure are quite interdependent. However, the Product Structure does not completely determine the item that should be master scheduled. Several other intervening variables (type of competition, demand response, and number of products or modules) also affect the decision. In closing, we provided a chart that suggested the Master Scheduling method for each logical permutation of the four variables above.

Our intent was to provide sufficient background in these very important topics so that you can understand both the power and constraints of M&CRP, because the MPS and BOM form the foundation for M&CRP operation.

Selected Bibliography

Berry, William L., Thomas E. Vollmann, and D. Clay Whybark: *Master Production Scheduling, Principles and Practice,* APICS, Falls Church, VA, 1979.

Deis, Paul: *Production and Inventory Management in the Technological Age,* Prentice-Hall, Englewood Cliffs, NJ, 1983, Chap 4.

Erhorn, Craig R.: "Multiple Configuration Bills of Material," *APICS 1985 Conference Proceedings,* APICS, Falls Church, VA, 1985, pp. 275–277.

Fogarty, Donald W., J. H. Blackstone, and T. R. Hoffmann: *Production and Inventory Management,* South-Western, Cincinnati, OH, 1991, Chaps 2 and 4.

———, T. R. Hoffmann, and P. W. Stonebraker: *Production and Operations Management,* South-Western, Cincinnati, OH, 1989, Chap 7.

Garwood, Dave: *Bills of Material: Structured for Excellence,* Oliver Wright Ltd., Essex Junction, VT, 1988.

Gessner, Robert A.: *Master Production Scheduling,* Wiley, New York, 1985.

Lunn, Terry with Susan A. Neff: *MRP: Integrating Material Requirements Planning and Modern Business,* Irwin, Homewood, IL, 1992, Chaps 4 and 6.

Mather, Hal: *Bills of Material,* Dow Jones-Irwin, Homewood, IL, 1987.

———: "Which Comes First, the Bill of Material or the Master Production Schedule," *MCRP Reprints,* APICS, Falls Church, VA, 1991, pp. 190–193.

Orlicky, Joseph: *Material Requirements Planning,* McGraw-Hill, New York, 1975, Chaps 10 and 11.

———, George W. Plossl, and Oliver W. Wight: "Structuring the Bill of Material for MRP," *MCRP Reprints,* APICS, Falls Church, VA, 1991, pp. 52–29.

Plossl, George W.: *Production and Inventory Control: Principles and Techniques,* 2d ed., Prentice-Hall, Englewood Cliffs, NJ, 1985, Chap 7.

Schultz, Terry: *Business Requirements Planning: The Journey to Excellence,* The Forum Ltd., Milwaukee, WI, 1984, Chaps 5 and 6.

Smith, Spencer B.: *Computer Based Production and Inventory Control,* Prentice-Hall, Englewood Cliffs, NJ, 1989, Chap 7.

Schwendinger, James R.: "Modular Planning Bills: They Explode to the Bottom Line," *MCRP Reprints,* APICS, Falls Church, VA, 1991, pp. 203–206.

Vollmann, Thomas E., W. L. Berry, and D. C. Whybark: *Manufacturing Planning and Control Systems,* 3d ed., Irwin, Homewood, IL, 1992, Chap 6.

4
How MRP Works

The Basic Mechanics

Introduction

Material Requirements Planning (MRP) is a computer-based information system designed to order and schedule dependent demand inventories (raw materials, component parts, and subassemblies) in a coordinated manner. Material Requirements Planning is as much a philosophy as it is a technique, and as much an approach to scheduling as it is an approach to inventory control. It views inventory from the vantage point of the stockroom, trying to insure that there will always be "just enough" on hand to meet projected demand.

Until the 1970s, the materials planning process in a manufacturing environment suffered from two problems. The first was the enormous task of setting up schedules, keeping track of large numbers of parts and components, and coping with schedule and order changes. The second was the perception that a company had to choose between investing in high quantities of inventory or having excessive stock-outs. Practitioners used inventory planning techniques that were designed for independent demand items, resulting in high inventories *and* frequent stock-outs. Starting in the late 1960s and early 1970s, manufacturers recognized that planning dependent items differently from independent items (using MRP) could produce lower inventories *and* lower stock-out rates. Additionally, they enlisted the power of the computer to handle much of the burden of keeping records and determining material requirements.

The main purposes of an MRP system are to control inventory levels and assign operating priorities for ordered items. These may be briefly expanded as follows:

- Inventory
  Order the right part
  Order in the right quantity
  Order at the right time (start date)
- Priorities
  Order with the right due date
  Keep the due date valid

The *motto* of MRP is "getting the right materials to the right place at the right time." The operating *philosophy* of MRP is that materials should be expedited when their lack would delay the overall production schedule, and de-expedited when a schedule change postpones their need. To this end, MRP logic will always plan inventory to the lowest possible amount, unless instructed otherwise by order modifiers. Order modifiers, including Safety Stock and lot sizes are discussed later in this chapter.

Based on a Master Production Schedule derived from a Production Plan, an MRP system identifies what to order, how much to order, and when to order. MRP systems usually assume infinite capacity and use computer programs to handle the many detailed calculations.

This chapter assumes that you understand the distinction between independent and dependent demand, and the basics of Bills of Materials and Master Production Scheduling, which were discussed in Chapters 2 and 3 respectively. You should also understand that Material and Capacity Planning is the third level of the priority and capacity planning hierarchy described in Chapter 2. The focus of this chapter is on *dependent* demand, which is derived from the demand of a higher level product, and we discuss some of the details regarding how to accomplish the philosophy of integrated priority and capacity planning.

Material Requirements Planning Inputs

Figure 4-1 illustrates the five major sources of information required for MRP to operate:

1. A *Master Production Schedule,* which tells how many finished products are desired and when
2. A *Bill of Materials,* which tells what parts are required to make each assembly and finished product

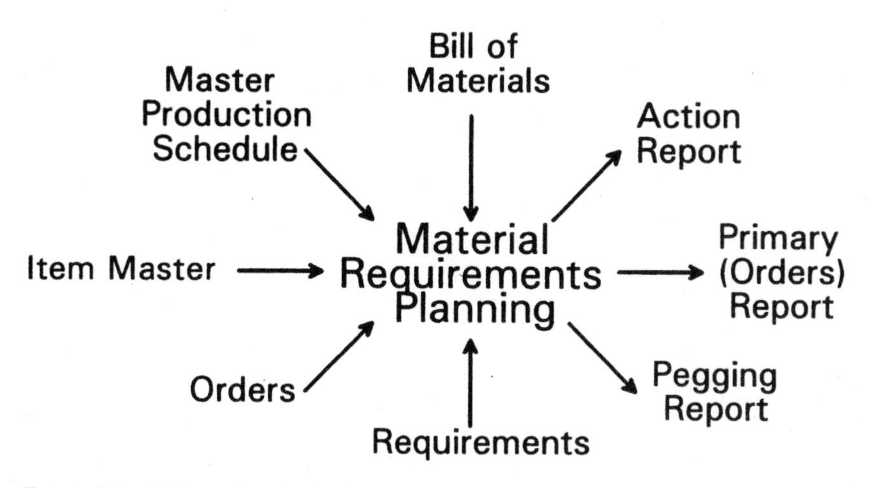

Figure 4-1. MRP inputs and outputs.

3. An *Item Master,* which tells how much inventory is on hand, the Lot Sizing factors, and ordering information for each part
4. *Orders,* which tells how much of each part is being acquired (will increase stock-on-hand in the future)
5. *Requirements,* which tells how much of each part is required (will decrease stock-on-hand in the future)

Each of these inputs are discussed in more detail in the following sections; the major outputs are discussed in Chapter 5.

Master Production Schedule

The *Master Production Schedule* states which end items (items that are sold to customers) are needed, in what quantities, on which specific dates, and when these items will be produced. The Master Production Schedule was discussed in detail in Chapter 3.

A Master Scheduler (a person) usually generates an initial Master Schedule in terms of what is needed and not what is possible, so it may not be feasible, given the limits of the production system. Moreover, what appears on the surface to be a feasible Master Schedule may not be so when end items are translated into detailed requirements of procurement, fabrication, and assembly. Unfortunately, the MRP system cannot distinguish between a feasible Master Schedule and an unfeasible one. Consequently, the Master Scheduler must sometimes process a proposed Master Schedule through MRP, in order to obtain a clearer picture of actual production requirements, which can then be compared to available capacity using the Capacity Requirements Planning technique described in Chapter 6.

The Bill of Materials (BOM)

A *Bill of Materials* (also called a product structure or parts list) is a list of all the materials, and the quantity of each, required to produce one unit of a manufactured product, or parent. MRP uses the bills of materials as the basis for calculating the amount of each raw material required for each time period. We have discussed Bills of Materials in some detail in Chapter 3.

In an MRP system, a *Bill of Materials File* is an up-to-date computerized file that contains a single record for each individual parent-component relationship. A Bill of Materials File resembles Figure 3-19 very closely, but with several additional fields. BOM files are discussed in greater detail in Chapter 8. MRP and MPS must assume that the BOMs they are using are indeed accurate and up-to-date, and that the BOMs represent the actual sequence of fabrication and assembly (e.g., that they are Manufacturing BOMs).

Item Master

In an MRP system, the *Item Master File* is a computerized file with a complete record for each item, or part. Because MRP systems are part-oriented, the Item Master File is the heart of the system. Each item, no matter at how many levels it is used in a product, or in many products, and no matter whether it currently has stock-on-hand or not, has one and only one record. The Item Master Record for a part contains many types of information, including: *static* data, such as part description, unit of measure, and MRP *planning factors* (lot sizes, lead times, safety stock, and scrap rates); plus *dynamic* data, such as various costs, current quantities on hand and on order. Chapter 8 contains more detail on the contents of a typical Item Master Record. An Item Master Record in a full-featured MRP II system might contain several hundred data elements for each part!

Requirements

A *requirement* is a computerized record of a future stockroom issue that will diminish stock-on-hand. There are two types of requirements:

1. *Internal,* which will be used within the plant to make other products
2. *External,* which will be sent outside the plant, such as customer orders, service parts, and forecasts. Note that a customer order is not really an

order in an MRP system, because it will not replenish the stock-on-hand in our inventory. This unfortunate terminology has confused practitioners for more than two decades, in spite of APICS' best attempts to revise the terminology.

A typical requirements record contains the item number of the part required, the quantity required, the date on which it is needed, and the quantity already issued from the stockroom. Customer orders also contain such additional information as the customer name and ship-to address, the date that the customer wants delivery, and the date we promised to ship. The record is explained in more detail in Chapter 8.

Orders

An *order* is a computerized record of a future stockroom receipt that will increase stock-on-hand. Just as there are two types of requirements, there are two types of orders:

1. Shop Orders (or Work Orders or Manufacturing Orders), which will be manufactured within our own plant. These are similar to our internal requirements, because they will be procured internally. These place a load on our plant and will appear in our CRP reports.
2. Purchase Orders, which will be procured from outside our plant. These are similar to our external requirements, because they will come into our plant from external sources. These place no load on our plant and will not appear in our CRP reports.

We can also categorize the incoming orders (both shop orders and purchase orders) in a different manner, which tells whether the order has been released, or whether it is still only planned. The categories are:

1. *Scheduled Receipt* (or Open Order, or Released Order), which is an order that has been officially released, either in the shop or to a supplier. A Scheduled Receipt commits our company to take action and spend money.
2. *Planned Order*, which exists only in the computer, and perhaps some printouts at this point. Our company has not yet been authorized to spend money; no supplier or shop has been authorized to start work on this order.

Planned Orders can become Scheduled Receipts *only* when a human expressly takes action. We are not aware of any MRP systems that perform this conversion automatically; this is one of the primary responsibilities of a materials planner.

An MRP order record contains considerable data, including item number being ordered, order quantity, original due date, actual received quantity, revised due date, quantities in MRB (Material Review Board) and scrap, supplier (if a purchase order), and other information. The record is explained in more detail in Chapter 8.

The Horizontal MRP Display

Because of its condensed graphical format, we will primarily use a horizontal MRP display, as shown in Figure 4-2. However, some practitioners utilize a vertical format, shown later in this chapter, when actually planning items.

The fields in the horizontal format include:

- *Item,* which is the item number of the item that we are planning.
- *LT,* which is the Lead Time of the item, in periods (days or weeks). Lead time is defined as the span of time required to perform an activity. This activity is normally the procurement of an item either from an outside supplier or from our manufacturing facility. An item's lead time can be comprised of any or all of the following: order preparation time; move or transportation time; manufacture, assembly or purchase time; receiving and inspection time; stocking time, and data transaction time. The item's lead time should include the total elapsed time from the day that MRP suggests an order until the item is ready for use.
- *OQ,* which is the Order Quantity of the item that tells MRP how many to order or which Lot-Sizing technique to use. We will use the simplest Lot-Sizing technique, called Lot-for-Lot (LFL) unless otherwise noted.

	PERIOD								
Item LT OQ SS	1	2	3	4	5	6	7	8	9
Gross Requirements									
Scheduled Receipts									
Projected Available									
Net Requirements									
Planned Order Receipts									
Planned Order Releases									

Figure 4-2. MRP horizontal display.

- *Safety Stock,* which is a planner-defined inventory target level to protect against fluctuations in demand and/or supply. When present, MRP plans to maintain stock levels at this level at all times.
- *Gross Requirements,* which is the total anticipated use, or stockroom withdrawals, during each time period. A given part can have Gross Requirements that include both dependent and independent demand. For example, if we build personal computers, we might also sell replacement disk drives, keyboards, and power supplies directly to the customer as replacement parts, as well as using them on our production lines for finished unit sales.
- *Scheduled Receipts,* which are replenishment orders that have already been placed, either in our own plant (shop orders) or at a supplier (purchase orders). These will increase the inventory on hand.
- *Projected Available,* which is the expected quantity in inventory at the *end* of the period, available for use in subsequent periods. This is calculated by adding the Scheduled Receipts and Planned Order Receipts for the current period to the Projected Available from the previous period, then subtracting the Gross Requirements for the current period. This equation assumes that all receipts will come into the stockroom before any requirements are issued. The larger the planning time periods, the greater the risk of this assumption.
- *Net Requirements,* which is the projected shortage for this period unless we take action. It is calculated by subtracting the Projected Available for the current period from the Safety Stock, which is specified for each part by a materials planner.
- *Planned Order Receipts,* which is the total quantity of items on replenishment orders (manufacturing orders and/or purchase orders) that MRP has planned to receive in a given period to cover the Net Requirements. With lot-for-lot ordering, the Planned Order Receipts in each period are always the same as the Net Requirements in the period. If Planned Orders are modified by a Lot-Sizing policy, the Planned Orders can exceed the Net Requirements. Any excess beyond the Net Requirements goes into Projected Available inventory for use in future periods.
- *Planned Order Releases,* which indicates the total quantity of items on Planned Orders to be placed (released) in a given period, so that the items will be available when needed. This is the Planned Order Receipts quantity, offset for lead time. Planned Order Releases for a manufactured part generate Gross Requirements for that part's direct components (children). When a material planner actually releases the Planned Order, the Planned Order Receipt and Planned Order Release quantities disappear, and that quantity appears in the Scheduled Receipts row.

Material Requirements Planning Logic

The MRP process is a combination of four very simple and logical processes, performed on each item in the database, one item at a time:

Determining Net Requirements, by period

Determining Planned Orders, by period

Lead time offsetting

Exploding Planned Orders

Each of these processes will be briefly described in this section and then combined in the next section to form the MRP process.

For easiest understanding, we begin by reviewing the simplest situation possible, On-Hand and Gross Requirements, and then add the following factors incrementally:

Scheduled Receipts

Planned Orders, with Requirements Explosion

Lot Sizing

Safety Stock

Multiple Products with Common Components

Scrap and Yield

Allocations

Remember that the tactical objective of MRP is to determine what items are required, how many of them are required, and when they are required. MRP's strategic objective is to minimize inventory without stocking out, ordering additional items only when all other avenues for filling demand have been exhausted.

On-Hand Quantity and Gross Requirements Only

We sell item *A*, a bicycle, to customers. Our marketing department forecasts that we will sell 25 units per week. Forecasts are requirements, because they will cause our stock-on-hand to decrease. We have 250 units physically on-hand. Figure 4-3 shows how MRP plans item *A*, starting from the first period, and ending when it completes the 9th period. To keep this example as simple as possible, we assume there is no Safety Stock planned, and use only those rows of the horizontal format that apply to this example. The MRP display shows that we start with 250 units in inventory, shown in the chart

		PERIOD								
Item A - Bicycle	SS:0	1	2	3	4	5	6	7	8	9
Gross Requirements		25	25	25	25	25	25	25	25	25
Projected Available	250	225	200	175	150	125	100	75	50	25

Figure 4-3. MRP display showing on-hand quantity and Gross Requirements only.

on the Projected Available line before period 1. During period 1 we have orders to sell 25, leaving us with a balance of 225 units. We repeat this process for eight more weeks, finishing with a projected ending inventory of 25 units.

Scheduled Receipts

To the very simple example included above as Figure 4-3, let us add Scheduled Receipts (also called released or open orders). These can be for either purchase or manufacturing orders. We will give these Scheduled Receipts their own row, right under Gross Requirements. Scheduled Receipts are added to the Projected Available from the previous period, because they will increase projected stock-on-hand. In the example shown in Figure 4-4, we have two Scheduled Receipts, one in period 1 and one in period 3, each for a quantity of 50.

Thus, the Projected Available at the end of period 1 is:

Projected Available from previous period	250
Minus Gross Requirements	–25
Plus Scheduled Receipts	50
Projected Available at end of period	275

The Projected Available at the end of each succeeding period is calculated using the same formula.

Rescheduling Scheduled Receipts

Figure 4-4 indicates that we have plenty of inventory to cover all our requirements. In fact, we have *too much* inventory. Remember that MRP's strategic objective is to minimize inventory while avoiding stock-outs. An MRP system will suggest that the two Scheduled Receipts be *canceled,* reducing the cash drain on the company. The actual rescheduling of the order would be done by the planner or planner/buyer.

Let us vary this example by reducing the starting on-hand quantity from 250 to 150 as indicated in Figure 4-5, to show how MRP would suggest *rescheduling out, or de-expediting* rather than canceling. MRP would suggest

Item A - Bicycle	SS:0	1	2	3	4	5	6	7	8	9
Gross Requirements		25	25	25	25	25	25	25	25	25
Scheduled Receipts		50		50						
Projected Available	250	275	250	275	250	225	200	175	150	125

Figure 4-4. MRP display showing Gross Requirements, Scheduled Receipts, and Projected Available.

moving the two Scheduled Receipts to periods 7 and 9, as shown by the italics.

Notice that MRP permits the Projected Available to reach zero, but not fall below it. Using the numbers from Figure 4-5 as a base, if our Gross Requirements change in several periods (owing to any number of reasons, including sales promotions, new dealers opening up, or changes in forecasts), MRP will then request that the Scheduled Receipts that are now in periods 7 and 9 be *rescheduled in,* or *expedited,* to periods 5 and 8, as shown in Figure 4-6.

To summarize, MRP suggests that Scheduled Receipts be rescheduled out or in to minimize inventory and avoid a stock-out.

Planned Orders for Purchased Items

We have shown how MRP calculates projected inventory balances period by period, using stock-on-hand, Gross Requirements, and Scheduled Receipts. In this next example, we will show what happens when stock-on-hand and Scheduled Receipts cannot fill all the Gross Requirements. This situation causes MRP to create planned replenishment orders, called *Planned Orders.*

MRP creates a Planned Order for each time period in which the Projected Available falls below the set Safety Stock amount and there are no Scheduled Receipts available to reschedule. Equation (4-2) defines Net Requirements as the amount required to bring Projected Available back up to Safety Stock level. In the most simple examples, the Planned Order quantity is the same as the Net Requirements. We will explain the reasons for the differences later under the heading Lot Sizing. The due date of the Planned Order is the period when Projected Available falls below Safety Stock level; the start date is the due date minus the lead time for the item.

Item A - Bicycle	SS:0	1	2	3	4	5	6	7	8	9
Gross Requirements		25	25	25	25	25	25	25	25	25
Scheduled Receipts								*50*		*50*
Projected Available	150	125	100	75	50	25	0	25	0	25

Figure 4-5. MRP display showing Scheduled Receipts being rescheduled.

PERIOD									
Item A - Bicycle SS:0	1	2	3	4	5	6	7	8	9
Gross Requirements	25	25	35	35	35	25	20	25	25
Scheduled Receipts					*50*			*50*	
Projected Available 150	125	100	65	30	45	20	0	25	0

Figure 4-6. MRP display showing Scheduled Receipts being expedited.

To show this in our horizontal display format, we add three rows: (1) Net Requirements, (2) Planned Order Receipts, and (3) Planned Order Releases. Because purchased items are easier to plan than manufactured items, we will explain purchased items in this section, and manufactured items in the next section. To create a situation in which stock-on-hand and Scheduled Receipts combined cannot fill all the Gross Requirements, let us change the starting stock-on-hand from 150 to 25, as shown in Figure 4-7. Note that MRP reschedules the Scheduled Receipts in again, because it meets demand with available resources before committing additional resources.

Determining Net Requirements, by Period

The basic objective of MRP is to have as little inventory as possible on hand, but still meet all requirements. In this respect, MRP is in total agreement with Just-in-Time. MRP uses the Safety Stock level for a given part, determined by the material planner, as the level at which it will maintain inventory.

MRP *plans* a part, period by period, starting with period 1. MRP first compares the current on-hand inventory to Safety Stock. Assuming that the current on-hand inventory is greater than or equal to Safety Stock, MRP then subtracts the Gross Requirements in period 1 and compares the resulting Projected Available to Safety Stock. If the Projected Available is greater than

PERIOD									
Item A - Bicycle SS=0; LT=2	1	2	3	4	5	6	7	8	9
Gross Requirements	25	25	35	35	35	25	20	25	25
Scheduled Receipts		*50*	*50*		0				
Projected Available 25	0	25	40	5	0	0	0	0	0
Net Requirements					30	25	20	25	25
Planned Order Receipts					30	25	20	25	25
Planned Order Releases			30	25	20	25	25		

Figure 4-7. Standard six-line MRP display.

or equal to Safety Stock, MRP repeats the process in period 2, then period 3, until there are no more periods to process for this part. This calculation can be expressed by Equation (4-1). If, during this process, MRP discovers Scheduled Receipts, it suggests rescheduling them to the period when they are needed to keep Projected Available from falling below Safety Stock.

$$\text{Projected Available at end of current period} = \text{Projected Available at end of previous period} + \text{Scheduled Receipts for current period} + \text{Planned Order Receipts for current period} - \text{Gross Requirements for Current period} \quad (4\text{-}1)$$

When the Projected Available falls below Safety Stock, MRP calculates the Net Requirements for this part in this period, as expressed in Equation (4-2).

$$\text{Net Requirements} = \text{Gross Requirements} + \text{Safety Stock} - \text{Scheduled Receipts} - \text{Projected Available at end of previous period} \quad (4\text{-}2)$$

MRP plans to fill the Gross Requirements (or projected shortage) by using existing orders before it suggests any new ones. In this respect, MRP is designed to conserve the resources of the company (and help controllers sleep well at night). It prints an action message which suggests that the material planner should reschedule the due date of Scheduled Receipts to the time period where the Net Requirements exist. MRP continues its calculations, assuming that the planner will indeed reschedule the Scheduled Receipt. In practice, however, planners actually reschedule far too few Scheduled Receipts, which causes the MRP plan to be unrealistic.

The calculations for periods 1 through 4 are the same as before: Starting with the Projected Available at the end of the previous period, subtract the Gross Requirements for this period, add the Scheduled Receipts for this period, and finish with the Projected Available for this period. We assume Safety Stock is zero and the lead time is two periods. However, in period 5, there is a Net Requirement of 30, and no more Scheduled Receipts or original stock-on-hand to use.

Determining Planned Orders, by Period

When MRP has no more Scheduled Receipts available to use, it turns to its last available tool, the Planned Order. In its simplest form, a Planned Order

Receipt is for the same quantity as the Net Requirements, due in the same time period as the Net Requirements. This obvious simplicity gives MRP tremendous power: it plans to receive material only when it is really needed, and in the minimum quantity which is really needed. This is very different from the Order Point systems discussed in Chapter 2.

A Planned Order Receipt answers the three basic questions of production control:

1. Which part(s) do I need? A Planned Order Receipt is for a specific part, and even answers the unasked question, "Why do I need this?" by allowing the planner to see the reasons for the requirement.
2. How many do I need? A Planned Order Receipt states a specific quantity.
3. When should I order them? A Planned Order Receipt defines the need date as the date on which stock-on-hand will fall below Safety Stock level. It defines the order date by subtracting the lead time from the need date, which we explain more fully in the section entitled Lead Time Offsetting.

In period 5 of Figure 4-7, the calculations add some more factors. In period 5, Projected Available becomes a –30, which is clearly unacceptable if we want to ship products to our customers in a timely manner. This causes the Net Requirements to also be 30. MRP knows that it has already used up all existing stock-on-hand and all existing Scheduled Receipts. It therefore must plan a new order, called a Planned Order. Because the Net Requirements is 30, MRP plans a Planned Order for 30, and makes it due for the period that had the Net Requirements (Period 5). A materials planner has already told MRP that the lead time for part *A* is 2 periods. So MRP subtracts two periods from the due date of period 5, creating a start date of period 3 for this Planned Order. MRP then adds the Planned Order Receipt quantity back into the Projected Available quantity for period 5, which brings the Projected Available quantity back to zero. MRP repeats this same process for periods 6 through 9. To recap the revised formula, in period 5 we computed as follows:

Projected Available from previous period	5
Minus Gross Requirements	–35
Plus Scheduled Receipts	0
Equals Initial Projected Available at end of period	–30
Plus Planned Order Receipts	30
Projected Available at end of period	0

Perhaps a simple personal example will help. Planned Order Receipts is a feature we would like to have with our checking accounts. It works like this: we can have as much money deposited into our checking accounts as we need, as long as we give our mysterious benefactor sufficient time to make the deposit (the lead time for the part). However, we cannot order more than we really need. This will cause our projected checking account balance to fall to zero, which is somewhat uncomfortable, but much better than negative.

Finally, one subtle but important point about Planned Orders must be made. The Planned Order is a *suggestion* by the MRP software that a real order be placed in that period by the planner. The MRP software will calculate the related components (children) required and make additional suggestions in the form of additional Planned Orders, but the Planned Order still exists only in the MRP logic. This suggestion is based upon the current set of circumstances. The same set of suggestions may or may not appear the next time MRP is processed.

Lead Time Offsetting

When MRP creates a new Planned Order Receipt, the item number, quantity required, and due date, are known. What is not known is the required start of the Planned Order. Timing the release of the Planned Orders to cover existing or future requirements is called offsetting for lead time, or simply, *lead time offset.* MRP computes the start date by using a process called *backward scheduling.*

Backward scheduling is the practice of computing the start date of an order by subtracting the order lead time from the due date, so that its completion occurs when needed to fill the Net Requirements. We do this all the time in our personal lives. To be at work at 8 a.m., we subtract the travel time (45 minutes) to compute the departure time (7:15 a.m.). Thus, if a Planned Order Receipt is due in period 5, and item *A* (the item MRP is planning) has a lead time of two periods, the Planned Order Release date would be 5 – 2 = period 3. Period 3 is the latest time period during which the order can be started and still be available in time to avert the projected shortage. So MRP writes the order quantity in period 3 in the Planned Order Release row.

If the Planned Order is for a purchased part, MRP is finished with this period and starts planning the next period. If, however, the part we are planning is manufactured or assembled, MRP displays another of its strengths by calculating the components that will be needed for that Planned Order. This is called "Exploding" the order; we explain it in greater detail in a following section.

Planned Orders for Manufactured Items

The MRP display for manufactured items looks absolutely identical to that for purchased items. So where is the difference? The difference is what MRP does with the Planned Order Releases. For purchased items, MRP assumes that a buyer or buyer-planner will create purchase orders, or releases against a blanket, and that our company has no obligation to worry about any components of those purchased parts. However, MRP assumes that our company will make its manufactured parts in-house, furnishing the components from our own stockrooms. Therefore, we must indeed focus on the components of manufactured parts.

Exploding Planned Orders. If our company is going to make 30 of item *A*, with a due date of period 5, we realize the shop order must begin in period 3. We also realize that we require 30 sets of components when we start the shop order. MRP can help us ensure that the components will be available when we need them in period 3. MRP does this by *exploding* the shop order into the components that will be used. This logic assumes that we will use the components in the shop order on that date, whether or not the sales for that end item ever materialize. As shown in Figure 4-8, if we are going to make 30 bicycles starting in period 3, we will need 30 frames and 60 wheels. Whether the bicycles will eventually sell in period 5, 6, and 7 is up to the marketing and sales group. Production's job is to build the bicycles; materials management's job is to have the proper components available on schedule (and no more components than absolutely necessary, according to our friends in finance and accounting) so production can build the bicycles.

Although this might look confusing at first, the calculations are quite simple. The Gross Requirement for a component is the Planned Order Release Quantity of the parent, times the Quantity Per (or usage) in the Bill of Materials. Again, to make 30 bicycles, we need 30 frames and 60 wheels.

Note that *at this time MRP does* not *check to see if the required material will be available for any component.* MRP instead *assumes* that it can plan to have enough material when it plans the component items, *B* and *C.* Thus, MRP is a "can-do" system; you can tell it what you want to ship, and it will tell you all the components required, by quantity and date.

One other concept is vital. When MRP explodes the planned shop order, it checks the effective dates in the BOM, and creates Gross Requirements for *only* those components that are *effective* on the planned shop order *start date.* Let us illustrate this with a simple example, with the help of Figure 4-9.

We started building the bicycle on April 15, 1988, with two wheels and a steel-tube frame assembly, comprised of the steel-tube frame and a seat. To

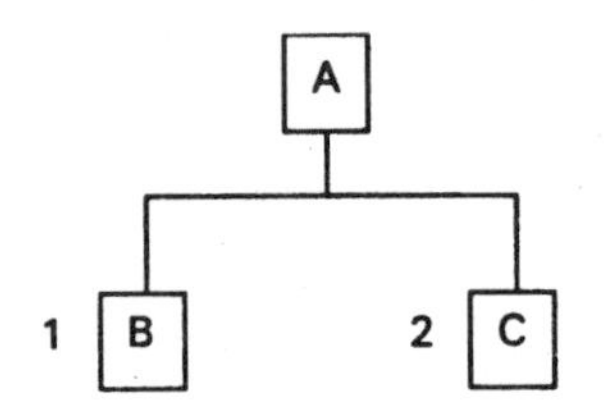

PERIOD

Item A - Bicycle SS:0; LT:2		1	2	3	4	5	6	7	8	9
Gross Requirements		25	25	35	35	35	25	20	25	25
Scheduled Receipts			*50*	*50*		0				
Projected Available	25	0	25	40	5	0	0	0	0	0
Net Requirements						30	25	20	25	25
Planned Order Receipts						30	25	20	25	25
Planned Order Releases				30	25	20	25	25		

x2 x1

PERIOD

Item B - Frame	SS:0	1	2	3	4	5	6	7	8	9
Gross Requirements				30	25	20	25	25		

PERIOD

Item C - Wheel	SS:0	1	2	3	4	5	6	7	8	9
Gross Requirements				60	50	40	50	50		

Figure 4-8. MRP display showing Planned Order Releases for a parent exploding into Gross Requirements of components.

keep our customers delighted, we have wanted for years to change to a titanium frame, because it is much lighter and does not rust. The breakthrough came when, in December 1992, Purchasing found a source that was not much more expensive than our current frames. We agreed on January 11 to change the BOM, so the engineer entered all the changes into the BOM at that time. We can have the new frames in-house on March 17, 1993.

Reviewing the concept of BOM effectivity that we discussed in Chapter 3, the engineer enters the changes into the BOM for the bike as follows:

- Create two new part numbers, C1 and D1.
- Create one single-level BOM relationships for C1 and each of its components, D1 and E, with immediate effectivity. This means that whenever we build C1, we will use D1 and E.
- Create a new BOM relationship between A and C1, with a starting effective date of March 17 (based on Purchasing's assurance of supply).

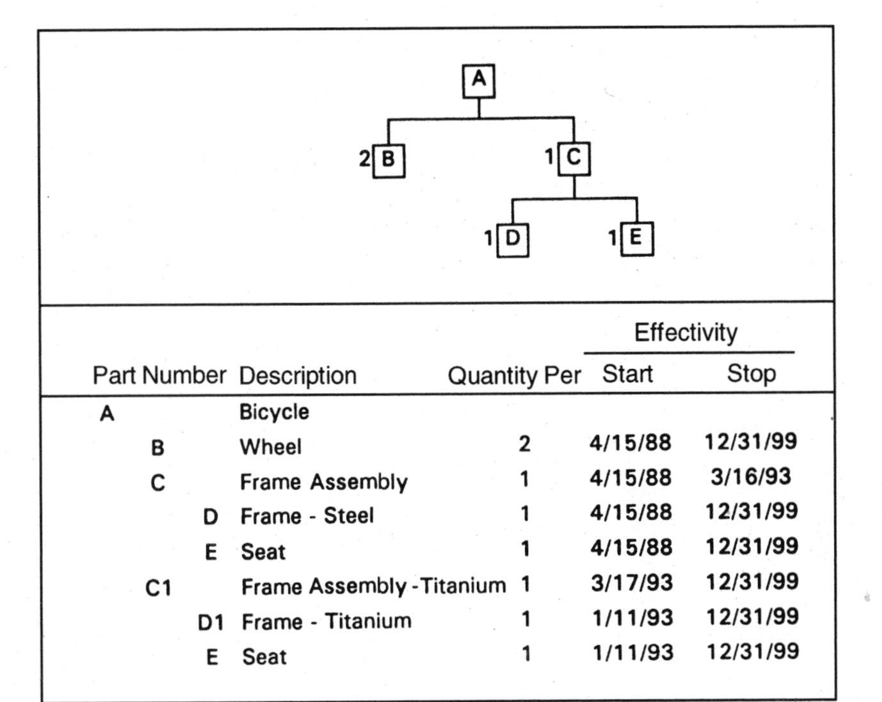

Part Number			Description	Quantity Per	Effectivity Start	Effectivity Stop
A			Bicycle			
	B		Wheel	2	4/15/88	12/31/99
	C		Frame Assembly	1	4/15/88	3/16/93
		D	Frame - Steel	1	4/15/88	12/31/99
		E	Seat	1	4/15/88	12/31/99
	C1		Frame Assembly-Titanium	1	3/17/93	12/31/99
		D1	Frame - Titanium	1	1/11/93	12/31/99
		E	Seat	1	1/11/93	12/31/99

Figure 4-9. MRP and BOM effectivity.

- Change the Effectivity Stop date for the A-C relationship, from 12/31/99 to 3/16/93.

When MRP runs in January, February, and the first half of March, it will continue to plan requirements to use the steel frame in our bike in all Planned Order Releases with start dates that are less than or equal to March 16. For all Planned Order Releases with start dates equal to or greater than March 17, MRP will plan to use the new titanium frame. For example, during the weekly MRP run on January 16, MRP plans a Planned Order Release for *A*, with a planned start date of March 16. Based on the effective dates in the BOM, MRP explodes into the old steel frame. MRP then plans a second Planned Order Release, with a planned start date of March 17. It explodes into the new titanium frame, again based on the BOM effective dates.

Lot Sizing

In this next example, we introduce the basics of Lot Sizing. We discuss Lot Sizing in greater detail in Chapter 5. Instead of Lot-for-Lot (LFL) Lot Sizing, Figure 4-10 shows a fixed Lot Size (OQ) of 50 for *A*. In our example, we

	PERIOD								
Item A - Bicycle; OQ=50	1	2	3	4	5	6	7	8	9
Gross Requirements	25	25	35	35	35	25	20	25	25
Scheduled Receipts		*50*	*50*		0				
Projected Available 25	0	25	40	5	20	45	25	0	0
Net Requirements					30	5			25
Planned Order Receipts					50	50			50
Planned Order Releases			50	50			50		

	PERIOD								
Item B - Frame	1	2	3	4	5	6	7	8	9
Gross Requirements			50	50			50		

	PERIOD								
Item C - Wheel	1	2	3	4	5	6	7	8	9
Gross Requirements			100	100			100		

Figure 4-10. MRP display showing Lot Size.

will convert the Net Requirements into Planned Order Receipts by using the Lot Size (OQ) and the following rule:

$$\text{Planned Order Release (Quantity)} = \text{the greater of Net Requirements or Lot Size (OQ)}$$

We can then compute the Projected Available to complete the calculations for the period by using Equation (4-2): subtract the Gross Requirements from the Scheduled Receipts, plus the Planned Order Receipts, plus the Projected Available from the previous time period.

Note that although the total quantity ordered stays essentially the same, the Lot Size rule causes more inventory to be on hand for the parent (increasing the overhead expenses associated with carrying inventory), and in addition, causes the component requirements to be concentrated in earlier periods. Additionally, larger Lot Sizes cause lumpy demand on the components. Companies that implement JIT deliberately reduce Lot Sizes, with the ultimate objective of having Lot Sizes of one. If we were to enter a Lot Size of 5000 for the bicycle, part *A*, MRP would assume that we knew what we were doing, accept the change, and respond by ordering enough components to make 5000 bicycles!! MRP is not designed to ask the very important question: "Are you out of your mind?"

Safety Stock

Ideally, Safety Stock should only be used at the end-item level, to cushion against uncertain customer demand, and at the lowest raw-material level if suppliers are not reliable. However, many companies who have implemented MRP have applied Safety Stock at many intermediate levels as well, because they are not yet comfortable in removing it. Safety Stock tends to be a permanent investment in inventory that does not turn over. One of the fastest ways to cut inventory levels is to reduce Safety Stocks.

For our example, we use the same MRP display of Figure 4-10, and add a Safety Stock level of 25 to create Figure 4-11. This means that MRP will now plan to maintain the Projected Available at 25 units.

By just adding a modest Safety Stock, we have caused MRP to pull the Planned Orders in by one week each, expedite the requirements on each component, and increase inventory. Safety Stock does not change the formulas used or the original requirements. In this case, it changes the suggested timing of MRP orders.

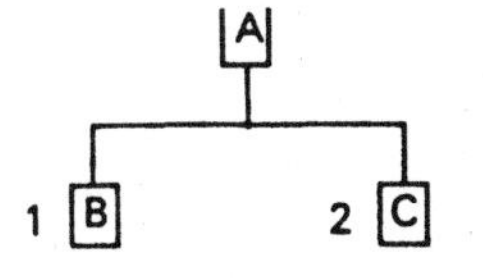

	PERIOD								
Item A - Bicycle; OQ=50; SS=25	1	2	3	4	5	6	7	8	9
Gross Requirements	25	25	35	35	35	25	20	25	25
Scheduled Receipts	*50*		*50*						
Projected Available 25	50	25	40	55	70	45	25	50	25
Net Requirements				20	5			25	
Planned Order Receipts				50	50			50	
Planned Order Releases		50	50			50			

	PERIOD								
Item B - Frame	1	2	3	4	5	6	7	8	9
Gross Requirements		50	50			50			

	PERIOD								
Item C - Wheel	1	2	3	4	5	6	7	8	9
Gross Requirements		100	100			100			

Figure 4-11. MRP display showing Safety Stock.

Summary of MRP Process

The MRP process can be summarized using two different views. The first is a macro, or overall, level, which shows an entire product planning cycle. The second is a micro, or more detailed, view of how the MRP logic works for one single part.

Macro Level. Figure 4-12 shows the entire processing cycle for product *A* and its components and their components. Time moves forward from left to right; "today" is March 1, which is at the leftmost margin of the chart. To simplify this example, our entire inventory has only the five parts shown. Our only stock-on-hand is for the raw materials, *D* and *E*, because we are a Make-to-Order shop.

When we run the MRP program "today," MRP plans a Planned Order for 10 *A*s, due on the specified ship date, March 18. MRP backs off two working days (the lead time to make *A*) to create the start date of the Planned Order, March 16. Because *A* is manufactured, MRP then explodes the Planned Order for *A* to create Gross Requirements for 20 *B*s and 10 *C*s, each due on the start date of the Planned Order for *A*. MRP has now finished the planning process for all parts in low-level code 0.

MRP then performs the gross-to-net calculations on *B* and *C*. It can do

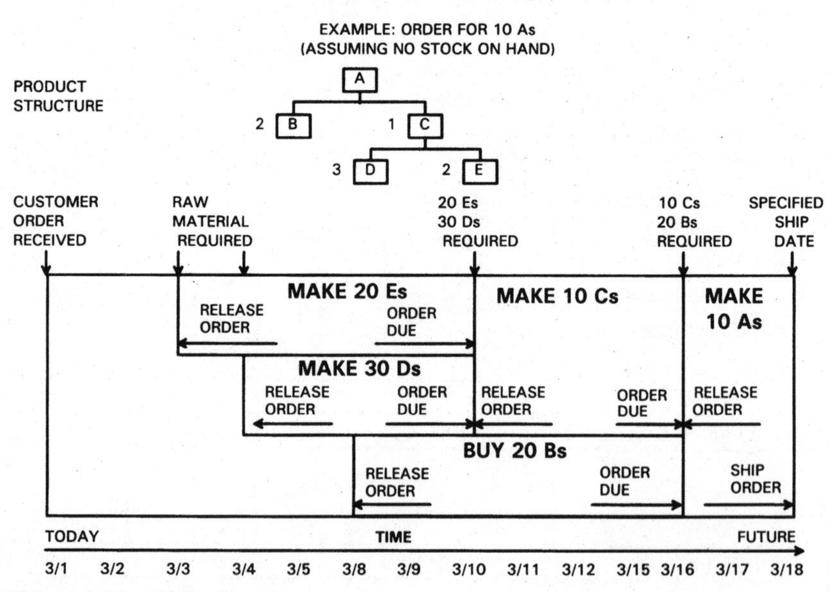

Figure 4-12. MRP processing summary.

either *B* or *C* first; they are both at the same low-level code 1. MRP processes *B* in the same way that it did *A*, creating a Planned Order for 20 due on March 16, and starting on March 8. Because *B* is purchased, MRP does not explode the Planned Order into any components. MRP processes *C* in a similar manner, creating a Planned Order for 10 due also on March 16 (the same due date as the order for *B*), but with a later start date of March 10, because the lead time for *C* is only four work days. MRP explodes the order for the 10 *Cs* into the effective components, *D* and *E*. MRP has now finished the planning process for all parts in low-level code 1.

MRP plans the two remaining parts, *D* and *E*, in the same manner. MRP creates the Gross Requirements for the raw materials as it plans *D* and *E*. Because we will have adequate raw materials on hand to build *D* and *E*, we do not need to create purchase orders to acquire the raw materials.

In the example, as in the real world, time marches on. It is easier to understand if you place a ruler on the left edge of the page (so that it goes from the top to the bottom of the page), then start moving the ruler slowly to the right. After a bit, the ruler should cross March 2, then March 3. On March 3, the material planner releases the shop order to make the 20 *Es*. One day later, right on schedule, the planner releases the shop order to make the 30 *Ds*. Two working days later (with a weekend in between), the buyer/planner releases the purchase order to buy the 20 *Bs*. So far, we have committed the minimum possible company resources to make the shipment of *As* exactly on schedule.

As the poet predicted, "the best-laid plans of mice and men oft go awry." On March 9, the customer calls to change the order ship date to March 25. When we run MRP that night, MRP suggests that we reschedule the due date on the *Ds* and *Es* to March 17, and the due date on the *Bs* to March 23. MRP plans the Planned Order Release for the 10 *Cs* to start on March 17 and be due on March 23. Our planner and buyer/planner change the scheduled receipts as MRP suggests.

Micro Level. MPS runs first, creating total Gross Requirements by period for the MRP items. If MRP encounters an MPS item, MRP does not perform grossing and netting, or revise the projected availability. MRP merely explodes the MPS Planned Orders into their appropriate components.

MRP processes items by starting with the end items at the top of the product structure, called Low-Level Code 0, and working down the low-level codes one at a time until all parts have been processed. Figure 4-13 shows the steps that MRP follows to plan, or process, a single item.

- For each MPS part, MRP explodes the MPS orders into the components.
- For each MRP-planned part, MRP calculates the Projected Available balances and Net Requirements for each period, starting with period 1. If

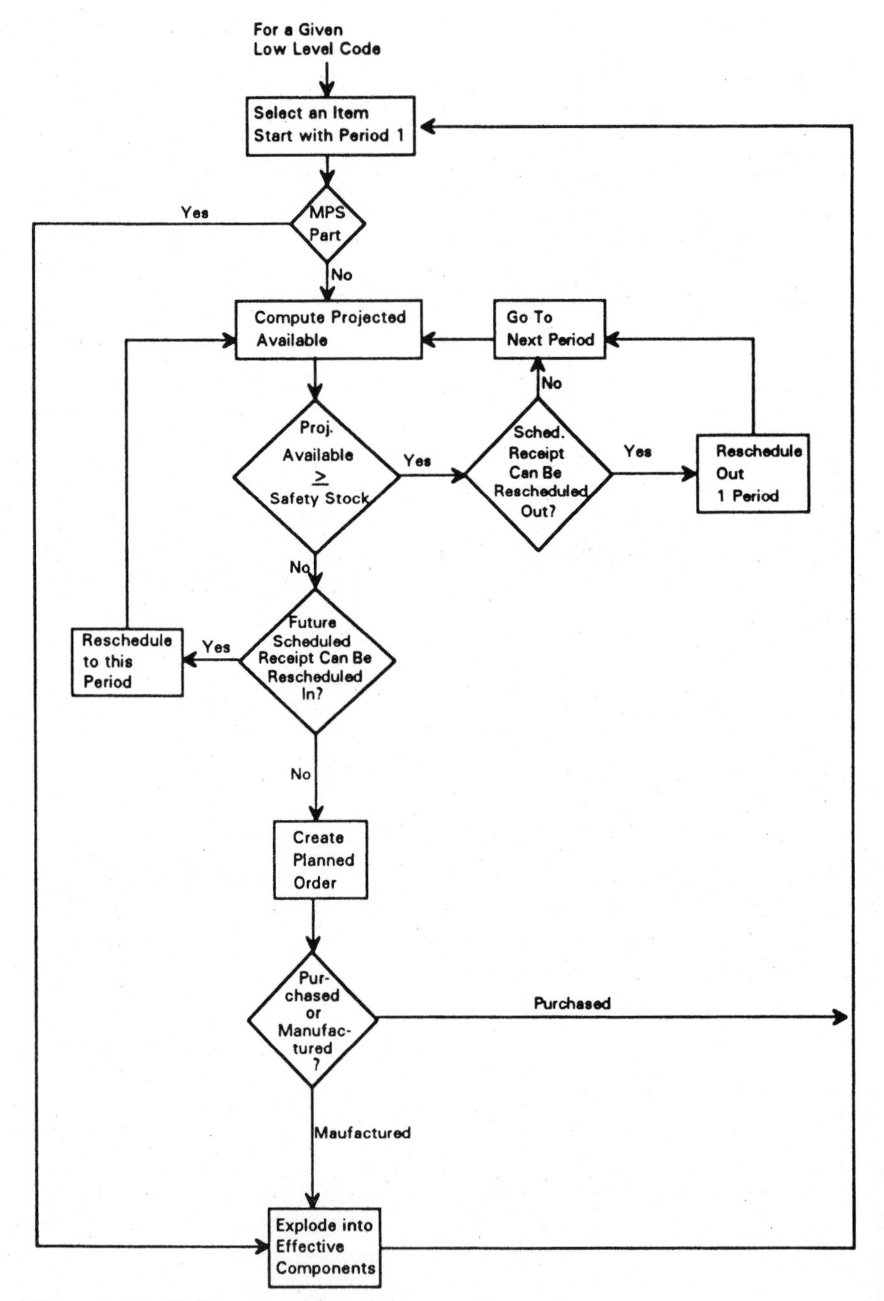

Figure 4-13. MRP processing flow chart.

Scheduled Receipts are in the wrong period, MRP suggests rescheduling them to the period of need, where they will prevent Projected Available from falling below Safety Stock level. To compute the Net Requirements, MRP compares the Projected Available balance at the end of the period to the Safety Stock level. If the Projected Available balance is less than the Safety Stock level, the difference is the Net Requirement.

- MRP plans an order to cover the Net Requirement in each period, using the Lot-Sizing technique specified by the planner for each part. MRP calculates the start date for each order by subtracting the lead time from its due date.
- For manufactured parts, MRP explodes each Planned Order into its components that are effective on the parent's Planned Order start date, using the quantity specified in the Bill of Materials. These dependent requirements form the bulk of the Gross Requirements for lower-level components. (The other Gross Requirements come from external, or independent, demand.)
- MRP plans (e.g., performs projected availability and gross-to-net calculations) each item only when all the Gross Requirements for that item have been computed, by using the item's low-level code.
- This process is continued, level-by-level, until all items controlled by MRP have been scheduled.

MRP for Multiple Products with Common Components

To show how MRP works when one component is used in more than one parent, we will use the same component *C* in two parents, *A* and *W*, as shown in Figure 4-14. *C* is still a wheel; *A* is our trusty bicycle; *W* is a unicycle. Usage quantities are shown in parentheses. For simplicity, all items use Lot-for-Lot lot sizing and zero Safety Stock.

To clarify which parent caused the Gross Requirements for component *C*, we will put all data for parent *W* in italics. For this illustration, we will only show the MRP displays for *A*, *W*, and *C*; the other components would be planned in a like manner.

In Figure 4-14 we added two extra lines for the Gross Requirements for *C*, so that the source of each Gross Requirement would be obvious. MRP programs do not have a separate line for each Gross Requirement, because the number of parents of a commonly used component, such as a fastener, is unlimited. Instead, bucketed MRP reports summarize all Gross Requirements into one line. Bucketless MRP reports use one line for each requirement, as we show in a following section.

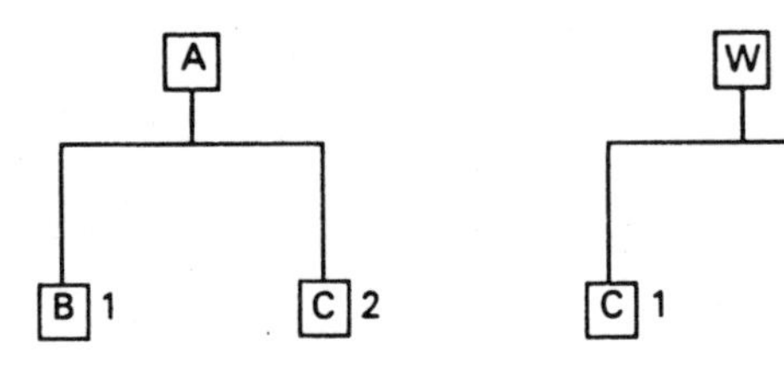

PERIOD

Item A - Bicycle LT 2	1	2	3	4	5	6	7	8	9
Gross Requirements	25	25	35	35	35	25	20	25	25
Scheduled Receipts		50	50		0				
Projected Available 25	0	25	40	5	0	0	0	0	0
Net Requirements					30	25	20	25	25
Planned Order Receipts					30	25	20	25	25
Planned Order Releases			30	25	20	25	25		

PERIOD

Item W - Unicycle LT 1	1	2	3	4	5	6	7	8	9
Gross Requirements	*10*	*10*	*10*	*10*	*10*	*10*	*10*	*10*	*10*
Scheduled Receipts			*20*						
Projected Available *25*	*15*	*5*	*15*	*5*	*0*	*0*	*0*	*0*	*0*
Net Requirements					*5*	*10*	*10*	*10*	*10*
Planned Order Receipts					*5*	*10*	*10*	*10*	*10*
Planned Order Releases				*5*	*10*	*10*	*10*	*10*	

PERIOD

Item C - Wheel	1	2	3	4	5	6	7	8	9
Gross Req'ts from A			60	50	40	50	50		
Gross Req'ts from W				*5*	*10*	*10*	*10*	*10*	
Total Gross Requirements			60	55	50	60	60	10	

Figure 4-14. MRP plan for two products.

Scrap and Yield

Where scrap is likely to result from a production process, or where the yield of a process is less than 100 percent, practitioners have traditionally increased the quantity of materials started through the process, in order to have enough good items at the end. MRP will automatically increase the Planned Order Release by the necessary amount. JIT, on the other hand, encourages the practitioner to improve the process in question, in order to eventually eliminate scrap or yield loss.

The scrap or yield allowance is applied to the Planned Order Releases, rather than to the Gross Requirements, because scrap predicts loss during manufacturing (e.g., a Planned Order), rather than storage. Gross Requirements can be filled from stock; they do not necessarily require Planned Orders.

Scrap calculations are somewhat deceptive. The formula for computing the Planned Order Release quantity is:

$$\text{Planned Order Release qty} = \frac{\text{Planned Order Receipt Quantity}}{(1.00 - \text{Scrap Rate})} \qquad (4\text{-}3)$$

Figure 4-15 shows item *A*, our bicycle, with a scrap allowance of 10 percent, which is applied by dividing the Planned Order Receipt quantity by (1.00 – .10, or .90). To understand this formula, try a scrap rate of 50 percent rather than 10 percent. With a scrap rate of 50 percent, we have to start twice as many items as we need, not just 50 percent more.

Scrap or yield adjustments can take many forms. A simple percentage is usually adequate, but sometimes rather complex formulas are needed. Also, if each setup produces a constant number of units of scrap, because of setup or destructive testing, then it would be best to use this constant per setup rather than a percentage. (However, most MRP packages do not yet allow for a constant scrap quantity per operation.)

Yield is the reverse of scrap. Yield is normally a very high percentage, rather than a low percentage like scrap, because yield represents the percentage of units that are good. Mathematically,

$$\text{Yield} = 1 - \text{scrap percentage} \qquad (4\text{-}4)$$

Allocations

Because MRP systems assume that all components are issued immediately after an order is released, they only explode Planned Orders. However, when a planner releases a Planned Order to become a Scheduled Receipt

		PERIOD								
Item A - Bicycle	Scrap 10%	1	2	3	4	5	6	7	8	9
Gross Requirements		25	25	35	35	35	25	20	25	25
Scheduled Receipts			50	50		0				
Projected Available	25	0	25	40	5	0	0	0	0	0
Net Requirements						30	25	20	25	25
Planned Order Receipts						30	25	20	25	25
Planned Order Releases				33	28	22	28	28		

Figure 4-15. MRP showing with a scrap allowance of 10 percent.

just prior to an MRP run, all components will probably *not* be issued from stock by the time MRP runs. The MRP program would then underestimate the requirements for the components, by the amount not yet issued, increasing the probability of material shortages. Bucketed MRP systems, therefore, track unissued released requirements, called "Allocations." When a planner releases an order, the MRP system changes each component requirement record to "Allocated," to show that the parent shop order has been released to the floor. MRP summarizes the total of all Allocations for a part and places them in the first period (or subtracts them from on-hand inventory, depending on the individual software package), because it assumes that the stock clerk is pulling the components even while MRP is running.

Low-Level Codes

Most companies have components that are common to several different Bills of Material, as well as being in various levels within Bills of Material. A Low-Level Code identifies the lowest level in any Bill of Material throughout the database at which the particular item appears, as illustrated in Figure 4-16. Although the convention is confusing to humans, the higher a part in the data structure, the smaller its low-level code. Final shippable end items that have no parents have a low-level code of zero.

In Figure 4-16, item *D* appears at three different levels in product *A*. The "Not Coded" BOM on the left side of Figure 4-16 has item *D* appearing at the three logical levels, while the "Low-Level Coded" BOM on the right side of Figure 4-16 shows all the uses of item *D* at its lowest level, level 4. In fact, the low-level code of *D* is 4.

Low-level coding is critical to the MRP software, to prevent MRP from

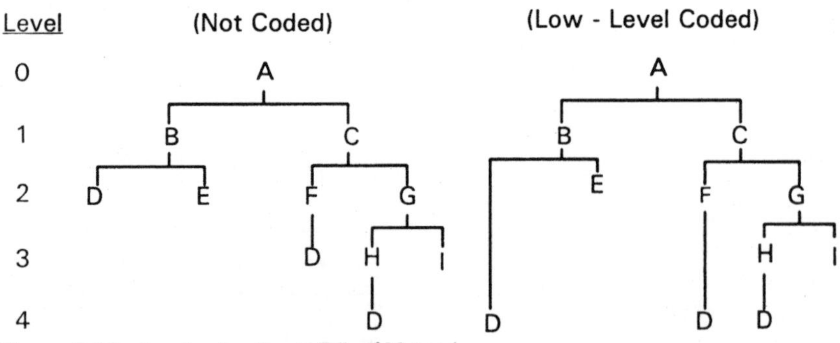

Figure 4-16. Low-level coding in Bills of Material.

calculating Net Requirements for a given item until all of that item's Gross Requirements have been computed. If another parent exploded more Gross Requirements into the component later, MRP would have to replan that component, dramatically increasing processing time. Because all parts that have low-level code 0 have no parents, by definition, MRP can plan them first. MRP can plan the parts of any low-level code in any order, because they are all the same low-level code. Once it finishes with low-level code 0, it can plan the low-level code 1 parts, because all Gross Requirements for them can come only from low-level code 0 parts, and they have already been computed.

The Bill of Material processing program automatically calculates and maintains the low-level code, while maintaining Bill of Material relationships.

Types of MRP Computer Processing

There are two alternative approaches for MRP computer processing: regeneration and net change.

Regeneration Processing

Regeneration is the method of planning used in the earliest MRP systems; it is still widely used today. When using regeneration, MRP replans, or "regenerates" the entire MRP plan for every item under its control. At the start of a regeneration (or "regen") MRP run, the MRP program first deletes all Planned Orders and all the Gross Requirements that those Planned Orders created. It then starts with a clean slate, using current stock-on-hand and MPS or customer orders. People enter maintenance transactions, including customer orders, shipments, shop orders, inventory transactions, and Bill of Material changes on a daily or continuous basis, so that they are working with timely data.

MRP is commonly regenerated once per week, with processing being carried out on the weekend. However, some companies run regenerative MRP on a daily basis. MRP processes in batch mode, and can require from less than an hour to more than 10 hours, depending on the number and complexity of products, the size and speed of the computer, and the design of the MRP program itself.

Some regenerative systems can perform the MRP process in *conversational* mode. As each low-level code of processing is completed, the human planner can review schedules for items on that level on a CRT. The planner may modify the plan, as he or she sees fit, by entering or modifying Planned

Orders. MRP includes these Planned Orders when it processes the items at the next lower level. Although this sounds interesting on paper, we know of no companies that actually use this technique.

Net Change Processing

Companies operating in an environment of frequent and rapid change find that updating the MRP once a week is not often enough, because the plan quickly becomes out-of-date during the week, possibly within the day. Because full regeneration MRP could take more than 24 to 72 hours in the 1970s, MRP package developers created *net change.*

Under net change, the MPS and MRP are considered to be continuous plans needing modifications from time to time. Net change only replans the items that need to be replanned, leaving everything else untouched (including Planned Orders). Transactions that will cause net change to replan a part include:

- A change or extension in the MPS
- A change in gross requirements for an item, resulting from a change in the schedule for a higher level item, service demand, or interplant shipments
- Unplanned inventory transactions such as
  - A correction in an inventory balance (e.g., cycle count or physical inventory transaction)
  - A lot being rejected in manufacturing
  - Any planned receipt or issue (in some systems)
- Engineering changes such as
  - Change in product structure
  - Change in effective dates
  - Change in usage
- Change in inventory planning parameters, such as
  - Lead Time
  - Safety Stock
  - Scrap Factor
  - Lot Sizing

MRP processes these "exception" parts in the same manner as a regenerative MRP run, proceeding level-by-level. Once an item has been replanned in a net change run, all its components will be replanned as well. Thus, net change does not always yield the dramatic reductions in processing time that we might initially expect. In fact, for one very commonly used

MRP package, the run time for net change was *greater* than for regenerative processing, if the number of parts that had changed exceeded 15 percent!

The most common way of implementing net change is to run net change MRP each night. Net change can also be carried out *on-line*, where MRP updates the plan continuously throughout the day as transactions are entered. One interesting variation of net change MRP is the ability to replan a single part, or family of parts, or plant, or location, on demand.

Comparing Regeneration and Net Change

Most MRP software packages include both regeneration and net change systems, each having advantages and disadvantages. Regeneration is more simple. Most companies that use net change for their standard MRP processing run regeneration software every few months. By generating a new plan with each regeneration, it purges errors that may have accumulated in prior runs. The major disadvantage apart from potentially long-run time, is that the MRP plan becomes increasingly out-of-date as time passes after a regeneration. This disadvantage can easily be overcome by running the regenerative MRP more frequently.

The principal advantage of net change is that the MRP is kept more current. Also, the processing speed of net change can be faster than regeneration, but not always. Net change systems tend to be more "nervous"—undergoing many changes, sometimes to schedules that were changed only yesterday.

Both regenerative and net change systems offer on-line transactions, which we strongly encourage practitioners to use. Batching might be suitable for accounting systems that summarize events, after the fact, into monthly buckets, but most operations systems require data to be much more timely than a classic batch system can support.

The increasing speed of computers has made regeneration sufficiently fast for many companies to run it as often as they want. As companies continue to make shorter and shorter runs of many complex products to serve increasingly volatile markets, MRP itself will continue to change. We discuss this more fully in Chapter 13.

Refinements of Material Requirements Planning

Firm Planned Orders

Planned Orders are computer suggestions based upon the information available to the computer at the time of MRP processing. Planned Orders are not "real," because they exist only in the computer. Unless acted upon by the planner, they may be changed in the next net change MRP run, and will be totally deleted and replanned in a regenerative MRP run. As the name implies, a Firm Planned Order (FPO) is a Planned Order that MRP can *not* change; only a human has the authority to change a Firm Planned Order.

The use of Firm Planned Orders provides a mechanism for human planners to override the replanning logic of MRP computer programs, by freezing Planned Orders in terms of quantity, release date, and due date. This is sometimes necessary to account for special situations that are temporary in nature, and do not call for permanent revision of input data or planning parameters. If a supplier is going to shut a plant for vacation, the planner will move the Planned Orders, which MRP plans for the vacation shut down period, to an earlier period, and then make them firm so that MRP will not reschedule them back to the actual date of need. If more rapid delivery than usual is required to meet a special need (e.g., using air freight), lead time can be reduced for that one order. If a work center will be overloaded for a week or two, some orders can be moved earlier or later.

However, Firm Planned Orders include a heavy administrative cost. Because the computer is prevented from performing its normal MRP plan for this order, a human must basically perform the entire MRP review process for this part. Thus, each Firm Planned Order adds a direct planning burden on a planner. If a planner is too busy to adequately review FPOs, the plan for the item in question, *and perhaps all of its components,* will become invalid. Because of this operational expense, we encourage practitioners to use FPOs sparingly.

As a concrete example of the use and effect of a Firm Planned Order, let us assume that our bicycle wheel, part *C* is a purchased part, and the supplier's sales representative has told us that the plant will be on vacation for the weeks of 6/30 and 7/7. If we group all three weeks of orders into 6/23, they will not be able to produce all of them in time. So we agreed to increase our order quantity to 100 per week for the next month, and to order enough to keep us in stock during 6/30 and 7/7. We can order the normal amount the week before the shut down, for delivery the week after the shutdown. We also know that our supplier does not always ship on time the first week after vacation, so we order enough before vacation to keep us supplied for 7/14.

In the first record of Figure 4-17, we have reproduced the original MRP plan for part *C*, which we computed back in Figure 4-14. Some additional gross requirements have come from our second line of bicycles. Our desired change is shown in the second record of Figure 4-17.

If we do not use Firm Planned Orders, the next MRP run will revert to the original MRP plan, as indicated in the top record in Figure 4-18, because the lot size of 75 and lead time of 1 are still in effect. (The MRP run

Original MRP Plan, run 5/26

Item C - Wheel OQ=75	1	2	3	4	5	6	7	8	9
	PERIOD								
Date	6/2	6/9	6/16	6/23	6/30	7/7	7/14	7/21	7/28
Gross Requirements	45	80	60	55	50	60	60	10	30
Scheduled Receipts	75								
Projected Available 40	70	65	5	25	50	65	5	70	40
Net Requirements		10		50	25	10		5	
Planned Order Receipts		75		75	75	75		75	
Planned Order Releases	75		75	75	75		75		

Desired Change, 5/26

Item C - Wheel OQ=75	1	2	3	4	5	6	7	8	9
	PERIOD								
Date	6/2	6/9	6/16	6/23	6/30	7/7	7/14	7/21	7/28
Gross Requirements	45	80	60	55	50	60	60	10	30
Scheduled Receipts	75								
Projected Available 40	70	90	130	175	125	65	5	70	40
Net Requirements		10						5	
Planned Order Receipts		100	100	100	0	0		75	
Planned Order Releases	100	100	100	0	0		75		

Figure 4-17. Need for Firm Planned Orders.

assumes that one week has passed; therefore the data that used to be in column 2 is now in column 1, and so on.) However, if we make the Planned Orders firm, the lead time and Lot Sizing rules will be overridden, and our desired changes will remain intact after the MRP run, as indicated in the bottom record of Figure 4-18.

There is a significant difference in the way Firm Planned Orders are used in MRP as opposed to MPS. Typically, all MPS Planned Orders are Firm Planned Orders. In MRP, Firm Planned Orders are used on an exception basis. The difference in the way Firm Planned Orders are used in MPS, as opposed to MRP, shows that MRP is carried out with greater reliance on the computer system than on human intervention. In part, this is because we assume that MRP is a logical extension to the MPS plan. Furthermore, MRP items are usually less costly and much more numerous than the MPS items, which means that heavy human intervention would be quite expensive.

Scrap in the BOM

Sometimes a particular item requires a special scrap rate in the Bill of Materials, rather than at the finished parent item. For example, a company

Next MRP Run Without Firm Planned Order

Item C - Wheel OQ=75	1	2	3	4	5	6	7	8	9
Date	6/9	6/16	6/23	6/30	7/7	7/14	7/21	7/28	8/4
Gross Requirements	80	60	55	50	60	60	10	30	
Scheduled Receipts	100								
Projected Available 70	90	30	50	0	15	30	20	65	
Net Requirements			25		60	45		10	
Planned Order Receipts			75		75	45		75	
Planned Order Releases		75		75	75		75		

Next MRP Run With Firm Planned Order

Item C - Wheel OQ=75	1	2	3	4	5	6	7	8	9
Date	6/9	6/16	6/23	6/30	7/7	7/14	7/21	7/28	8/4
Gross Requirements	80	60	55	50	60	60	10	30	
Scheduled Receipts	100								
Projected Available 70	90	130	175	125	65	5	70	40	
Net Requirements							5		
Planned Order Receipts		100	100	0	0		75		
Planned Order Releases	100	100	0	0		75			

Figure 4-18. Effect of Firm Planned Order.

builds china cabinets that have sliding glass doors. The company breaks many more glass doors than the wood cabinets during the assembly process.

Some MRP systems allow scrap to be specified at the Bill of Material relationship. In the example above, the company set the scrap factor 5 percent for the glass doors as a component of the wood case. Then, when MRP plans an order for 100 finished china cabinets, it explodes into 100 wood cases and 105 sets of glass doors.

Summary

We have shown the basic logic of MRP processing and how MRP uses the basic inputs, including: Bills of Materials, Master Production Schedule, Inventory, Orders, and Requirements. We discussed the four basic processes of MRP: determining Net Requirements by period, determining Planned Orders by period, Lead Time Offsetting, and Exploding Planned Orders.

We also showed how MRP performs its calculations by using a working example, starting with a very simple form of MRP, and covering all the major factors:

- On-hand quantity and Gross Requirements. We showed how on-hand quantity is a foundation for all further MRP calculations for a part (thus requiring very high accuracy).
- Scheduled Receipts, including both manufacturing and purchase orders. MRP frequently suggests reschedules so that inventory will be minimized. We stated that most planners do not have the time to reschedule orders out (or de-expedite orders). This totally undermines the credibility and relative priority of these orders.
- Planned Orders, with requirements explosion. We showed how some seemingly simple Lot Sizing can increase inventory investment and cause the shop to produce orders sooner.
- Safety Stock. We showed how MRP plans to never, ever, use Safety Stock. We strongly encourage you to use Safety Stock sparingly, and preferably only on end items and purchased parts.
- Multiple Products with common components. We showed how multiple parents can explode into a common component.
- Scrap and Yield. We showed the effect of scrap and yield on MRP plans. If you have a 5 percent scrap rate at each level of a 6-level BOM, MRP will plan almost 35 percent more components at the bottom level, owing to the effect of the compounding. One of MRP's weaknesses is that it does not present this waste forcefully to management so they will take action. MRP assumes that management and planners know what they are doing.
- Allocations. These cover a potential problem, in which MRP would lose track of components that had not yet been issued to released manufacturing orders (Scheduled Receipts).

We covered the two types of computer processing: regeneration and net change, concluding that although net change might still be appropriate, computers have become so fast that regenerative is a very viable alternative. Finally, we discussed Firm Planned Orders and scrap factors in the BOM.

In short, we attempted to provide you with enough depth of understanding about how MRP actually functions, that you can effectively use it to manage your company toward increased competitiveness.

Selected Bibliography

Browne, Jimmie, J. Harhen, and J. Shivnan: *Production Management Systems: A CIM Perspective,* Addison-Wesley, Reading, MA, 1988, Chap 5.

Deis, Paul: *Production and Inventory Management in the Technological Age,* Prentice-Hall, Englewood Cliffs, NJ, 1983, Chap 7.

Fogarty, Donald W., J. H. Blackstone, and T. R. Hoffmann: *Production and Inventory Management,* South-Western, Cincinnati, OH, 1991, Chap 10.

Lunn, Terry with Susan A. Neff: *MRP: Integrating Material Requirements Planning and Modern Business,* Irwin, Homewood, IL, 1992, Chap 3,

Orlicky, Joseph: *Material Requirements Planning,* McGraw-Hill, New York, 1975, Chaps 2, 3, 4, and 5.

Plossl, George W.: *Production and Inventory Control: Principles and Techniques,* 2d ed., Prentice-Hall, Englewood Cliffs, NJ, 1985, Chap 6.

Schultz, Terry: *Business Requirements Planning: The Journey to Excellence,* The Forum Ltd., Milwaukee, WI, 1984, Chap 8.

Smith, Spencer B.: *Computer Based Production and Inventory Control,* Prentice-Hall, Englewood Cliffs, NJ, 1989, Chap 9.

Vollmann, Thomas E., W. L. Berry, and D. C. Whybark: *Manufacturing Planning and Control Systems,* 3d ed., Irwin, Homewood, IL, 1992, Chap 2.

5
Using MRP Effectively

"You want to use this tool to your benefit? Then listen to what it is telling you!"
(AN ANONYMOUS CARPENTER, CIRCA 1950)

Introduction

Success in using MRP depends upon actions taken in response to the system feedback and outputs. The MRP planner/buyer must respond correctly and expeditiously to the outputs provided by both MRP, CRP, and all other modules of the MRP II system, as well as information external to the system. However, to respond appropriately, the MRP planner/buyer must first understand the reasons why a response is required in the first place, the alternatives available, and the costs and benefits of each alternative. For example, although MRP might suggest a reschedule of a Scheduled Receipt, the MRP planner/buyer must insure that such a reschedule is, in fact, the proper course of action. It is in making and executing these decisions that MRP eventually succeeds or fails.

This chapter is a continuation of Chapter 4. As such, it will primarily describe the outputs of the MRP process, and discuss how to use these outputs in a closed-loop mode. We will also discuss when and how the planning data of the MRP system should be revised.

Material Requirements Planning Outputs

As shown in Figure 4-1, MRP produces three principal output reports:

MRP Primary Report

MRP Action Report

MRP Pegging Report

The content and format of these reports and inquiries vary considerably from one MRP computer package to another. These reports are available on paper or on a computer screen. For each report, we describe the common elements that are contained in most of the MRP computer systems.

MRP Primary Report

The *MRP Primary Report,* often called simply the *MRP Report,* normally uses either the horizontal format, with time in buckets (usually days and weeks), or the vertical format, with time in days. Figure 5-1 shows a "standard" horizontal format, which is the same format that we used in Chapter 4.

The information included for each item generally includes at least the following:

- Header (planning) information: item number; item description; order quantity policy; order quantity minimum, maximum, or multiple; shrinkage factor; lead time; Safety Stock.
- Periods: the planning periods (usually days and weeks) are normally designated by their starting date or ending date. In some reports the periods are numbered.
- Gross Requirements: the Gross Requirements summarized for a period. To learn what caused these requirements, you can review the Pegging

		PERIOD								
Item A - Bicycle		1	2	3	4	5	6	7	8	9
Gross Requirements		25	25	25	25	25	25	25	25	25
Scheduled Receipts			50		50					
Projected Available	25	0	25	0	25	0	0	0	0	0
Net Requirements							25	25	25	25
Planned Order Receipts							25	25	25	25
Planned Order Releases					25	25	25	25		

Figure 5-1. MRP report in the standard horizontal format.

Report, which is discussed later in this chapter. These requirements can include both shipments to customers and usage on the shop floor.

- Scheduled Receipts: the Scheduled Receipts summarized for each period. Scheduled Receipts can be either purchased or manufactured.
- Projected Available Balance: the Projected Available Balance at the end of each period. This is the quantity that should be in the stockroom, available for use, at the end of each period.
- Net Requirements: the Net Requirements for each period, which MRP computes by comparing the Projected Available Balance to the Safety Stock Level.
- Planned Order Receipts: the Planned Order Receipts for each period, which MRP computes by adjusting the Net Requirements upward by a Lot-Sizing factor. This is the expected receipt quantity of the Planned Order. It can differ from the order start quantity because of shrinkage.
- Planned Order Releases: the Planned Order Releases summarized for each period. This is the start date and order quantity of each Planned Order. MRP computes Planned Order Releases by subtracting the lead time from the Planned Order Receipt date.

The MRP Primary Report contains virtually all the data used for MRP processing, so that if a planner wants to understand how the MRP program calculated a given value, he or she has the requisite data to manually replicate the program's calculation. The most important information contained in the Primary Report is the Planned Order Releases. These are suggested replenishment orders that MRP has created; each Planned Order is for a specific quantity of a specific item, with specific start and finish dates. The MRP planner must take action to convert these "suggested" orders into

		PERIOD								
Item A - Bicycle		1	2	3	4	5	6	7	8	9
Gross Requirements		25	25	25	25	25	25	25	25	25
Scheduled Receipts			50		50					
Projected On-Hand	25	0	25	0	25	0	-25	-50	-75	-100
Projected Available	25	0	25	0	25	0	0	0	0	0
Net Requirements							25	25	25	25
Planned Order Receipts							25	25	25	25
Planned Order Releases					25	25	25	25		

Figure 5-2. MRP report in the expanded horizontal format.

"real" orders. MRP can only plan orders; it takes a human being to release them and make them "real."

Alternative MRP Display Formats

The horizontal MRP display format that we have been using in the previous section is widely recognized. However, you should be aware of two other horizontal formats, expanded and abbreviated, as well as the vertical format.

Expanded Horizontal Format. The expanded horizontal format adds an additional row for Projected On-Hand. Projected On-Hand is the same as Projected Available, except that Projected On-Hand includes only starting physical Stock-on-Hand and Scheduled Receipts. *Projected On-Hand excludes Planned Order Receipts.* This row is usually provided in addition to the Projected Available row, but in some of the older literature it is used in lieu of Projected Available. Figure 5-2 illustrates this point.

Abbreviated Horizontal MRP Display. Because of the limited space available on screens, many MRP systems use the abbreviated horizontal MRP display. Some older systems also use it for reports. It also appears in the older APICS certification literature. Its brevity causes it to be somewhat difficult to use for manual MRP processing. With the Net Requirements and Planned Order Receipts rows missing, a person performing manual MRP calculations will have to do these calculations in his or her head, on scratch paper, or somewhere on the form. One alternative for manual processing is to use the Scheduled Receipts row for Planned Order Receipts, and the lower half of the projected availability row for the Net Requirements. Figure 5-3 demonstrates this method by showing the Planned Order Receipts and Net Requirements entries in italics.

Vertical Format. The horizontal format used in this chapter, up to this point, is common in textbooks, articles, and instructional materials, be-

		PERIOD								
Item A - Bicycle		1	2	3	4	5	6	7	8	9
Gross Requirements		25	25	25	25	25	25	25	25	25
Scheduled Receipts			50		50		*25*	*25*	*25*	*25*
Projected Available	25	0	25	0	25	0	0/*25*	0/*25*	0/*25*	0/*25*
Planned Order Releases					25	25	25	25		

Figure 5-3. MRP report in the abbreviated horizontal format.

cause it is a good visual tool. Figure 5-4 shows the same information as Figure 5-1, only in vertical format. We will make the following changes, however, so that we can illustrate the automatic pegging capabilities of this format.

- We will assume that our bicycles can be used as components in a Complete Bicycle Assortment (item number CBA) that we sell to department stores.
- We also need to purchase the Scheduled Receipt due on July 6, because our shop will be closed for our annual vacation during that week. The Gross Requirements, starting on June 22, are due to Planned Orders for the parent, rather than released work orders.

The vertical format contains all this information in one place, rather than requiring the planner to scroll through multiple screen programs or multiple reports. The vertical format is usually available for screen inquiries, as well as in printed report form.

The vertical, or list method, permits much greater system flexibility. This is especially useful when lead times need to be expressed in days, rather than in weeks. Some manufacturing orders, and even some purchases, can be accomplished in very short times, but artificially long-lead times caused by the use of weekly buckets can result in an unrealistically long cumulative lead time.

ITEM = A Bicycle, LT = 8 days, OQ = LFL, SS = 0, ON HAND = 25						
Date	Reference	Parent	Start	Recpt	Reqt	Avail
6/15						25
6/15	WO3519	CBA			25	0
6/22	WO3618		6/12	50		50
6/22	PL3622	CBA			25	25
6/29	PL3707	CBA			25	0
7/6	PO2298		6/14	50		50
7/6	PL3889	CBA			25	25
7/13	PL3945	CBA			25	0
7/20	PL4478		7/6	25		25
7/20	PL4389	CBA			25	0
7/27	PL4621		7/13	25		25
7/27	PL4599	CBQ			25	0
etc.						

Figure 5-4. MRP report in the vertical format.

Vertical or Horizontal? Which format should you use? Each has its advantages and disadvantages, as listed below.

- Horizontal

  Advantages
  - Intuitive, easy to understand
  - Condenses considerable information in a small space
  - Has been the industry standard

  Disadvantages
  - Buckets are generally in weekly increments. A weekly bucket hides the problem that the incoming purchase order will arrive at the stockroom on Friday, but the requirement to use the item on the floor must be pulled on the previous Monday.
  - Not easy to see any details which would tell the causes of the requirements.
- Vertical

  Advantages
  - All pegging detail available in the one report
  - Uses dates to show expected stock transactions; can spot impending shortages inside a given week.

  Disadvantages
  - Can be far too long, owing to the detail, especially for common fasteners and raw materials.

Selecting the Period Size

In the horizontal format, each period is commonly called a "bucket." MRP planners specify the sizes of these "buckets," and MRP simply arranges the data (requirements, receipts, projected available, and so on) into the buckets based on dates. Traditionally, planners have used weekly buckets for the first half (or more) of the report or screen and monthly buckets for the remaining report or screen.

To avoid problems with "surprises" (a requirement on Tuesday, but the covering receipt not arriving until Friday), we strongly encourage you to use:

- *Daily* buckets for the first 10 days or so,
- *Weekly* buckets for the next several periods, and
- *Monthly* buckets for the final few months. These monthly buckets should be out beyond the cumulative lead times of the products, to insure that all plans are visible.

Manufacturing Calendars

When scheduling, many planners prefer to use the common Gregorian calendar (e.g. 6/21/93), skipping over holidays, weekends and other nonworking days. Planners define these nonworking days in the MRP shop calendar.

Some companies use manufacturing calendar days, or "M-days," rather than standard calendar dates. Manufacturing calendars are developed from standard calendars by consecutively numbering only those days of the year when production is actually planned, excluding holidays, periods of plant shutdowns, and possibly Saturdays and Sundays. Humans find calculating lead-time offsets and other time-related functions much easier with manufacturing calendars. With the proliferation of affordable computers, manufacturing calendars are less necessary.

MRP Action Report

The *MRP Action Report,* often called the *MRP Exception Report,* focuses the MRP planner on those items needing immediate attention, and recommends a course of action. The system can, and does, automatically reschedule or replan Planned Orders. However, MRP cannot change the due date or quantity of Firm Planned Orders and Scheduled Receipts; MRP can only *suggest* changes. The Action Report is how MRP suggests those changes. The report (see Figure 5-5) typically includes the following information:

Release an order

Release order with insufficient lead time

Reschedule in (expedite)

Reschedule out (de-expedite)

Cancel an order

Review order past due

MRP ACTION REPORT						
Planner: WJB					Run Date: 8/15	
Item	Description	Action	Order	Quantity	Date from	Date to
B21	Mountain Bike	Rel/Exp	P2469	200	8/15	8/8
		Release	P2475	200	8/15	
B23	Mountain Bike—Special	Resched	W3321	100	8/31	9/22
S445	Spokes	Expedite	S8293	1000	9/15	8/28
S466	Seat—Vinyl	Cancel	S7321	50	9/12	

Figure 5-5. MRP Action (Exception) Report.

Later in this chapter we discuss more fully how a planner can respond to each of these situations.

The Action Report provides the planner with an effective and efficient method of prioritizing where to spend his or her time. If an item does not need attention, it is not included in the report. Before deciding what action, if any, to take with respect to an item on the Action Report, the planner will normally also review the MRP Primary Report and the Pegging Report for that item.

Some MRP systems can organize these exceptions according to a priority system. For example, orders to be canceled may be listed in declining dollar value. Orders past due may be listed in declining days past due.

"You know you're going to have a bad day when . . . " the MRP Action Report is over two feet high. We actually saw one that high at a large, high-tech manufacturing company.

MRP Pegging Report

When using the horizontal format in the MRP Primary Report, the *MRP Pegging Report,* shown in Figure 5-6, provides the source of requirement at the next higher level in the Bill of Materials, for example, each parent order that exploded directly into this item to cause each gross requirement. The vertical format MRP report automatically includes single-level pegging.

An MRP system can create Pegging Reports in two different ways:

1. Record the pegs as the MRP explosion process is being carried out. As Gross Requirements for an item are being calculated based on needs of the parents, and orders planned to cover those needs, the needs of the parents entering into those calculations are retained and form the Pegging Report.

ITEM: 4327	DESCRIPTION: Bearing		DATE: 6/11
Required date	**Quantity**	**Source of requirements**	**Order number**
6/15	11	Item 7653	W5473
6/15	17	Item 9365	W7754
6/22	30	Item 4768	W4752
6/22	5	Portland Plant	W1732
6/22	10	Service Forecast	
6/29	17	Item 9365	W7758
6/29	15	Item 3472	W4471

Figure 5-6. A single-level Pegging Report.

2. Use the where-used logic associated with the BOM files whenever a query is received regarding pegging. The system determines what parents the item has, searches the MRP to determine what orders are planned for the parents, and then shows how these orders created the requirement.

These two approaches provide similar, but not identical, results. The second approach indicates the part numbers of *all* items on which that part is used. The first approach is more selective; it shows only the specific part numbers that produce the specific Gross Requirements in each time period.

The principal use of pegging arises when, for some reason, a component will have a shortage, owing perhaps to unexpected sales, anticipated late delivery, poor quality, or an inventory adjustment. Using the Pegging Report, the planner determines the requirements that caused the order. With this information, he or she can explore alternatives at this level and at higher levels in the Bill of Materials. Can a Lot Size be changed? Can Safety Stock be used? Can another component be substituted to meet certain of the requirements? Can some of the needs be satisfied by rescheduling earlier completion of another order for the item that is already in process? Which higher level items can be delayed? Which can not?

Additionally, planners can use pegging to validate the MRP Planned Orders. Also, pegging can be used if a sudden unexpected demand occurs, and management asks the age-old questions:

- Can the components and materials be made available quickly enough to meet the new demand
- What disruption would this cause in meeting other commitments?

Some systems also offer full Pegging Reports. Full pegging shows the requirement all the way up to the MPS, or possibly to customer orders. Although full pegging is intended to determine which customer orders will be late, owing to some problem with materials or the plant, the answer is often not appropriate. Usually, the more appropriate response to the same situation is to figure out how to overcome the difficulty and ship the order to the customer on time. For example, we could use substitute materials, or work some overtime, or buy a "manufactured" part from a supplier (or even a competitor). Only as a last resort should we call customers to tell them that we will ship late, or partial. Full pegging is also complex, and even misleading, because it must deal with scrap allowances and Lot-Sizing rules. For example, how does the pegging program decide which of four potential end-item orders will receive the shortage amount, inasmuch as all four end-item orders have been combined into a single lot by an intermediate part? Because of these considerations, single-level pegging is, by far, more widely used. A planner can trace requirements all the way to the MPS, or customer order, by following the single-level pegging upward, one level at a time. Although this can be a lengthy job, the planner can review the alternatives and make decisions at each level.

Using Material Requirements Planning System

The outputs of Material Requirements Planning are used primarily by planners in production control, inventory control, or purchasing, who are usually called Materials Planners, Planner/Buyers or Buyer/Planners. Computerized MRP systems often encompass thousands of part numbers. To handle this volume, planners are generally organized around logical groupings of parts. Even so, planners should use the Action Report to quickly determine which items need action. The following situations require action on the part of the MRP planner:

- *Releasing Orders:* Releasing, or opening, shop or purchase orders when indicated by the system
- *Priority Planning:* Rescheduling the due dates of open orders when desirable
- *Responding to Changes:* Reacting to changes in MPS, BOM, ECO, Lot Sizes, orders, deliveries, and other factors
- *Bottom-up Replanning:* Using pegging data to solve material shortages and to develop new schedules
- *Revising Planning Data:* Analyzing and revising planning data, such as lot-sizing technique, lead time, Safety Stock and/or safety lead time, and scrap allowances.

Releasing Orders

Figure 5-7 shows the process of releasing orders in an MRP system. The process is initiated when a Planned Order Release shows up in the current time period on the MRP Action Report. Although it sounds obvious, orders should be released when MRP suggests, unless there are extenuating circumstances, such as bottlenecks in the shop.

Releasing Orders is the process of releasing orders to the shop, for make items, or to suppliers, for buy items. MRP suggests that Planned Orders be released just in time to be in the stockroom on their due dates, and prints action messages for each order that is due for release within a user-specified

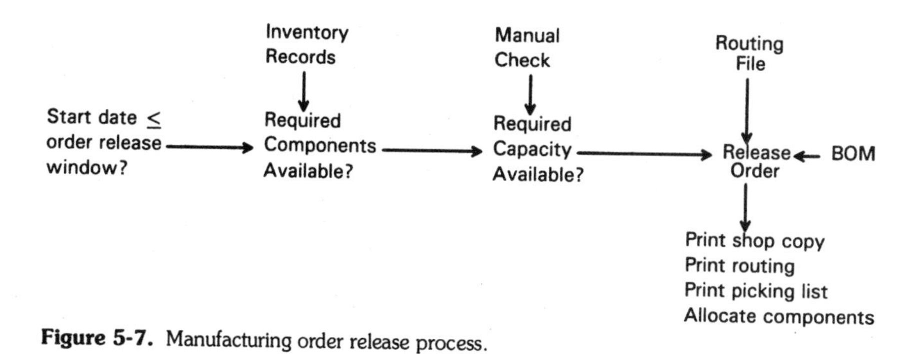

Figure 5-7. Manufacturing order release process.

window. Order release converts the Planned Order Release or Firm Planned Order into a Scheduled Receipt, as either a shop order or purchase order. Most MRP systems provide a list of orders to be released as part of the Action Report; this greatly simplifies the work of the planner. For those systems that do not, the planners must scan the entire MRP Primary Report to find the Planned Order Releases in the current time period.

Although Planned Orders are controlled by the computer, Scheduled Receipts (open orders) are not. Once a Planned Order has been changed to an open order, only people can change it, because material and capacity are allocated to it.

When a purchase order is released, Purchasing confirms the details with the supplier, and enters the additional data and changed data into the system. Manufacturing orders involve several more steps. Before releasing a manufacturing order, planners generally check the availability of required material and capacity. Most MRP systems check component availability for any order that is scheduled to be released, and report any discrepancies in the Action Report. Planners can release the order if the system shows a material shortage.

Planners should also perform a manual capacity check to ensure that capacity is available, prior to releasing the order. Presumably, we have already performed Capacity Requirements Planning, which indicated that the necessary capacity will probably be available under normal conditions. It would be nice if we could do a quick capacity check using finite loading to ensure that capacity is really available. However, most CRP systems do not have this capability. All we can do is manually determine if there are any abnormal conditions that would change the available capacity, such as machines down or undergoing maintenance, operators out, or a large backlog. There is little point in releasing an order if capacity is not available; the new order will only increase the backlog. Once a planner releases an order to the shop, the planner has lost much control over that order. If an order will languish in a queue, the planner is better advised to hold it in Production Control, until the backlog situation has been resolved.

The system automatically allocates the proper quantities of components to that order. The allocation means that this quantity of components is reserved for this specific shop order, and therefore is not available for any other use.

During the order release process, the MRP system prints *pick lists* for the stockroom, unless the system is truly on-line and paperless. In many companies, a pick list is like a check at a bank; no list, no parts. The pick list calls for a specific quantity of a part number to be removed from one or more inventory locations, and to be delivered to a particular department or location for a specific shop order. When the stock clerk notifies the computer that a part has been picked, the computer removes the allocation and reduces the on-hand balance for that part. Because of the time involved in picking, the amounts shown as Projected Available balance may be lower than the physical inventory balance, by an amount equal to the pick list. For example, a large kit might include 75 different items and require two hours to pick. In most companies, the stock clerk reports the pick only once, when the picking is complete. Just before the kit is completely picked, the stock-on-hand balances for the parts that have been picked will be lower than the computer expects, by the amount that has been picked.

The system can also create job or traveler documents. These can include:

- A copy of this work order's routing and Bill of Materials
- Turnaround documents (not needed for companies that use bar coding for shop floor feedback)
- Copies of the engineering specifications, if these are accessible by the MRP system

Priority Planning (Expediting and De-expediting)

Priority planning refers to maintaining correct due dates by constantly evaluating the true date of need for Scheduled Receipts, and *expediting* and *de-expediting* orders as necessary. The due date determined by MRP is the most important control variable in Work-Center Dispatch lists, which are, in turn, the primary priority-control devices on the shop floor. Given the importance of these due dates, we cannot overemphasize the need for the MRP planner to review and reschedule *all* orders that MRP suggests, both expediting and de-expediting.

To further emphasize the importance of priority planning, let's explore the relationship between priority planning and priority control. Priority planning is the process of determining what is needed and when it is needed. Priority control is the process of ensuring that planned priorities are achieved. Priority planning is performed by the MRP system. Priority control is performed by the Shop Floor Control system primarily by using Work-Center Dispatch lists. Priority planning by MRP establishes the due date for an order, which becomes the controlling factor of the work-center dispatch list.

A due date is established for a job when the shop order is released. If the due date is several weeks into the future, the requirement that caused the order will quite possibly change before the original due date is reached. The MRP system keeps track of these changes, computes the real due date (called the *date of need*) for each order each time that MRP runs, and issues an action notice for the order to be rescheduled when the date of need differs from the published due date.

Figure 5-8 provides two examples of priority planning. The top record, which depicts the situation before priority planning is required, shows Gross Requirements of 50 in period 1 and 150 in period 2. It also shows a Scheduled Receipt of 200 in period 3. Processing the inventory record with those inputs results in Planned Order Releases of 200 in periods 1, 3, 4, 5, and 6.

In the second record, the requirement in period 2 has changed from 150 to 50 (owing to a customer cancellation). As a result, the Scheduled Receipt of 200 is no longer needed in period 3; the Projected Available inventory in period 3 is sufficient to satisfy the requirement. MRP will generate an action notice to de-expedite the Scheduled Receipt. The logical thing to do is to delay the Scheduled Receipt by one period. Only a human can do this; the computer can not reschedule open orders. Whether or not we delay the Scheduled Receipt until period 4, MRP will create new Planned Order Releases of 200, in periods 2, 3, 4, and 6.

The bottom record shows a different situation. In this record, the Gross Requirement is now 150 in periods 1 and 2. MRP generates an action notice to expedite the Scheduled Receipt. The logical way to resolve this problem is to expedite the Scheduled Receipt by one period. This would result in Planned Order Releases in periods 1, 2, 3, 4, and 6. On paper, this is no harder than delaying it one period. However, from a practical standpoint, it is considerably more difficult. The shop can delay a job with very little problem. However, expediting one job usually means that another job will be delayed. One of the most difficult tasks facing the MRP planner is determining what shop orders can be delayed that will release capacity in the right work centers, with little or no disruption to customer commitments and the shop floor.

If item *S* is the lowest level component, then we need worry only about expediting one Scheduled Receipt. However, if *S* has components, then the

	PERIOD								
Item A - Bicycle; OQ=200 SS=0; LT=3	1	2	3	4	5	6	7	8	9
Gross Requirements	50	150	100	150	150	200	150	150	200
Scheduled Receipts			200						
Projected Available 200	150	0	100	150	0	0	50	100	100
Net Requirements				50		200	150	100	100
Planned Order Receipts				200		200	200	200	200
Planned Order Releases	200		200	200	200	200			

	PERIOD								
Item A - Bicycle; OQ=200 SS=0; LT=3	1	2	3	4	5	6	7	8	9
Gross Requirements	50	50	100	150	150	200	150	150	200
Scheduled Receipts			200 →	*200*					
Projected Available 200	150	100	0	50	100	100	150	0	0
Net Requirements					100	100	50		200
Planned Order Receipts					200	200	200		200
Planned Order Releases		200	200	200		200			

	PERIOD								
Item A - Bicycle; OQ=200 SS=0; LT=3	1	2	3	4	5	6	7	8	9
Gross Requirements	150	150	100	150	150	200	150	150	200
Scheduled Receipts		*200*	← 200						
Projected Available 200	50	100	0	50	100	100	150	0	0
Net Requirements				150	100	100	50		200
Planned Order Receipts				200	200	200	200		200
Planned Order Releases	200	200	200	200		200			

Figure 5-8. Examples of priority planning in MRP.

change in the Planned Order Release schedule of *S* may result in the need to expedite open orders for lower level components.

Responding to Changes

MRP is a planning system. MRP's plans are based on certain assumptions, or predictions, regarding what will happen in the real world. When some of these assumptions prove incorrect, MRP has to develop new plans that take these changes into account. This is an important aspect of being a truly closed-loop system. The essence of MRP is the ability to react to change, which is one of the primary benefits of using the system. Once MRP is replanned, it will advise planners of the individual items that need attention through the use of action messages in the Action Report. The Action Report lists, or flags, only those items that require some sort of action. Changes in the MRP may cause changes to CRP, which may result in CRP action messages too. As in the reschedules in the previous section, the MRP planner must keep MRP totally up-to-date with respect to the effects of these changes. If the MRP system does not contain current data, the people in the company will start using informal systems again, thus slowly undermining the validity and worth of the MRP system.

In the follow paragraphs, we discuss some of the more common changes that cause MRP to generate action messages.

MPS Revisions. Sometimes the MPS is revised within the time fences, affecting near-term requirements. Time fences are designed to inform management that MPS changes will cause additional expense by requiring extraordinary actions, as discussed in Chapter 3. But time fences are guidelines, rather than hard and fast rules. Very few companies will continue to manufacture a product (including those orders within the time fence) when sales have collapsed, or refuse to increase production to take advantage of a blossoming market, or tell a valued customer that they must wait the full standard lead time if that customer needs a product quickly.

Engineering Changes. Some engineering changes to the Bill of Materials must be implemented immediately. As discussed in Chapter 3, good engineering practice tries to phase in changes by planning them well ahead of time, establishing an *effectivity date* that takes into account existing stocks of material and incoming shop orders and purchase orders. Some situations, however, call for immediate revision, especially when safety and quality are involved. When a Bill of Materials changes suddenly, the lower level manufactured and purchase orders, many of which may be Scheduled Receipts, will be affected.

Inventory Corrections. When errors are discovered in inventory data, often as a result of cycle counting, the cycle counter changes the current stock-on-hand quantity, which is the basis of the MRP plan for a part. Because MRP requires highly accurate inventory balances, companies often start cycle count programs that continually verify on-hand information, and eliminate the underlying causes of the errors. The adjustment to on-hand balances can cause MRP to reschedule orders.

Excessive Scrap. If actual scrap exceeds estimated quantities, less material will be completed than needed for requirements. Planners can include scrap allowances in the MRP planning parameters, as discussed in Chapter

4. However, even these allowances may be exceeded sometimes. When they are, the deficiency can be rectified by increasing the size of Scheduled Receipts, or creating new Planned Orders to meet requirements. Companies who are implementing JIT view the removal of scrap factors as an obvious way to save money. Rather than planning for scrap losses through MRP, they identify and implement ways to eliminate them completely. To a JIT-oriented company, *any* scrap is excessive.

On the other hand, a Scheduled Receipt might have less scrap than planned. In this case, the system will simply accept the excess quantity. The next MRP run might reduce the quantity of, or reschedule out, some Planned Order Releases, or suggest de-expediting the next scheduled delivery.

Customer Order Revisions. Occasionally, customers will request changes in their orders. These changes can be to the requested delivery date, the product or quantity ordered, or even an order cancellation. Most customer-initiated changes will be reflected in the MPS.

In a Make-to-Stock environment, these changes are relatively less serious. In all other environments, we may have already started purchasing and/or manufacturing customer-specific items. The planner can utilize the Pegging Report to analyze the impact of the customer-requested change, and to decide how best to use any existing stocks if the customer does not need them now.

Supplier or Production Problems. Many manufacturers somehow believe that their suppliers can walk on water, leap tall buildings with a single bound, and otherwise perform miracles. However, suppliers are manufacturers, just like our own plant. Both suppliers and our plant can fail to produce on time, fail to deliver correct quantities, and/or fail to meet our quality standards. All of these problems cause a shortage of expected Scheduled Receipts.

Bottom-Up Replanning

When discussing priority planning, we mentioned that expediting Scheduled Receipts was one means of solving material shortage problems. Expediting is not always possible and, even when it is possible, it is not always desirable, because it can be disruptive to the planning routines as well as the shop floor. By using pegging data, we can often solve the material shortage problem, without resorting to expediting. The process of using pegging to solve material shortages is called *Bottom-up replanning*, which we will explain by using an example.

Figure 5-9 shows the past MRP plan of component *S*, and the Bill of Ma-

P

Q 1 R 2

S 1

Item S - Rim; OQ=200 SS=0; LT=3		1	2	3	4	5	6	7	8	9
Gross Requirements				200			200			
Scheduled Receipts				200						
Projected Available	50	50	50	50	50	50	50	50	50	50
Net Requirements							150			
Planned Order Receipts							200			
Planned Order Releases				200						

(Column headers 1–9 are under PERIOD.)

Figure 5-9. Past MRP Plan of component *S*.

terials for component *S* and its parents. The requirements of 200 in periods 3 and 6 are satisfied by a Scheduled Receipt in period 3, and a Planned Order Release in period 3.

However, things have changed, as shown in Figure 5-10. Now there are requirements of *300* in periods 3 and 7. The Scheduled Receipt of 200 in period 3 cannot completely satisfy the requirement of 300 in period 3. We have a net requirement of 50 in period 3. We could try to solve the problem by releasing a Planned Order, but that would be difficult, because the release date is already past.

Instead, we generate a pegging report on component *S*, and learn that its parent is subassembly *R*. We review subassembly *R* (shown in Figure 5-11), finding that the reason for the change in requirements was caused by an additional customer order in period 7. Subassembly *R* uses a Fixed Order Interval of four weeks for determining Lot Size.

Item S - Rim; OQ=200 SS=0; LT=3		1	2	3	4	5	6	7	8	9
Gross Requirements				300				300		
Scheduled Receipts				200						
Projected Available	50	50	50	150	150	150	150	50	50	50
Net Requirements				50				150		
Planned Order Receipts				200				200		
Planned Order Releases	200				200					

(Column headers 1–9 are under PERIOD.)

Figure 5-10. Current MRP plan of component *S*.

Item R - Wheel; FOI=4 SS=0; LT=1		1	2	3	4	5	6	7	8	9
						PERIOD				
Gross Requirements			100	100	100	50	50	100	50	100
Scheduled Receipts			200							
Projected Available	0	0	100	0	200	150	100	0	250	150
Net Requirements					100				50	
Planned Order Receipts					300				300	
Planned Order Releases				300				300		

Figure 5-11. Current MRP plan of parent *R*.

After looking at various possibilities for action, we decide to change the Fixed Order Interval Lot Size of subassembly *R* to two weeks (see Figure 5-12). This will eliminate the requirement to expedite component *S*.

For ease of understanding, we have presented a very simple problem with a very simple solution. However, pegging can be used to solve some very complex problems. The solution to complex material shortages may involve compressing lead times throughout the product structure, using pegging and bottom-up replanning.

Revising MRP Planning Data

Managing Lead Times

Our increasingly competitive environment has defined time as a key competitive weapon, such that long lead times are no longer tolerable. Additionally, long MRP lead times increase the work-in-process, because MRP uses the lead time to schedule the order release date. Long lead times also cause additional on-hand inventories, in the form of Safety Stocks. Finally,

Item R - Wheel; FOI=2 SS=0		1	2	3	4	5	6	7	8	9
						PERIOD				
Gross Requirements			100	100	100	50	50	100	50	100
Scheduled Receipts			200							
Projected Available	0	0	100	0	100	50	0	50	0	100
Net Requirements					100			100		100
Planned Order Receipts					F200			150		200
Planned Order Releases				F200			150		200	

Figure 5-12. Revised MRP plan of parent *R*.

long lead times force us to forecast further out, thus reducing our forecast accuracy. The ideal situation would be to have lead times so short that we could manufacture customer orders and ship them so quickly that the customers think we stock their product.

Most MRP systems use a fixed lead time, which should be in days rather than weeks. (In the future lead times will be calculated in hours, rather than days.) We discuss the problems of these fixed lead times and potential solutions in greater detail in Chapter 13. MRP breaks down the Bill of Materials into many levels, and adds the lead time for each part at each level together to establish the cumulative lead time for ordering. In other words, it accumulates the data we provide. If the cumulative lead times are too long, we need to provide different data.

Most systems define manufacturing lead times as the sum of queue, setup, run, wait, and move. The productive elements of manufacturing lead time are setup time and run time. All the other elements, queue, wait, and move, are nonproductive or wasted time, and should be reduced, in actual practice, as close to zero as possible. In fact, even setup should be reduced to zero as quickly as possible. This enables a manufacturer to respond quickly with small Lot Sizes, thus dramatically reducing both WIP inventories and lead times.

There are several ways that users of MRP systems can reduce their lead times. The vast majority (often over 90 percent) of total manufacturing lead time is queue time. Thus, merely cutting queues in half should effectively reduce our lead times, and our WIP, by half.

Lead times are actually self-fulfilling prophecies. Lead times can be directly controlled by production planners, manufacturing engineers, and the people who make things happen on the shop floor. The most effective lead-time reduction method is to implement a planned program to progressively reduce the lead time of all products over a certain time period. The quickest and easiest way to do this is to simply measure existing queues for several days or weeks, find the lowest amount, and cut all queues to just above that level. The changes in queue time *must* be reflected in the lead-time field of the Item Master record of each affected item. We have focused on queue times for the same reason that Willie Sutton robbed banks: "That's where the money is." While we do not want to discourage improving run speeds or move times, we strongly encourage practitioners to focus their efforts in the area with the largest payback.

For a lead-time reduction program to be effective, personnel on the shop floor must be involved in the initial planning, and must wholeheartedly support the program. To do this, shop personnel must realize that a reduction in lead times will result in a reduction of inventory company wide. This inventory reduction equates to a similar *increase* of available capital that can be used by the company. The floor personnel must also be assured that faster throughput, brought about by an overall reduction in lead times, will not result in an overall reduction of employment.

The same approach can also work for purchased parts and suppliers. We can work with them to reduce the total lead time. In the long run, this should lower their costs, and will certainly help them become more competitive.

Using Safety Stock or Safety Time

There are two ways to buffer uncertainty in an MRP system. One is to specify a quantity of Safety Stock in much the same manner as for independent demand inventory planning (as in the order point systems discussed in Chapter 2). The other method, *safety lead time,* plans order releases earlier than indicated by the requirements plan, and schedules their receipt earlier than the required due date. Both approaches increase inventory levels to provide a buffer against uncertainty, but the techniques operate quite differently.

In the top record of Figure 5-13, there is no buffering. MRP generates net requirements in periods 5, 7, and 9 and plans order releases in periods 3, 5, and 7, because the lead time is two periods. The second record is identical, except that we have specified a Safety Stock of 30. This means that each period the Projected Available inventory drops below 30, MRP will generate a Planned Order Receipt. This happens in periods 3, 6, 8, and 9, with order releases planned for periods 1, 4, 6, and 7.

If we add the units on hand for each period, we discover the true cost of the Safety Stock: a total of 590 units of inventory (66 units per period) was carried in the Safety Stock case, but only 190 units (21 units per period) in the case with no Safety Stock.

In the last record we use a safety-lead time of one period. The Planned Order Receipt dates remain the same as the no buffering case, but the Planned Order Release dates will move to the left by one period, to periods 2, 4, and 6, and the due dates will be one week earlier than the actual need dates, that is, in periods 4, 6, and 8. When implementing safety lead time in packaged systems, we must do more than merely add safety lead time to the regular lead time; we must take steps to change the due dates also. Safety lead times have two major disadvantages:

1. They put false due dates on orders. When people discover that the dates are false (and they *will* find out!), the credibility of the system will be destroyed. A supplier who expends considerable extra effort to ensure that we receive a shipment on time, and then discovers that we didn't need it for another week or two, will remember that experience for a long, long time.

Item A - Bicycle; OQ=80 No Buffering	PERIOD 1	2	3	4	5	6	7	8	9
Gross Requirements	30	60	50		40	30	50	40	60
Scheduled Receipts		80							
Projected Available 60	30	50			40	10	40		20
Net Requirements					40		40		30
Planned Order Receipts					80		80		80
Planned Order Releases			80		80		80		

Item A - Bicycle; OQ=80 Safety Stock = 30	PERIOD 1	2	3	4	5	6	7	8	9
Gross Requirements	30	60	50		40	30	50	40	60
Scheduled Receipts		80							
Projected Available 60	30	50	80	80	40	90	40	80	100
Net Requirements			30			20		30	10
Planned Order Receipts			80			80		80	80
Planned Order Releases	80			80		80	80		

Item A - Bicycle; OQ=80 Safety Lead Time = 1	PERIOD 1	2	3	4	5	6	7	8	9
Gross Requirements	30	60	50		40	30	50	40	60
Scheduled Receipts		80							
Projected Available 60	30	50	0	80	40	90	40	80	20
Net Requirements					40		40		30
Planned Order Receipts				80		80		80	
Planned Order Releases		80		80		80			

Figure 5-13. Using Safety Stock and safety lead time.

2. They increase WIP inventories, if used for shop orders. This increases queues at the various work centers. This action is totally antithetical to our attempts to *reduce* lead times.

Theoretically, Safety Stock is the best buffering technique where there is an uncertainty in either demand or supply *quantity*, while safety lead time is best in cases where demand or supply *timing* uncertainty exists.[1] The *type* of

[1] D. C. Whybark and J. G. Williams, "Material Requirements Planning Under Uncertainty," *Decision Sciences*, vol 7, October 1976, pp. 595–606.

uncertainty, and not the *source,* determines the best buffering technique. These guidelines have important practical implications. Supply timing and demand quantity are the two categories that have the highest levels of uncertainty. Orders from suppliers are subject primarily to timing uncertainty, owing to variability in both production and transportation times. Demand quantity uncertainty exists primarily at the MPS level, because we are dealing with independent, rather than dependent, demand at this level. However, quantity uncertainties can exist at other levels of the Bill of Materials as well, owing to uncertainty of yields in some types of manufacturing processes (e.g., semiconductors and other high-tech products), or suppliers whose processes are also somewhat variable.

Thus, Safety Stock buffering should be used:

Primarily at the MPS level

To the extent necessary at intermediate levels, to buffer against yield uncertainties

To the extent necessary at the purchased level, to buffer against supplier quantity delivery uncertainties

Safety lead time should only be used to buffer the *timing* uncertainty of supplier deliveries, assuming that we can control the timing of deliveries in our own facilities. In spite of the theory, we cannot recommend that a company mislead its suppliers, because excellent supplier-customer relationships require a foundation of trust. Thus, we recommend using Safety Stock rather than safety lead time for all buffering, including raw materials. One other possibility is hedging, which we discuss in Chapter 9.

The theory discussed above is based on a reactive approach to the problems. Instead, we recommend a pro-active, and more productive approach; we should work closely with our suppliers to improve their reliability and timeliness. If our current suppliers are unreliable, we need to develop suppliers who are reliable, even if the out-of-pocket costs appear greater. Current accounting systems cannot measure, and thus do not accurately reflect, the true total cost of the confusion and management attention required by delivery problems. For intermediate levels, we can work on our own processes to increase the predictability of quantities. On the customer demand side, we can work more closely with marketing and our customers to generate better forecasts.

Selecting Lot-Sizing Technique

M&CRP uses Lot-Sizing techniques to recommend how much material to order, for each Planned Order. Lot-Sizing techniques can be classified a number of ways for purposes of explanation. Figure 5-14 lists nine Lot-

	Cost Tradeoff No	Cost Tradeoff Yes	Dynamic No	Dynamic Yes
Lot for Lot (LFL)	X			X
Fixed Order Quantity (FOQ)	X		X	
Fixed Order Interval (FOI)	X			X
Economic Order Quantity (EOQ)		X	X	
Period Order Quantity (POQ)		X		X
Least Unit Cost (LUC)		X		X
Least Period Cost (LPC)		X		X
Least Total Cost (LTC)		X		X
Wagner-Whitin Algorithm (WW)		X		X

Figure 5-14. Lot-Sizing techniques.

Sizing techniques in order of complexity, with Lot-for-Lot being the simplest technique, and the Wagner-Whitin Algorithm being the most complex. This is the order in which we describe the techniques. Also, we can classify them as to whether they are cost-trade-off, or dynamic. A *cost-trade-off* technique trades off the costs of carrying and ordering inventory, to try to achieve a minimum combined cost. A *dynamic* Lot-Sizing technique recalculates quantities each time MRP is run, whereas a nondynamic or *static* Lot-Sizing technique is calculated once and never changes. If a dynamic Lot-Sizing method is used, a Planned Order may be for one quantity during one planning cycle but, because of small changes in requirements, may be for a completely different quantity in the next cycle. Under the same conditions, a static Lot-Sizing technique will plan an order for the same quantity, but for a different period in the next cycle.

Lot-for-Lot (LFL). The Planned Order quantity suggested by the computer system is equal to the exact amount required. Planned Orders are suggested for each and every requirement. This is the simplest and most straightforward of all the exact ordering techniques. It provides period by period coverage of Net Requirements, and the Planned Order quantity always equals the total amount of Net Requirements per period. These order quantities are recomputed whenever their respective Net Requirements change.

- Advantages:

  Easiest of all techniques to use.

  Minimizes inventory carrying costs, because it orders exactly what is required each period.

Generates a steadier flow of orders and work than other techniques that generate fewer and larger orders.

- Disadvantages:

  It ignores the costs related to set up and reordering.

  In a vertical display format, it can generate pages of Planned Orders for a commonly used component.

- When to use:

  For items that are expensive.

  Where demand is sporadic.

  Where the ordering or setup cost is relatively low, and the carrying cost is very high.

  Where demand is so large that some quantity will be ordered every period.

- Final recommendation: Use LFL as the Lot-Sizing technique of choice, if ordering costs are low or can be controlled.

Fixed Order Quantity (FOQ). The FOQ Lot-Sizing technique causes Planned Orders to be generated for a predetermined fixed quantity at all times.

- Advantages:

  Extremely easy to use, because all Planned Orders are of fixed quantities set by the planner.

- Disadvantages:

  Ignores the visibility of future demand available in MRP. Will often cause remnants of inventory to remain once the original demand is satisfied. This will increase overall inventory carrying cost.

- When to use:

  When production or purchasing quantities must be set to specific amounts, or where there is a particular price break or break-even amount the company wants to achieve. However, the same result can be accomplished by using a minimum order quantity, freeing the Lot Size to increase above the Fixed-Order Quantity, if requirements warrant.

  Where some physical characteristic of the manufacturing process requires a fixed Lot Size, for example, an oven with a limited capacity. This situation could, alternatively, be managed by the order multiple parameter.

- Recommendation: Use other techniques unless the physical characteristics of the manufacturing system demand FOQ. Even then, other techniques, such as using order modifiers, can accomplish the same objectives without the rigidity.

Fixed Order Interval (FOI). FOI causes MRP to create replenishment orders that are equal to the Net Requirements for a fixed interval (a given number of periods, or buckets). It resembles the Fixed Interval method of ordering, discussed in Chapter 2.

- Advantages:

  Easy to use.

  Uses future visibility of demand provided by MRP; more responsive than FOQ or EOQ.

  Reduces ordering costs, compared to LFL (useful in "bucketless" environments, especially for low-value parts that have many requirements, such as fasteners).

  Can be used to implement corporate inventory turnover objectives relatively easily. Six turns per year would equate to a two-month FOI.

- Disadvantages:

  Less responsive to changes in demand than Lot-for-Lot.

- When to use:

  Where external circumstances dictate ordering be done on a periodic basis. For example, when purchasing from a distant supplier, it might be cost-effective to order on a periodic basis, to minimize ordering and transportation costs.

- Recommendation: This is an alternative to LFL as the technique of choice.

Economic Order Quantity (EOQ). EOQ is a cost-trade-off technique that attempts to minimize the total costs by trading off the sum of the period carrying costs and ordering costs (assuming them to be fixed, rather than reducible). The following formula for EOQ was developed in Chapter 2 as Equation (2-4):

$$EOQ = \sqrt{\frac{2DS}{Ci}} \qquad (2\text{-}4)$$

where D = Demand, in units per period
EOQ = Economic Order quantity, in units
S = Order cost, in dollars
c = Unit cost
i = Interest, or carrying cost, in percent per period

As an example to demonstrate EOQ, let's assume we have a component with a unit cost of $200, carrying cost of .5 percent per week, an ordering cost of $100 per order, and Net Requirements over the next nine weeks, as shown in Figure 5-15. Note, we are only using the three rows of the inventory record that we need to demonstrate Lot-Sizing behavior. To solve Equation (2-4), we need to determine *D*, the average demand, by summing all Net Requirements and dividing by the number of periods:

$$D = (40 + 15 + 35 + 50 + 30 + 60 + 40)/9 = 30$$

$$\text{EOQ} = \sqrt{2(100)(30)/(200)(.005)} = 77.46 = 77 \text{ (rounded)}$$

- Advantages:

  Is the simplest of the cost trade-off techniques.

  Is the most widely known of the cost trade-off techniques.

  Brings attention to the cost factors of inventory, including ordering, setup, and carrying costs.

- Disadvantages:

  Ignores the visibility of future demand available in MRP.

  Is not an exact technique. It will often cause remnants of inventory to remain once the original demand is satisfied. This will increase overall inventory carrying cost.

  Ignores the extreme subjectivity of the cost factors of inventory, including ordering, carrying, and setup. These can vary, depending on the method of calculating, by factors of 1-5! If the data on which an EOQ is based is so subjective, the resultant EOQ must be equally imprecise.

  Assumes setup and ordering costs are fixed, rather than reducible.

- Recommendation: EOQ should not be the primary technique of choice. However, EOQ might be an entirely suitable choice for Lot Sizing those small inexpensive items where the EOQ is three or six months' supply. Other techniques can accomplish the same objectives, without requiring the square root formula. You can calculate EOQs, from time to time, as an incentive to reduce setup times and ordering costs.

	PERIOD								
	1	2	3	4	5	6	7	8	9
Projected Available	37	22	22	64	14	61	61	1	43
Net Requirements	40	*15*		35	*50*	30		*60*	40
Planned Order Receipts	77			77		77			77

Figure 5-15. Lot-Sizing using EOQ technique.

Period Order Quantity (POQ). The Period Order Quantity uses the same type of economic reasoning as the EOQ, but it determines the number of periods to be covered by each order rather than the number of units to order. We do this by dividing the annual demand rate D by the Economic Order Quantity, EOQ. Using Equation (2-4), we can develop an equation for POQ as follows:

$$POQ = D/EOQ$$

This results in a Fixed Order Interval rather than a Fixed Order Quantity, as in EOQ. Using the normal units for the variables, the answer will be in fractions of a year, which must be converted to periods, and then to the nearest whole period selected as the POQ. However, in our example, both the demand D and the carrying cost H are in weeks, so we will obtain POQ in weeks. To find POQ for our example:

$$POQ = EOQ/D = 77/30 = 2.56 = 3 \text{ weeks (to nearest whole)}$$

In applying POQ, as shown in Figure 5-16, we want to schedule an order every three weeks that should be just large enough to cover the requirements for the next three weeks.

Given that the POQ is based on EOQ, our comments about EOQ also hold true here, except that POQ uses an MRP projection of future requirements to order exactly needed quantities. Because POQ is a dynamic technique, we prefer POQ to EOQ and FOQ.

Incremental Cost-Trade-off Techniques

In this category we group the last four techniques that we will review: Least-Unit Cost (LUC); Least-Period Cost (LPC); Least-Total Cost (LTC); and Wagner-Whitin Algorithm (WW). LUC, LPC, and LTC provide a good solution, although not necessarily optimal. Wagner-Whitin computes an optimal solution, based on its enabling assumptions.

The first three techniques use stopping rules. That is, they start from the first period and test prospective orders covering the first period, then the

	PERIOD								
	1	2	3	4	5	6	7	8	9
Projected Available	15	0	0	80	30	0	0	40	0
Net Requirements	40	15		35	50	30		60	40
Planned Order Receipts	55			115				100	

Figure 5-16. Lot Sizing using POQ technique.

first and second period, then the first, second, and third periods, and so forth, until a stopping criterion is met. They create an order quantity that covers demands in all periods, through the stopping period. Then the process is repeated, starting at the next period after the last stopping period.

The basic cost calculations for the first three methods are the same. The only difference is the criterion used to determine the stopping point. The purpose of all three techniques is to minimize the sum of the carrying costs and the ordering cost. In rows *C, D,* and *E* of Figure 5-17, we have calculated the carrying costs, ordering costs, and total cost for an order that would cover one week, two weeks, three weeks, and four weeks, in columns 1, 2, 3, and 4 respectively. This is the data needed to make the first decision. We will use this data to demonstrate all three techniques. The stopping criterion for making the first decision for each method is identified in Figure 5-17 by the following symbols:

$$(U) = \text{LUC}, (P) = \text{LPC}, \text{ and } (T) = \text{LTC}$$

- Advantages: The cost-trade-off techniques.

  Uses the visibility of future demand in MRP to compute exact quantities.
  Attempts to minimize costs.

- Disadvantages:

  Too complex for most practitioners to easily understand.
  Solves the wrong problem; trades off setup and carrying costs, rather than reducing them.
  Can cause nervousness, changing quantities and periods with each MRP run.

- Where to use:

  Where the minimization of the combined carrying and setup costs are very important, and demand is very volatile and sporadic.

			Week		
Row	Variable	1	2	3	4
A.	Net Requirements (Units)	40	15	0	35
B.	Weeks Held in Inventory	0	1	2	3
C.	Carrying Cost for Lot ($)	0	15	15	(T) 120
D.	Ordering Cost for Lot ($)	100	100	100	(T) 100
E.	Total Cost for Lot ($)	100	115	115	220
F.	Prospective Lot Size (Units)	40	55	55	90
G.	Total Cost per Unit ($)	2.5	2.09	(U) 2.09	2.44
H.	Total Cost per Period ($)	100	57.5	(P) 38.33	55

Figure 5-17. Data for LUC, LPC and LTC Lot-Sizing techniques.

- Recommendation: With many firms moving toward greatly reduced setup costs, these techniques will be less and less relevant. In spite of considerable research, there is insufficient difference in the performance of the four techniques to provide a basis for selection. If the demand is stable and cost minimizing techniques are needed, we prefer the POQ technique, because it is much easier to apply and does not create nervousness in the system.

Least-Unit Cost. This is a dynamic Lot-Sizing technique that adds the ordering cost and inventory carrying cost for each trial lot size, and divides by the number of units in the lot size, picking the lot-size quantity with the lowest unit cost. In Figure 5-17 we divide Total Cost for Lot (row E) by the Prospective Lot Size (row F) to obtain the Total Cost per Unit (row G). The stopping rule is that the first time the Total Cost per Unit goes up, the previous period is the stopping period. Looking at the LUC stopping criterion in row G, we see that the stopping point is reached in period 3. This means our first order should cover the first three weeks. Looking at Figure 5-18, we see there are Net Requirements of 40 and 15 in the first three weeks. Therefore our first Planned Order would be for 55 units due in week one.

Least-Period Cost. The Least-Period Cost was developed by Edward Silver and Harlan Meal, and is often referred to as the *Silver-Meal method.* The procedure determines the total costs of ordering and carrying for lots covering a successively greater number of periods into the future, and then selects the lot with the Least-Total Cost per period covered. The LPC criterion is calculated in row H, and indicates a stopping point is reached in period 3.

Least-Total Cost. Also called *Part Period Balancing,* this method is a dynamic Lot-Sizing technique. The system calculates the order quantity by comparing the carrying cost and the set up, or ordering costs, for various trial lot sizes, and selects the lot size where these costs are most nearly equal. If we look at rows C and D of our data, we can see that the carrying costs and ordering costs are most nearly equal in period 4. Thus our first order of 90 should cover the requirements of the first four periods.

	PERIOD								
	1	2	3	4	5	6	7	8	9
Projected Available	15	0	0	50	0	60	60	0	0
Net Requirements	40	15		35	50	30		60	40
Planned Order Receipts	55			85		90			40

Figure 5-18. Lot Sizing using least-unit cost.

	PERIOD								
	1	2	3	4	5	6	7	8	9
Projected Available	15	0	0	80	30	0	0	40	0
Net Requirements	40	15		35	50	30		60	40
Planned Order Receipts	55			115				100	

Figure 5-19. Lot Sizing using Least-Period Cost.

Look Ahead-Look Back is a technique used to adjust a schedule of orders already obtained, using some other technique. It was originally proposed as refinement of Least-Total Cost, but it can be used to improve the schedules produced by other heuristics as well.

Look Ahead is applied first. The Look-Ahead technique tests an order to see if moving it to the next following period would reduce the number of part-periods in the schedule. (A part-period is the number of parts held for one period.) If the answer is yes, the order is moved ahead one period and the test is repeated. This process continues until the test fails. Then Look Ahead tests the next order, and continues until all orders have been adjusted.

Look Back tests orders that were not moved ahead, to see if moving it to the previous period would reduce the number of part-periods. If the answer is yes, the order is moved back one period and the test is repeated. This process continues until the test fails.

Wagner-Whitin Algorithm. The Lot-Sizing techniques discussed so far are all heuristic in that they attempt to provide near optimal solutions, using simple rules and a modest amount of computation. The Wagner-Whitin Algorithm, however, employs a mathematical-optimization technique, called dynamic programming, and guarantees an optimal solution. This mathematical equation was offered in several software packages in the 1960's, but required massive computer power and time. Because of its complexity, and the lack of any demonstrated superiority of results in the real world, it is used very rarely.

	PERIOD								
	1	2	3	4	5	6	7	8	9
Projected Available	50	35	35	0	90	60	60	0	0
Net Requirements	40	15		35	50	30		60	40
Planned Order Receipts	90				140				40

Figure 5-20. Lot Sizing using Least-Total Cost.

Minimum, Multiple, and Maximum Quantities. Once MRP has determined the basic order quantity, by using the chosen Lot-Sizing Technique, it applies the following finishing touches to the order quantity, which can be used individually or in concert:

- *Minimum,* which defines a floor, or minimum, order quantity. This can be used for purchased parts, when the supplier has specified a minimum, or to offset long setups in our own plant.
- *Maximum* can be caused by limits of production, distribution, or storage space. For example, if we use pre-built shipping containers, the maximum would prevent us from ordering more than we can store. As another example, if we can run a die only so many impressions before it needs to be reground or replaced, the maximum would prevent us from ordering more than a single die can make.
- *Multiple,* perhaps the most useful of the three, allows a planner to specify the multiple of the order, totally independent of any other Lot-Sizing rule. Pencils, for example, come by the dozen or the gross. Parts are often shipped in containers that hold specific quantities. Ovens and injection molding dies have a fixed number of units per cycle.
- Recommendation: We encourage you to utilize any combination of these three order finishers on top of very simple Lot-Sizing rules, such as LFL, FOI, or POQ.

Lot-Sizing Policy

In selecting a Lot-Sizing technique, the planner must understand what is the most important objective for the company and for that particular product. If a planner selects one of the cost-trade-off techniques, the planner is implicitly assuming that the carrying and ordering costs must be traded off. We encourage the practitioner, instead, to reduce the underlying costs.

Cost-trade-off techniques should only be used if setup costs are a major element in manufacturing costs. If setup costs are not a major factor, Lot-for-Lot will minimize inventory carrying costs more than a cost minimization technique will. The smaller the setup times, the less inventory throughout the manufacturing facility. This approach attacks the root of the problem, setup times, rather than trying to minimize their effects by increasing inventories.

With the current trend toward the continual reduction od setup time, and the movement tmward smaller Lot Sizes overall (i.e. JIT), most practical uses of cost minimizing techniques have disappeared. Some process industries must still use cost-trade-off techniques.

If a fixed-Lot-Sizing technique (either FOQ or EOQ) is used, the planner must create a policy regarding what to do if the Net Requirement is greater than the fixed-Lot-Size. For EOQ, if the Net Requirement is larger than the EOQ, the planner should order the Net Requirement, because the EOQ curve tends to be relatively flat at the lowest point, suggesting that a quantity close to the EOQ will still have reasonably minimal total costs. The same is true for FOQ, except that when the FOQ size is controlled by a physical constraint of the manufacturing system (oven size, and so on), the planner should order the multiple of the FOQ that minimally exceeds the Net Requirement.

Correcting Nervousness of MRP System

Changes in Planned Order Releases at the higher levels produce chain reactions downward through the product structures, and hence throughout the Material Requirements Plan. Changes for upper level inventory items may cause other lower level items to have insufficient, or excessive immediate coverage, or to require an additional immediate order release. The degree to which upper level change causes lower level action messages is called *MRP nervousness.*

Nervousness in an MRP system is usually measured by the number of messages on the action reports, which indicate that the previous MRP schedule has changed. Nervousness usually generates the following types of action: the need for expediting, the need for de-expediting, canceling Scheduled Receipts, or releasing Planned Orders in expedite mode.

Causes of Nervousness

Action messages result from quantity and/or timing imbalances, caused by one or more of the following reasons.

Imperfect End-Item Forecasts. One reason for an imbalance may be imperfect end-item forecasts. Customers often change the quantity and/or timing of their orders, which results in a change in the projected requirement and on-hand balances of the end items affected. Because forecasts are updated as time elapses and current sales data are received, forecast demand for specific end items can increase or decrease several times prior to actual shipment.

Lot-Sizing Side Effects. Another reason for the imbalance may be Lot-Sizing side effects, particularly by dynamic Lot-Sizing techniques. Because dynamic Lot-Sizing techniques recalculate the Lot Size every MRP run, a small change in the requirements may result in the new planned orders being in different periods, and for significantly different quantities. A borderline situation may result in the quantity and timing changing back and forth; triggered by only small changes in the requirements.

Long-Lead Times. If the products have long-lead times, the planning horizon must also be "long." A stable and realistic plan far out on the planning horizon may become totally different as time progresses. The shorter the distance that we have to forecast, the more accurate our forecast will be. Long-lead times, like long setup times, cost much more than most practitioners realize, because they reduce flexibility and increase administrative effort.

Scheduled Order Receipt Changes. Changes may occur in Scheduled Order Receipts for a number of reasons. For example, a supplier may be unable to meet the scheduled delivery date, or deliver a quantity different than the scheduled quantity. The same problems may occur internally with manufacturing Scheduled Order Receipts.

Methods for Reducing MRP Nervousness

We discuss the six most common methods for reducing MRP nervousness. Usually no single method will reduce the nervousness to a satisfactory level; two or more methods may need to be used together.

Stabilize the MPS. By improving demand forecasting and strictly enforcing time fences, the MPS can be greatly stabilized. Also, developing a good spare-parts forecast, as an input to MRP Gross Requirements, will reduce nervousness from spare parts.

Stabilize the Planning Data. Refrain from changing the planning data, unless a change is truly needed.

Reduce Lead Times. The smaller the lead times throughout the product lines, the smaller the upheaval that upper level demand changes will cause. Long-range plans would not have to be created, and more time could be spent servicing the customer. Of all the actions, this will be the most beneficial in the long term, because it eliminates the basic cause of nervousness. Imagine, for example, that the time from customer order receipt to fabrication, assembly, and shipment can be shrunk to two days. MRP nervousness would basically disappear, as would time fences, action messages, and Firm Planned Orders.

Constrain Action Messages. A constraint, or damper, can be imposed that restricts reschedule action messages to those exceeding the damper (e.g., do not print any reschedule messages for one week or less). The same type of constraints can be imposed on quantity of either Scheduled Order Receipts or on-hand inventory. For example, if the requirement for an item increased, an action message would not be generated unless the previous requirement is exceeded by a specified quantity. Although these dampers reduce the action messages, they do not reduce the underlying causes, and can cause operational "surprises." This is a bit like disconnecting the temperature gauge in a car so that we won't see that the engine is running hot.

Use Firm Planned Orders. MRP cannot automatically change Firm Planned Orders in future schedules. Changed requirements within the time vicinity of the FPO must be handled by other orders. This method will not decrease the number of action messages, because MRP will still suggest rescheduling the FPO to the date of actual need. This practice may, however, reduce schedule nervousness at levels below the FPO. Its impact on other performance variables is unclear. MRP will not plan inside the FPO, and thus will reduce the number of MRP suggested changes there. However, as we discussed in Chapter 4, FPOs put the entire burden of material planning for that part on the planner. This is a high price, indeed.

Use Lot-Sizing Techniques. Quantity nervousness may be reduced by employing a fixed-Lot-Sizing technique (either FOQ or EOQ), particularly at the upper levels. However, this will have other undesirable effects, such as creating inventory remnants, and inhibiting the reaction to customer demands.

Summary

In Chapter 4, we studied the fundamentals of Material Requirements Planning. In this chapter, we discussed how to use the three MRP reports: primary, action, and pegging. We encouraged you to use the Action Report as a prioritizing tool and strongly recommended that if you use the horizontal MRP report format, you should set the first 10 periods, or buckets, to days, to minimize surprises. We provided examples in which using the single-level Pegging Report avoided the need to expedite, and still provided the shipment to the customer on time. We discouraged the use of the full Pegging Report.

We recommended the actions that an MRP planner should take with respect to these reports, including releasing orders, performing priority planning, responding to changes, and performing bottom-up planning. We insisted that the MRP planner keep MRP totally informed of all changes, and that the planner review and act on each MRP exception message.

The nonroutine actions expected of a M&CRP planner were also studied, including entering or revising the planning data in the M&CRP system. These nonroutine actions involved:

- Selecting lead times (we recommended that you dramatically cut lead times)
- Using Safety Stock or safety time (we recommended that you use Safety Stock sparingly, and that you avoid using safety lead time)
- Selecting Lot-Sizing technique and policy (we recommended using LFL, FOI, and POQ as the best policies, modified by order minimum, multiple, and maximum)
- Correcting nervousness of M&CRP system (we recommended reducing lead times, and stabilizing the Master Schedule and the planning parameters as the best alternatives)

Selected Bibliography

Browne, Jimmie, J. Harhen, and J. Shivnan: *Production Management Systems: A CIM Perspective,* Addison-Wesley, Reading, MA, 1988, Chap 6 and 9.

Haddock, Jorge, and Donald E. Hubicki: "Which Lot-Sizing Techniques Are Used in MRP," *MCRP Reprints,* APICS, Falls Church, VA, 1991, pp. 247–251.

Kanet, John J.: "Toward A Better Understanding of Lead Times in MRP Systems," *Journal of Operations Management,* vol 6, no 3, May 1986, pp. 305–315.

Kropp, D.H., and R. C. Carlson: "A Lot-Sizing Algorithm for Reducing Nervousness in MRP Systems," *Management Science,* vol 30, February 1984, pp. 240–244.

Lunn, Terry with Susan A. Neff: *MRP: Integrating Material Requirements Planning and Modern Business,* Irwin, Homewood, IL, 1992, Chap 7, 8, and 10.

Minifie, J. Roberta, and Robert A. Davis: "Survey of MRP Nervousness Issues," *Production and Inventory Management,* 3d Qtr., APICS, Falls Church, VA, 1986, pp. 111–120.

Orlicky, Joseph: *Material Requirements Planning,* McGraw-Hill, New York, 1975, Chap 6 and 7.

Plossl, George W.: *Production and Inventory Control: Principles and Techniques* 2d ed., Prentice-Hall, Englewood Cliffs, NJ, 1985, Chap 3.

Proud, John F.: "Using the MRP Output," *MCRP Reprints,* APICS, Falls Church, VA, 1991, pp. 101–103.

Smith, Spencer B.: *Computer Based Production and Inventory Control,* Prentice-Hall, Englewood Cliffs, NJ, 1989.

St. John, Ralph: "The Evils of Lot Sizing in MRP," *MCRP Reprints,* Falls Church, VA, 1991, pp. 263–266.

6

Fundamentals of Capacity Requirements Planning

"The most basic assumptions can cause the greatest disasters!" (AUTHORS)

Introduction

MRP assumes that whatever we can schedule, we can make. Sometimes this assumption is valid, sometimes it is questionably valid, and sometimes it is just plain wrong. Capacity Requirements Planning (CRP) tests the assumption and identifies areas of overload and underload, so that planners can take appropriate action. CRP compares the load imposed on each work center by the open and planned orders generated by the Material Requirements Planning system, with the available capacity of each work center, for each time period of the planning horizon. Unlike MRP, which creates new Planned Orders to avoid future shortages, CRP does not create, reschedule, or delete any orders. Instead, it provides a very powerful simulation, which results in a feedback report.

We are aware that, in the early 1990s, most companies do not use CRP. Admittedly, some companies do not need to use it; their products and processes are sufficiently simple that they can load the shop from experience, or by using the Rough-Cut Resources Planning capability in Master Production Scheduling. This fact in no way diminishes the power and capability of CRP; it merely means that CRP remains an underutilized tool that can provide a competitive edge to the companies that utilize it intelligently.

With the assistance of Capacity Requirements Planning, management endeavors to regulate both the arrival of orders, and/or the capacity of a work center to achieve a steady flow of orders through the shop, with minimum queues. Longer queues cause longer shop-order lead times and contribute to inefficiency. Conversely, when the load is reduced, the queues are shortened, resulting in shorter shop-order lead times and greater efficiency.

Like MRP, CRP is used primarily by firms with traditional manufacturing environments: those manufacturing environments found in the fabrication and assembly industries that manufacture discrete products, using a job shop to fabricate the parts, and a small batch-flow line to assemble the final products. They use both Make-to-Order and Make-to-Stock Demand Response Strategies, depending upon the nature of the product and the competition. With modifications, CRP can be used in other manufacturing environments, which will be discussed in Chapter 9. In this chapter we will confine our discussion to CRP operating in a traditional manufacturing environment.

CRP, together with MRP, comprises the third level of the priority-capacity planning hierarchy, as we discussed in Chapter 2 and illustrated in Figure 2-1. Highest on the capacity side of the hierarchy is Resource Requirements Planning, which is a top-management responsibility that determines long-range, overall levels of capacity, regarding work force, and plant and facilities constraints. Second is Rough-Cut Capacity Planning, which is used to determine if sufficient resources are available to execute the Master Production Schedule, concentrating on those resources that are considered potential bottlenecks. Third in the hierarchy is Capacity Requirements Planning, which provides a detailed assessment of the resources needed to execute the manufacturing orders generated by the Material Requirement Planning process. Finally, Operation Sequencing and Input/Output Control provide concise lists of tasks to be accomplished, and detailed assessments of the planned and actual outputs of the shop floor.

This chapter reviews the fundamentals of Capacity Planning and the interactions with MRP and begins by looking at CRP as an input/output system. After describing the inputs of CRP, we discuss the logic and mechanics of the process in order, as follows:

- Identifying and calculating the capacity of the work centers
- Obtaining the necessary information on the orders and routings of the products
- Determining the load on each work center for each time period, including backward scheduling and infinite loading.

We close the chapter with a comprehensive example of CRP that demonstrates the concepts. This example also indicates how the CRP process ties in with the MRP process.

Capacity Concepts

Measures of Capacity

To compare load and capacity, you must use the same unit of measure, for example, pieces, tons, feet, or standard work hours per time period. Additionally, you should select a unit of measure that assists in bottleneck identification, which is a primary purpose of capacity analysis. For example, the process industries use such measures as tons or gallons per unit of time. Repetitive manufacturers can use physical units of product per unit time.

In the intermittent manufacture of discrete parts, where many items may be processed by the same work center during the course of a month, the most common unit of measure used is the *standard hour.* Standard hours produced are equal to the *standard time* per piece, multiplied by the number of pieces produced. An industrial engineer sets the standard time to represent the time required by a qualified and well-trained operator, working at a normal pace. We will be using standard hours as our measure of capacity throughout this chapter.

Capacity Definitions

Because different writers have used different terms for the same capacity concepts, we present the following definitions that will be used throughout the book. In case of duplicate terms, we list the preferred term first, with the alternates following in parentheses.

- *Capacity* (*or Available Capacity*) is the *rate* at which a productive system (worker, machine, work center, department, plant) can produce. This is defined in terms of units of output per unit of time (e.g., pieces per hour).
- *Required Capacity* is the capacity needed to achieve a production schedule.
- *Theoretical* (*or Maximum or Design*) *Capacity* is the maximum possible capacity of the productive system. This is based on the assumption of ideal conditions, such as three shifts, seven days per week, and no downtime.
- *Demonstrated* (*or Actual or Effective*) *Capacity* is the rate of output that can be expected based on experience, taking into account the current product mix and the current and planned levels of resources, such as manpower, overtime, and the number of shifts.

- *Available Work Time* (*or Productive or Scheduled Capacity*) is the number of work hours actually scheduled, or available, at a work center during a specified period.
- *Calculated* (*or Rated or Nominal*) *Capacity* is calculated by multiplying the Available Work Time times the Utilization and Efficiency. The formula is:

  Calculated Capacity = Available Work Time × Utilization × Efficiency
- *Output* is the amount of work that has been produced. Output is usually stated in terms of hours of work or units of production completed through a specified step in the process.
- *Backlog* (*or Queue*) is the amount of work waiting to be released, as well as the amount of work on the shop floor to be performed by a productive system.
- *Load* is the amount of work scheduled to be performed by a productive system in a specified time period.

These concepts are illustrated in Figure 6-1.

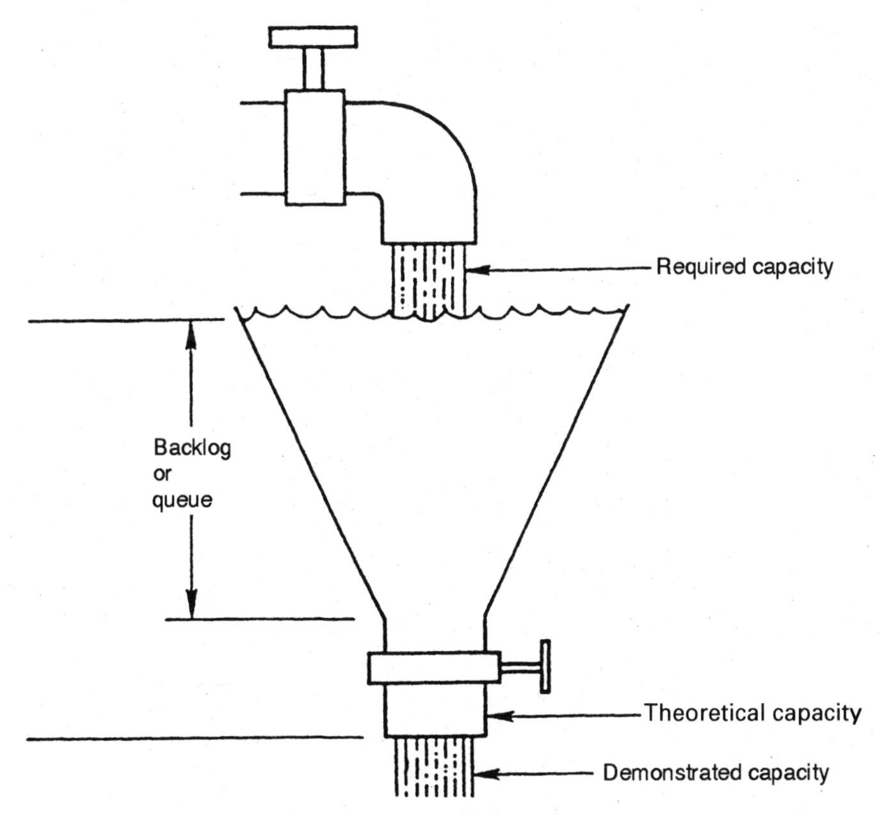

Figure 6-1. Pictorial representation of capacity concepts.

The difference between capacity and load is important. *Capacity* is the rate at which work can be *accomplished by* the system. *Load* is the rate at which work is being *imposed upon* the system. Load is an input rate; capacity is an output rate of a system. If the incoming flow of orders exceeds the capacity, then the load builds up in the form of queues of unprocessed work in front of the work center. If the flow of orders is less than the capacity, the work center is idle part of the time, because it has no work to do. This is called *idle time.*

Types of Load

We can classify load into the following types:

- Planned, which consists of all the Scheduled Receipts and Planned Order Releases that are generated by the Material Requirements Planning System. This type of load is known some time in advance, and capacity for meeting these planned loads can be arranged long before it is required. There are two types of planned load:

  Scheduled Receipts (released load). The released load consists of those orders that have been released to the shop by the MRP planner. These released orders appear on the MRP display as Scheduled Receipts, and on the shop dispatch lists as shop orders.

  Planned Order Releases (unreleased load). The unreleased orders, or Planned Order Releases, are orders in the MRP system for planning purposes only; they do not show up in the shop at all.

- Unplanned, which includes all load that was not generated by MRP. This can include emergencies, engineering requests, unplanned responses to very important customers, personal work, and various other situations. We strongly encourage you to minimize the unplanned load.

To obtain an accurate picture of capacity requirements in the future, both released and unreleased orders *must* be included when running CRP. For this reason, the unplanned load should be also entered into the MRP system as soon as it is known.

Types of Capacity

Capacity Requirements Planning programs determine the level of capacity (resources or inputs) that is needed to achieve the production schedule developed by MRP, and to compare this with available capacities. To perform this function, CRP must be able to identify, measure, and analyze all the resources needed to carry out the production schedule. These resources have historically included direct manufacturing labor and machine hours, but can also include tools, fixtures, material handling equipment, warehouse space, NC tapes, quality measuring equipment, and indirect labor for both shop (setup, engineering assistance, and quality) and office (design engineering).

Currently, most CRP systems only plan for the availability of two resources: direct labor and machine hours. Some CRP systems can only analyze the most critical of these two resources. However, CRP can potentially be used to analyze any one, several, or all of the manufacturing resources.

Input/Output Model of CRP

As shown in Figure 6-2, the Capacity Requirements Planning process transforms the inputs of open orders, Planned Orders, work-center data, and routing data into the output of the Capacity Requirements Plan, or Work-Center Load Reports. In the following paragraphs we describe each of these inputs and outputs as a prelude to a detailed description of the Capacity Requirements Planning process.

Work-Center Data

The work-center record contains many items of information, which are outlined in more detail in Chapter 8. The primary items of information in the work-center data file that directly pertain to capacity management and the manufacturing cycle include:

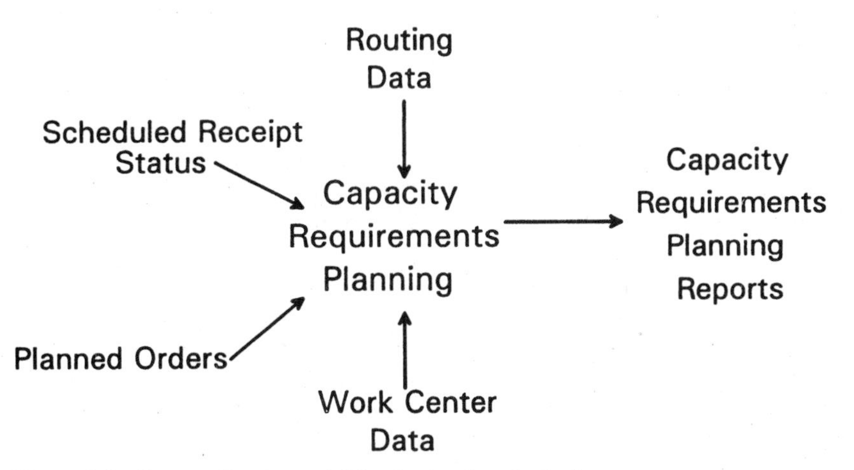

Figure 6-2. Capacity Requirements Planning inputs and outputs.

Identification and description

Number of machines or work stations

Number of workdays per period

Number of shifts per workday

Number of hours per shift

Utilization factor

Efficiency factor

Average queue time

Average wait and move times

Scheduled Receipt and Planned Orders (Load Data)

CRP has two primary sources of load data: Scheduled Receipts and Planned Orders. Other sources, such as product rework, quality recalls, engineering prototypes, and so on, must be translated into one of these two types of orders to be used by CRP. The Order file contains data on both planned and open orders. CRP uses the following data in order to compute the load for each of these sources:

Order quantity

Order due date

Operations completed

Operations remaining

Planned Order release date

Routing Data

A *routing* is the path that an item follows from work center to work center, as it is completed. Every manufactured part, assembly, and product has a unique routing, consisting of one or more operations. Each operation is a single step, or function, which is performed in a specific area or work center. The operations are sequenced, often numbered 10, 20, 30, and so on, to permit insertion of new operations without changing old operations numbers. Figure 6-3 shows an example of a routing for an item. The number of operations in a routing will vary considerably, depending on the nature of the product and the manufacturing process. We include a more complete description of the data elements in the routing file in Chapter 8. For capacity planning, we will concentrate on the following information:

Part Number: 27593 Description: Shaft Drawing No: D 2759

Operation Number	Work Center	Setup Time (hr)	Run Time per Pc (hr)	Operation Description	Tooling
10	Lathe	1.50	.20	Turn Shaft	Chisel
20	Mill	0.50	.25	Mill Slot	
30	Drill	0.30	.05	Drill Hole	
40	Grind	0.45	.10	Grind	
50	Grind	0.00	.05	Polish	

Figure 6-3. Routing for a shaft.

Operation number

Operation

Planned work center

Possible alternate work center

Standard setup time

Standard run time per unit

Tooling needed at each work center

Capacity Requirements Plan Report

The primary output of the CRP process is the Capacity Requirements Plan, or work-center load profile, shown in the form of a bar chart in Figure 6-4. This graphic representation enables a planner to easily see the relationships between projected load and capacity for each period, and readily

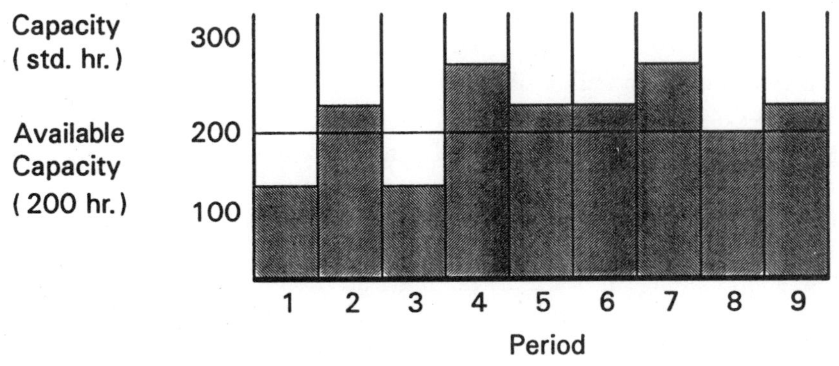

Figure 6-4. Work center Profile for work center *B*.

identify overloads and underloads. CRP produces a Capacity Requirements Plan report, such as this, for each work center in the factory. CRP can also compare the load versus capacity, in the form of columnar data.

CRP Logic and Mechanics

Overview

Capacity Requirements Planning is the process of determining, in detail, the labor and machine resources required to accomplish all manufacturing orders. We describe this procedure in detail in the following paragraphs. The CRP process consists of the following major tasks:

- Determining capacities of work centers
  Identifying and defining work centers
  Calculating capacities of work centers
- Obtaining information on orders and routings
  Obtaining the list of Scheduled Receipts and Planned Orders
  Obtaining routings of items on order
- Calculating load on each work center, in each time period by:
  Using backward scheduling
  Using infinite loading
  Multiplying load for each item by quantity of items scheduled in a time period
- Displaying (on paper or a screen) the results in a Capacity Requirements Plan, showing load vs. capacity for each time period
- Comparing work-center loads and capacities, and taking appropriate action

Determining Capacities of Work Centers

Identifying and Describing Work Centers. A *work center* is a set of one or more machines, and/or a number of workers, that can be considered as one unit for the purpose of capacity planning and detailed scheduling. Each work center has a unique ID number, or name. Production scheduling, shop loading and capacity planning all use the work-center capacity data that resides in the work-center master file. A work center does not have to be a machine, per se. It can be a work area, such as an assembly bench.

Resources can be organized into work centers in several ways:

- By groups of *similar machines*, physically located together, with a range of slightly varying capacities and capabilities. (This is the method normally used by a traditional job shop.)
- An entire *assembly line*, or flow line, can be considered as a single work center having a single capacity, because the entire sequence is run as a unit, or not at all. (This is the approach normally used by a flow shop.)
- Arranging several dissimilar machines in a *cell* to produce a group of products, and considering each cell a work center. (This is the approach taken by JIT work cells and *cellular* [group technology] *manufacturing*.)
- According to the *cost of operations* of the machines, such as those with high or low setup costs, to reduce the problem of small work orders being scheduled into work centers with high setup costs.
- By groups of *people* working on similar, or closely related tasks or products. (This method could be used by a firm that is very labor intensive.)

Figure 6-5 is an example of how work centers might be formed in a small job shop, using the first criterion. The job shop is first divided into three departments, according to similar machines and tasks, and then each department is divided into work centers using the same criterion.

Calculating Work-Center Capacity. Before we calculate capacity, we must make sure that we know which version we want to calculate. The three versions of capacity in use today were defined earlier in this chapter and are now discussed in more detail.

Theoretical capacity is the maximum amount of time, or possible capacity, that could be made available, working all people and/or machines all available hours, each day of the week, with no rest breaks, or machine downtime. Theoretical capacity can rarely be achieved. We will not use it except as an upper bound to test the reasonableness of our capacity calculations.

Demonstrated capacity is a measure of the work that has *actually* been produced by a work center in the past. It is usually obtained by measuring

Dept. 10 Fabrication	Dept. 20 Machining	Dept. 30 Assembly
WC101 Shear	WC201 Turn	WC301 Subassembly
WC102 Form	WC202 Drill	WC302 Final Assembly
WC103 Weld	WC203 Mill	WC303 Insp. and Test

Figure 6-5. Work center definition in a small job shop.

the output of a work center during a period of normal operations. However, this can sometimes be misleading. If a work center runs out of work, or has machine problems, for a period of time, the demonstrated capacity, as measured for that period, is less than it should have been. Practitioners normally use demonstrated capacity only if the data needed to determine calculated capacity is not available.

Calculated capacity is the version of capacity that is most frequently used in CRP and other capacity calculations, particularly when dealing with a job shop. As indicated by the definition and Equation (6-1), calculated capacity (usually stated in hours for a job shop) is the product of three factors.

$$\text{Calculated capacity per period} = \text{available worktime} \times \text{util} \times \text{eff} \tag{6-1}$$

where util = utilization
eff = efficiency

In a job shop, the work time available at a work center for a period is the product of four factors, as indicated in Equation (6-2)

$$\text{Available work time per period} = \text{number of workers or machines} \times \frac{\text{hrs}}{\text{shift}} \times \frac{\text{shifts}}{\text{workday}} \times \frac{\text{workdays}}{\text{period}} \tag{6-2}$$

The Available Work Time is the time that is actually scheduled; it is usually considerably less than the maximum time that could be available. If a work center has both workers and machines, the calculated capacity should be based on the smaller of the two capacities. In some cases, the planner may wish to calculate both, particularly if the human capacity is less than the machine capacity. If needed, additional human capacity can be added by using overtime or temporary workers. By combining Equations (6-1) and (6-2), we can state a more detailed equation for the calculated capacity as follows:

$$\text{Calculated capacity per period} = \text{number of workers or machines} \times \frac{\text{hrs}}{\text{shift}} \times \frac{\text{shifts}}{\text{workday}} \times \frac{\text{workdays}}{\text{period}} \times (\text{util}) \times (\text{eff}) \tag{6-3}$$

The number of hours actually worked may be less than the available hours, because of such factors as lack of materials, machine downtime, absenteeism, or the lack of other critical resources (such as skilled setup operators or first-piece inspection approval). The ratio of hours worked to hours available is called *utilization,* and is given by:

$$\text{Utilization} = \frac{\text{number of hours worked}}{\text{number of hours available}} \tag{6-4}$$

Utilization can be determined for either labor or equipment, or both, depending on which is more appropriate in a given situation.

Efficiency is a measure of the productivity of a worker, work center, department, or plant, and is measured by the ratio of standard hours produced to number of hours worked. That is:

$$\text{Efficiency} = \frac{\text{number of standard hours produced}}{\text{number of hours worked}} \tag{6-5}$$

Utilization and efficiency are monitored by the Production Activity Control system; planners often use the actual results for planning purposes. Smoothing techniques can summarize past data, putting greater weight on more recent experience. However, poor performance should be a signal for corrective action, not a basis for planning the future.

In some plants, especially those implementing Activity-Based Costing, setup time is considered direct work (that is, it will be charged directly to the production order involved), and so it is included in the hours worked and standard hours produced in figuring capacity. Elsewhere, particularly where the setup is performed by a setup crew and not by the machine operator, setup is considered indirect work and is charged to overhead. Under this arrangement, setup time is treated as machine downtime, and thus contributes to the number of scheduled hours not worked in calculating utilization. This practice increases the overhead percentage and decreases the machine utilization percentage.

An Example. To better understand the meaning and use of these concepts, let's look at an example. The ABC Company has a work center made up of 4 milling machines and 4 operators, each operating one 8-hour shift, 5 days a week. During the past four weeks they have averaged 16 hours of machine downtime per week. The measured efficiency of the operators is 95 percent. While producing a product that required one-fifth hour of milling, they produced quantities of 600, 620, 610, and 590 in each of the last four weeks. Using this data, we can answer the following questions.

1. What is the theoretical capacity?
2. What is the demonstrated capacity?
3. What is the calculated capacity?

1. Because machines and operators are equal, we can use either for our calculations in the following equation:

$$\begin{aligned}\text{Theoretical capacity per period} &= (\text{max. no. of machines})\\ &\quad\times(\text{max. no. of hours per shift})\\ &\quad\times(\text{max. no. of shifts per workday})\\ &\quad\times(\text{max. no. of workdays per period})\\ &= (4)\times(8)\times(3)\times(7)\\ &= 672\ \text{hrs/week}\end{aligned}$$

2. To find the demonstrated capacity, we will average the weekly output over the four weeks.

$$(600 + 620 + 610 + 590)/4 = 605\ \text{units/wk}$$
$$605\ \text{units/wk} \times 1/5\ \text{hrs/unit} = 121\ \text{hours/wk}$$

3. To find the calculated capacity, we must first find the utilization. To do so, we will use Equations (6-2) and (6-4).

$$\begin{aligned}\text{Available work time per period} &= (\text{no. of workers and/or machines}) \quad (6\text{-}2)\\ &\quad\times(\text{no. of hours per shift})\\ &\quad\times(\text{no. of shifts per workday})\\ &\quad\times(\text{no. of workdays per period})\\ &= (4)\times(8)\times(1)\times(5)\\ &= 160\ \text{hours/week}\\ \text{No. of hours worked} &= 160 - 16 = 144\end{aligned}$$

$$\text{Utilization} = \frac{\text{number of hours worked}}{\text{number of hours available}} = 144/160 = .90 = 90\%$$

$$\begin{aligned}\text{Calculated capacity per period} &= (\text{available work time}) \quad (6\text{-}1)\\ &\quad\times(\text{utilization})\\ &\quad\times(\text{efficiency})\\ &= (144)\ \text{hrs/wk}\times(.9)\times(.95)\\ &= 123\ \text{hours/week}\end{aligned}$$

In the above example, the work center actually performed (demonstrated capacity) very close to its calculated capacity.

Obtaining Information on Orders and Routings

Determining the Schedule of Open and Planned Orders. In M&CRP systems, Planned Orders and open orders are both in the Order file. Open orders (Scheduled Receipts) contain the current status, including the number of operations completed, the number of hours worked to date, and the number of operations remaining.

Preparing and Using Routings. A routing (also called a Bill of Operations, Operations Chart, Operations List, or Routing Sheet) is a set of information that details the method of manufacturing of a particular item. It includes the operations to be performed, their sequence, the various work centers to be involved, and the standards for setup and run. In some companies, the routing also includes information on tooling, operator skill level, inspection operations, drawings and testing requirements. We include sample data from a typical routing record in Chapter 8.

Once the work centers have been defined, Manufacturing Engineering can route each of the parts through them, and develop the step-by-step *routing* for manufacturing each part and assembly. Some systems use a unique identifying number for the routing, distinct from the number of the item being manufactured. This allows different items to share the same routing. For example, two assemblies with different item numbers, differing only in color, could share the same routing.

The standard routing for an item is its *primary routing*. An item can also have one or more *alternate routings*, which describe additional methods of manufacturing the same item. Alternate routings are used when the primary routing is unavailable, owing to capacity constraints or equipment unavailability. Additionally, a routing can include one or more *alternate operations*, which provide options for manufacturing the item when a given operation has capacity constraints or is otherwise unavailable. The standard operations are called *primary operations*.

Figure 6-6 illustrates an example of a routing sheet for a turned wood product, such as a shelf pole, or spacer. The operation numbers are assigned with intervals of ten so that the routing can be revised in the future, without having to renumber all the subsequent steps.

The routing identifies the work center for each step, but the machine in the work center is normally selected by the work-center foreman. Setup time refers to the length of time required to get ready to perform the task, such as positioning the saw blade, setting up a template for the lathe, or

Item Number 321 Wooden Shelf Pole				Drawing Number D1102		
Operation number	WC number	Operation description	Setup time (hrs)	Prod. rate (pieces/hr)	No. of people	No. of mach
10	102	Cut raw stock	0.1	350	2	1
20	201	Turn to shape	0.2	175	2	2
30	203	Sand	0.1	225	2	2
40	302	Varnish	0.2	250	1	1
50	401	Package	0.3	700	2	1

Figure 6-6. Routing for wooden shelf pole.

positioning the part for sanding. The production rate refers to the expected number of pieces per hour that should be produced, once the machine is set up.

The Production Activity Control system maintains the data, operation by operation, in the order-specific routing for each Scheduled Receipt. From this file, CRP learns the status of each open order, so CRP can compute the amount of work left. For planned shop orders, CRP uses the standard routing file for the part.

Understanding and Using the Manufacturing Cycle. In order to understand how certain additional information is used to schedule orders through the work centers, it is first necessary to understand the manufacturing cycle, and how load and nonload activities are distributed in this manufacturing cycle. The manufacturing cycle consists of all the activities that occur to an item between the release of an order to the shop and the receipt into stores, or the shipment to the final customer.

The manufacturing cycle in a job shop typically consists of the following activities: order preparation, queue, setup, run, wait, move, inspection, and store. Inspection can occur in the work center, or at designated areas, but is not discussed here. We will ignore order preparation and storing, because they usually do not occur in the work centers, and concentrate on those activities that do occur in the work centers, as shown in Figure 6-7.

Load activity is any activity that requires the resources or capacity of a work center. Of the five activities shown in Figure 6-7, only setup and run are load activities. The others, queue, wait, and move, are nonload activities, because they occur without using work-center resources. Setup can be divided into two types: *internal,* which absorbs machine capacity, and *external,* which allows the machine to perform other work.

- *Queue time is* usually an average time associated with a work center, and resides in the work-center files.
- *Setup time* and *run time* are usually associated with each operation on the routing sheet for a specific item, are unique to that item, and can be found in that item's routing file.

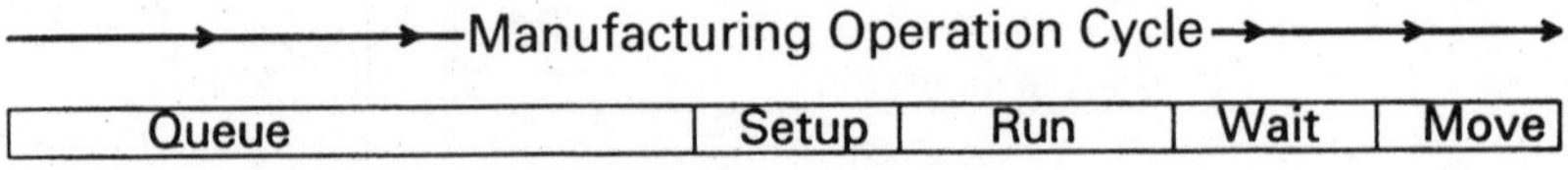

—Manufacturing Operation Cycle→

Queue	Setup	Run	Wait	Move

- Queue = time waiting before the operation begins
- Setup = time readying the machine for operation
- Run = time performing the operation
- Wait = time waiting after the operation ends
- Move = time physically moving between operations

Figure 6-7. The manufacturing cycle in a job shop.

- *Wait time* and *move time* are normally more a function of the material control and handling system in a particular shop or plant, and are often common for the whole shop. However, they are often found in the work center and routing files, respectively.

All activities, both load and nonload, must occur in the sequence indicated in Figure 6-7, and they all require time. Thus, we must know the times of all these activities in order to schedule manufacturing orders through a job shop. Times associated with load activities are sometimes called operation times, whereas times associated with nonload activities are sometimes referred to as *interoperation times*. JIT views these activities differently. In JIT terms, *value-added* activities are those that add value in the eyes of the customer. JIT attempts to identify and eliminate all nonvalue-added activities. Of the five activities in Figure 6-7, only run is truly a value-added activity.

Calculating Loads and Distributing to Work Centers by Time Period

Backward Scheduling of Scheduled Receipts and Planned Orders. CRP uses two approaches to scheduling: forward scheduling and backward scheduling.

1. Forward scheduling begins today with a Scheduled Receipt, or the start date of a Planned Order, and schedules the earliest start date for each operation in a forward direction (from start date to finish date). It then uses that finish date of the scheduled operation as the earliest start date of the next operation.
2. Backward scheduling begins with the scheduled due date and backtracks, using the routing to determine the latest start date for each operation. It then uses that latest start date as the scheduled due date for the previous operation, and repeats the backward scheduling until it has scheduled all operations for a given order.

These concepts can best be understood by an example.

In Figure 6-8, we will schedule two jobs *A* and *B* through a single work center, using forward and backward scheduling. The work center operates five days a week, using a single shift from 8 a.m. to 4 p.m. The release date, due date, and all the routing information on jobs *A* and *B* are provided at the top of Figure 6-8. To simplify the charting of the schedule, we have combined the setup time and the run time into a single load time. We use the following symbols in charting the activities: *Q* = Queue, *L* = Load, *W* = Wait, and *M* = Move. The symbols *A* and *B* preceding the activity symbol denote the job being scheduled.

(All times in hours)

Job	Release Date	Queue Time	Setup Time	Run Time	Load Time	Wait Time	Move Time	Due Date
A	1	16	1	6	7	8	4	15
B	4	20	1	10	11	8	4	19

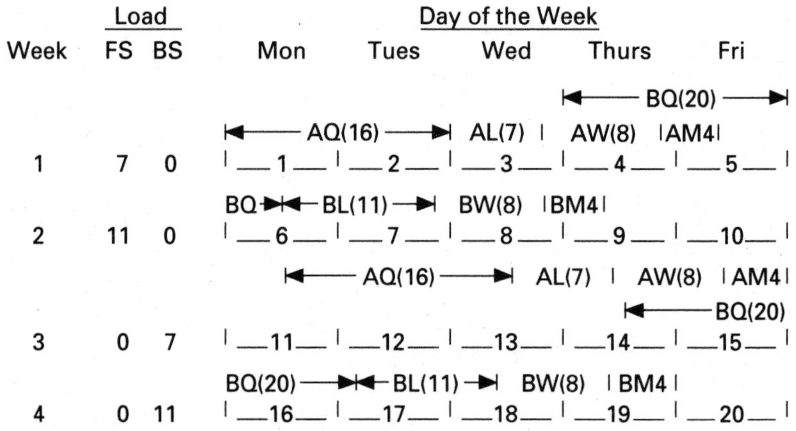

Week	Load FS	Load BS	Mon	Tues	Wed	Thurs	Fri
						←BQ(20)	→
			←AQ(16)	→	AL(7)	AW(8)	AM4
1	7	0	1	2	3	4	5
			BQ→←BL(11)	→	BW(8)	BM4	
2	11	0	6	7	8	9	10
			←AQ(16)	→	AL(7)	AW(8)	AM4
						←	BQ(20)
3	0	7	11	12	13	14	15
			BQ(20)	→←BL(11)→	BW(8)	BM4	
4	0	11	16	17	18	19	20

Figure 6-8. Forward and backward scheduling.

Using forward scheduling, we start job *A* on its release date of 1; it finishes on the 5th day, in the first week. We start job *B* on its release date of 4, and it completes on the 9th day, in the second week. Using forward scheduling, the two jobs are completed far in advance of their due dates.

Using backward scheduling, we finish job *B* on its due date. Thus we finish job *B* at the end of the 19th day (due date is 19), and backward schedule the activities of job *B*, from right to left, starting with the Move activity. From our schedule, we find out that the latest time that job *B* can start, and still finish on the due date, is at 1 p.m. on the 14th day. To schedule job *A*, which has a due date of 15, we find it must start not later than 1 p.m. on the 11th day. Thus, both jobs can start much later than their release dates, and still meet their due dates.

Although this example appears unrealistic, this situation is actually quite common in manufacturing companies. This graphically illustrates the difference between the static MRP lead times (15 days, or 3 weeks, for parts *A* and *B*), and the projected actual lead times, based on routing and quantity. To correct this obvious imbalance, you can use the steps suggested in Chapter 5, under the heading *Lead Time.*

Practitioners have debated the merits of backward versus forward scheduling in CRP for years. Backward scheduling is currently the accepted practice, because it comes closer to the way that a loaded plant actually operates. However, in an underloaded plant, the assumption that jobs will start on the last possible date is not valid, so the CRP load profile, produced by backward scheduling, is invalid. The converse is true; forward scheduling is equally invalid in a plant that is filled to capacity (and beyond) with orders. MRP uses backward scheduling, but fixed lead times, so that order lead times are averages at best. When CRP uses backward scheduling, employing *detailed scheduling* (using the precision of actual routing data), it assists the planners to release orders to start just in time to meet their due dates, and therefore minimizes the work-in-process inventory. However, if this backward scheduling step results in some orders with past due start dates, these orders must be forward scheduled, starting on the current date.

Utilize Infinite Loading to Generate Load Profiles. CRP uses *infinite* loading, in conjunction with backward scheduling, to generate the Capacity Requirements Plan, or Work-Center Load Reports. The difference between finite and infinite loading is illustrated in Figures 6-9*a* and 6-9*b*. Figure 6-9*a* shows the results of infinite loading as it might be projected by Capacity Requirements Planning. Figure 6-9*b* shows the load profile that would result from finite loading. Infinite loading distributes load into the work center's time buckets, without regard to the maximum capacity of the work center. The resulting work-center load profiles reveal both underloaded periods and overloaded periods, with the available capacity usually indicated by a horizontal line across the profile.

Finite loading starts with a schedule of work orders, as does infinite loading. Before finite loading can begin, however, priorities must be established for individual orders, generally by Sales and Marketing prioritizing customer orders. The highest priority orders get first claim on available capacity in each work center. Finite loading calculates the limiting capacities in each work center. The jobs are then loaded into the work centers in priority sequence. As soon as a work center is filled to its limiting capacity, the "overflow" jobs are scheduled where unused capacity exists, either earlier or later, depending on the rescheduling algorithm and the planning parameters. Finite loading generally utilizes backward, forward, and other alternative scheduling methods, to distribute the load evenly among the periods. Finite loading also utilizes alternate work centers and routings; infinite loading only uses the primary work centers and routings, because it would not know when to switch to the alternate.

The finite-loading approach will only schedule work in a work center up to its capacity. Thus, the 75 hours of work shown as past due in Figure 6-9*a* would be scheduled in week 1, under finite-loading techniques. Finite loading does not solve the undercapacity problem illustrated in Figure 6-9*a*. If

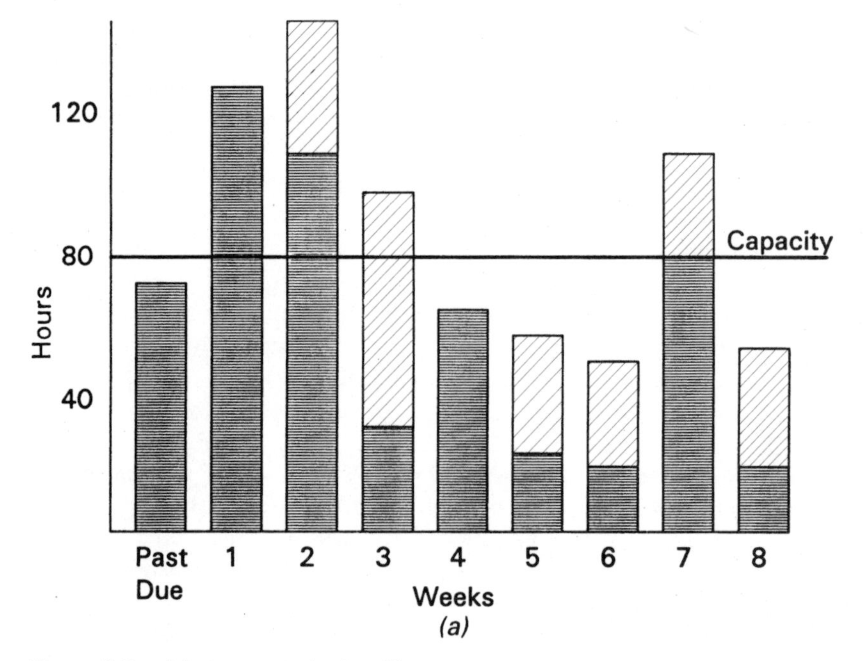

Figure 6-9*a*. Infinite capacity load profile.

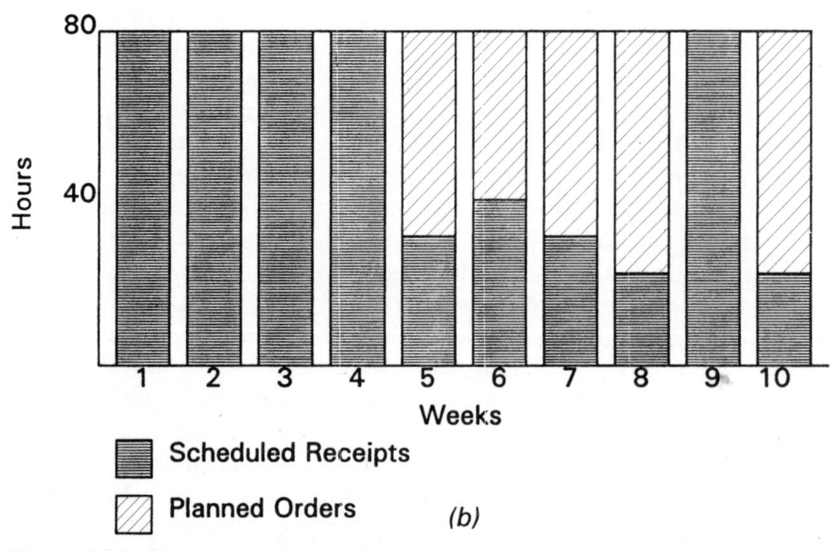

Figure 6-9*b*. Finite capacity load profile.

the capacity is not increased, only 80 hours of work will come out of this work center each week, regardless of the scheduling procedure. Finite loading will determine *which* jobs will come out, based upon priorities.

Finite-loading and infinite-loading techniques each have drawbacks:

- Loading work into a plant without regard for its capacity (as infinite loading does) is unrealistic. If a plant is overloaded, the schedules and expectations that infinite loading creates are invalid.
- If each order is fit into available capacity (finite loading), and nothing is ever done to change capacity, customer service will be severely affected. Further, finite loading masks the need for additional capacity, because it does not indicate periods when capacity overloads exist. It simply loads to capacity and carries all unscheduled work forward.

CRP normally uses infinite loading and backward scheduling, whereas Operations Sequencing, which is part of Production Activity Control module, uses finite loading and various forms of scheduling. The CRP process involves:

Performing infinite loading to determine potential problems

Modifying the available capacity

Modifying the MPS schedule, as a last resort.

This approach is more useful and appropriate at the M&CRP level, than either finite loading or infinite loading used alone.

A Comprehensive Example of Capacity Requirements Planning

Introduction

Conceptually, the mechanics of Capacity Requirements Planning are quite simple. CRP receives the Scheduled Receipts and Planned Order Releases from the Material Requirements Planning system, and simulates the path of each job through the plant, by work center and time period. From this simulation, CRP prepares a report (or inquiry screen) that compares the load in each work center to the available capacity.

However, actually performing the mechanics of capacity requirements can become quite tedious, even for small shops. To demonstrate the mechanics, we will use another simple bicycle *P*, similar to the item *A* used in

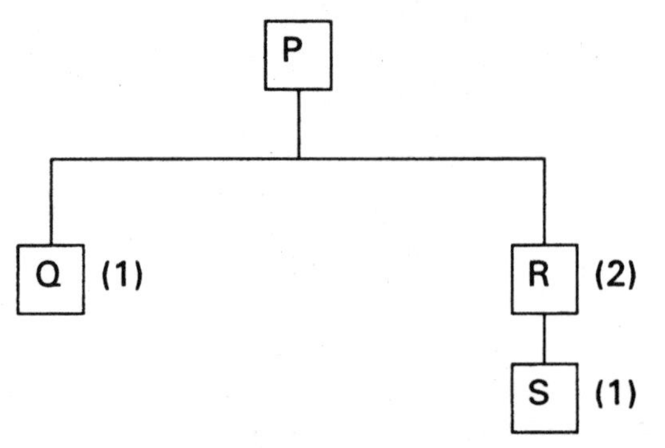

Figure 6-10. Bill of Materials of product *P.*

Chapter 4. In the Bill of Materials for product *P,* shown in Figure 6-10, the usage quantities are given in parentheses to the right of each item.

We will also use a very simple factory, consisting of only three work centers, as shown in Figure 6-11.

Determining the Capacities of the Work Centers

From the files on Work Centers A, B, and C, we find that each work center has one machine and one operator working a single 8-hour shift, 5 days a week. The utilization of all work centers is 90 percent, and the efficiency of all operators is 95 percent. We can use Equation (6-1) to calculate the Available Work Time for each work center:

$$\text{Available work time (in min.)} = 1 \times 8 \times 1 \times 5 \times 60 = 2400 \text{ min./week}$$

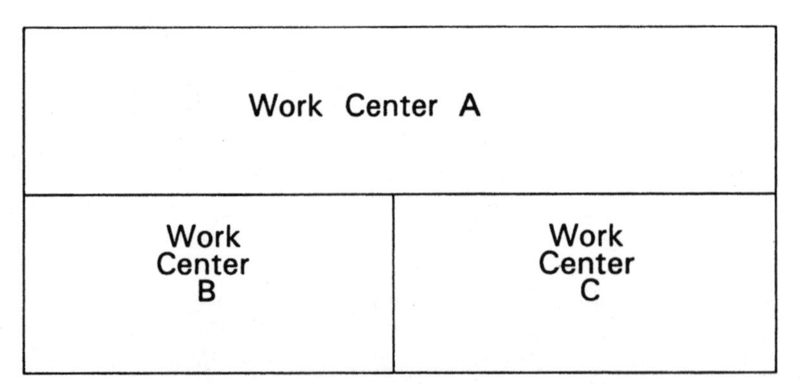

Figure 6-11. Plant layout to produce product *P.*

Work Center	Available Work Time	Utilization	Efficiency	Calculated Capacity
A	2400 min	90%	95%	2052 min
B	2400 min	90%	95%	2052 min
C	2400 min	90%	95%	2052 min

Figure 6-12. Capacities of Work Centers A, B, and C.

We elected to measure capacity in minutes to avoid fractions and to make the arithmetic easier to follow; normally, capacity is measured in hours. We can then use Equation (6-2), based on Figure 6-12, to determine the calculated capacity:

$$\text{Calculated capacity} = (2400 \text{ min.}) \times (.90) \times (.95) = 2052 \text{ min./week}$$

Determining the Schedule of Open and Planned Orders

Figure 6-13 displays pertinent planning information from the Item Master file on items *P*, *Q*, *R*, and *S*. There are no open orders (Scheduled Receipts).

From the Master Production Schedule, we obtain the information that item *P* has requirements in periods 6, 7, 8, and 9, as indicated in Figure 6-14.

We then enter the information contained in Figures 6-10, 6-13, and 6-14 from the Bill of Materials file, Item Master file, and the Master Production Schedule into the Inventory Record and process it manually, as described in Chapter 4, and shown in Figure 6-15.

From the data in Figure 6-15, we can summarize the Planned Order Releases for the four parts in Figure 6-16.

From the routing file we can obtain the routing information on items *P*, *Q*, *R*, and *S*, as shown in Figure 6-17*a*, which indicates item *P* has one operation, items *Q* and *R* have two operations each, and item *S* has three operations. To keep this initial example simple, we assume that each operation

Item	Lead Time	Lot Size	Safety Stock	Quantity On Hand
P	1	LFL	0	50
Q	2	LFL	0	50
R	2	LFL	0	100
S	3	LFL	0	600

Figure 6-13. Planning data on items *P*, *Q*, *R*, and *S*.

Period

Item P	1	2	3	4	5	6	7	8	9
Gross Req'ts						400	200	300	400

Figure 6-14. Master Production Schedule for item *P.*

Period

Item P	1	2	3	4	5	6	7	8	9
Gross Req'ts.						400	200	300	400
Sched. Receipts									
Proj. Avail. 50	50	50	50	50	50	0	0	0	0
Net Req'ts.						350	200	300	400
Pln. Ord. Receipt						350	200	300	400
Pln. Ord. Release					350	200	300	400	

Period

Item Q	1	2	3	4	5	6	7	8	9
Gross Req'ts.					350	200	300	400	
Sched. Receipts									
Proj. Avail. 50	50	50	50	50	0	0	0	0	0
Net Req'ts.					300	200	300	400	
Pln. Ord. Receipt					300	200	300	400	
Pln. Ord. Release			300	200	300	400			

Period

Item R	1	2	3	4	5	6	7	8	9
Gross Req'ts.					700	400	600	800	
Sched. Receipts									
Proj. Avail. 100	100	100	100	100	0	0	0	0	0
Net Req'ts.					600	400	600	800	
Pln. Ord. Receipt					600	400	600	800	
Pln. Ord. Release			600	400	600	800			

Period

Item S	1	2	3	4	5	6	7	8	9
Gross Req'ts.			600	400	600	800			
Sched. Receipts									
Proj. Avail. 600	600	600	0	0	0	0	0	0	0
Net Req'ts.				400	600	800			
Pln. Ord. Receipt				400	600	800			
Pln. Ord. Release	400	600	800						

Figure 6-15. MRP plan for items *P, Q, R,* and *S.*

	Period								
Pln. Ord. Receipt	1	2	3	4	5	6	7	8	9
Prod. P						350	200	300	400
Component Q					300	200	300	400	
Assembly R					600	400	600	800	
Component S				400	600	800			

Figure 6-16. Planned Order Receipts for items *P*, *Q*, *R*, and *S*.

requires exactly one period, including queue, setup, run, wait, and move. Figure 6-17*a* includes run and setup times for each operation, so that we can load the work centers.

Using Backward Scheduling to Obtain Schedule of Operations

When we combine the routing information in Figure 6-17*a* with the quantities and due dates provided by the Planned Order Receipts in Figure 6-16, we can develop the Schedule of Operations for items *P*, *Q*, *R*, and *S* as shown in Figure 6-17*b*. We use one line for each operation for each part. We put all the orders that pass through that operation on the same line. So, if we have four separate orders for item *R*, we will have four entries for *R*, operation 1 (abbreviated *R*1), and 4 entries for *R*, operation 2 (abbreviated *R*2). Each entry is for the order quantity. Taking the entry in line 1, period 6 as an example, 350*P*1 indicates that 350 units of item *P* should start and complete its 1st (and only) operation during period 6, where the entry is located. Interpreting the entry in line 8, period 2, 400*S*1 indicates that 400 units of item *S* should start and complete its 1st operation during period 2. Please remember that we have restricted each operation to exactly one period for simplicity of calculation in this example.

The Schedule of Operations in Figure 6-17*b* is developed *using backward scheduling* as follows. Because we are loading to infinite capacity, we can start with any order for any part; it makes no difference. So let us start by scheduling the first order of the end item *P*. We start by scheduling the last operation of the first order for the end item (350*P*1, in period 6). Because the last operation is the only operation, we schedule the operations for the other orders for the end item, in periods 7 through 9. We then schedule the last operations, 300*Q*2, 200*Q*2, 300*Q*2, and 400*Q*2 in periods 5 through 8. We schedule the first operations for *Q* in periods 4 through 7 at the same time. We repeat the process for *R*. Finally, we can backward schedule item *S*, with the last operation for the first order being in period 4, the middle operation in period 3, and the first operation in period 2. We can then pro-

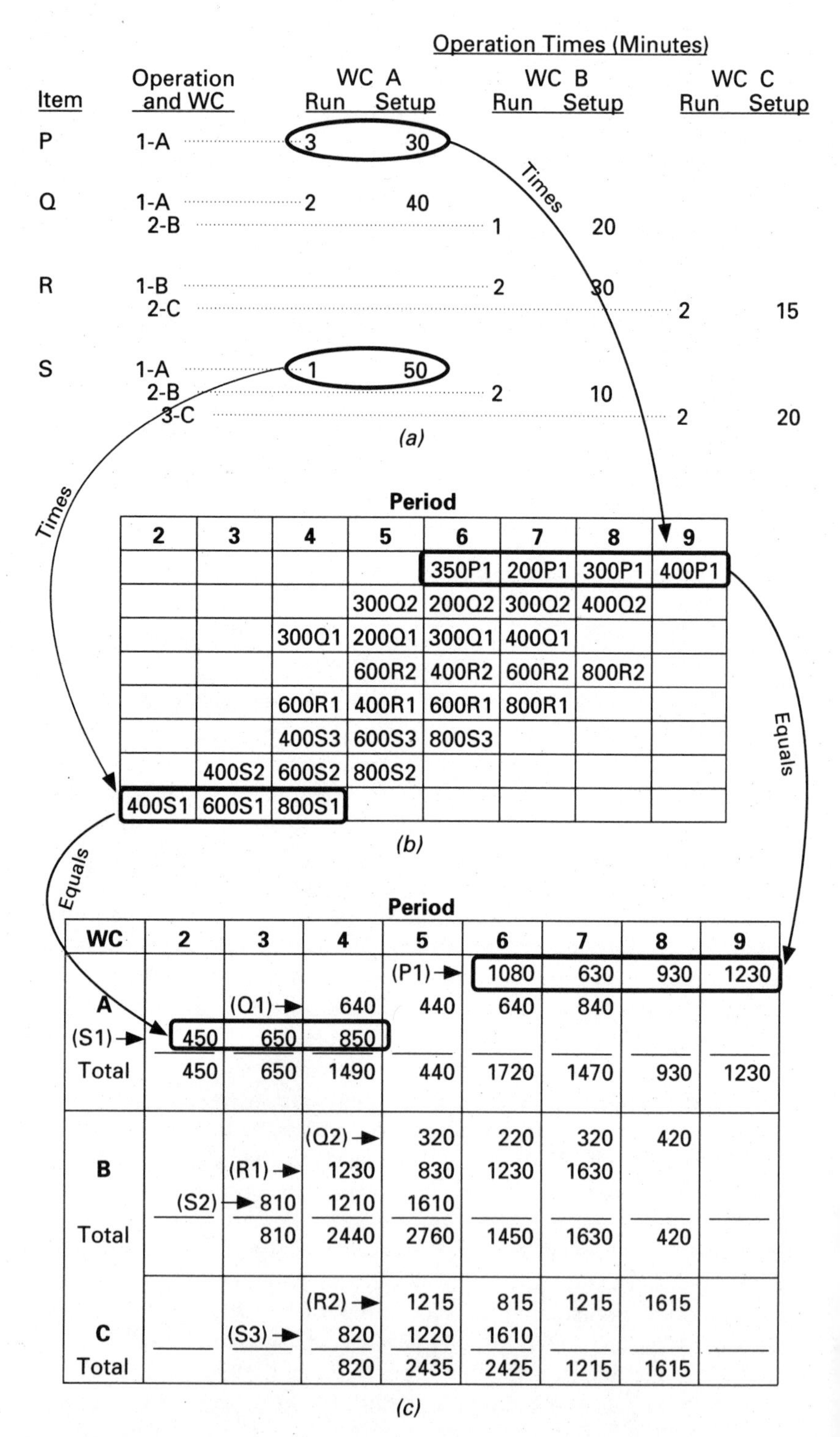

		Operation Times (Minutes)					
		WC A		WC B		WC C	
Item	Operation and WC	Run	Setup	Run	Setup	Run	Setup
P	1-A	3	30				
Q	1-A	2	40				
	2-B			1	20		
R	1-B			2	30		
	2-C					2	15
S	1-A	1	50				
	2-B			2	10		
	3-C					2	20

(a)

Period

2	3	4	5	6	7	8	9
				350P1	200P1	300P1	400P1
			300Q2	200Q2	300Q2	400Q2	
		300Q1	200Q1	300Q1	400Q1		
			600R2	400R2	600R2	800R2	
		600R1	400R1	600R1	800R1		
		400S3	600S3	800S3			
	400S2	600S2	800S2				
400S1	600S1	800S1					

(b)

Period

WC	2	3	4	5	6	7	8	9
A				(P1)→	1080	630	930	1230
		(Q1)→	640	440	640	840		
(S1)→	450	650	850					
Total	450	650	1490	440	1720	1470	930	1230
B			(Q2)→	320	220	320	420	
		(R1)→	1230	830	1230	1630		
	(S2)→	810	1210	1610				
Total		810	2440	2760	1450	1630	420	
C			(R2)→	1215	815	1215	1615	
		(S3)→	820	1220	1610			
Total			820	2435	2425	1215	1615	

(c)

Figure 6-17. *(a)* Routing information for items *P, Q, R,* and *S*; *(b)* Schedule of Operations for items *P, Q, R,* and *S*; *(c)* Accumulating loads for work centers by time periods.

ceed to schedule the other lots (600 in periods 5, 4, and 3, and 800 in periods 6, 5, and 4).

Calculating Loads and Distributing Using Infinite Loading

With the routing information available from Figure 6-17*a* and the Schedule of Operations available from Figure 6-17*b*, we can now manually perform the Capacity Requirements Planning process, and compute the load on each work center caused by the schedule for items *P*, *Q*, *R*, and *S*. To simplify the arithmetic, we will calculate the load in minutes; normally the load is computed in hours. The load resulting from each operation can be calculated by the following equation, with the run time and setup time available from Figure 6-17*a*, and the order size available from Figure 6-17*b*.

$$\text{Load} = \text{lot size} \times \text{run time} + \text{setup time}$$

This equation calculates the load imposed upon a specific work center, in a specific time period, by a specific operation. We deposit the results of these calculations into a matrix of work-center load versus time period, as shown in Figure 6-17*c*. Figure 6-17*a*, *b*, and *c* are positioned together to better enable you to follow the explanation. We show below the calculations for the first operation of *P* and *S*, which are also shown diagrammatically on Figures 6-17*a*, *b*, and *c*. The item and operation that is the source of the entries is shown in parentheses to the left of the entries in Figure 6-17*c*.

$$\text{Load}\ (P, 1,6) = (350 \times 3) + 30 = 1080$$
$$\text{Load}\ (P, 1,7) = (200 \times 3) + 30 = 630$$
$$\text{Load}\ (P, 1,8) = (300 \times 3) + 30 = 930$$
$$\text{Load}\ (P, 1,9) = (400 \times 3) + 30 = 1230$$

$$\text{Load}\ (S, 1,2) = (400 \times 1) + 50 = 450$$
$$\text{Load}\ (S, 1,3) = (600 \times 1) + 50 = 650$$
$$\text{Load}\ (S, 1,4) = (800 \times 1) + 50 = 850$$

Note: Load (x, y, z) = Load (item, operation, period).

Comparing Work-Center Loads and Capacities

We can obtain a good understanding of the capacity problems facing the managers of Work Centers A, B, and C, by plotting the planned load versus the available capacity for the time periods of interest, as shown in Figure 6-18. By looking at the graph, we can see that we have capacity problems at Work Center B, during periods 4 and 5, and at Work Center C, during peri-

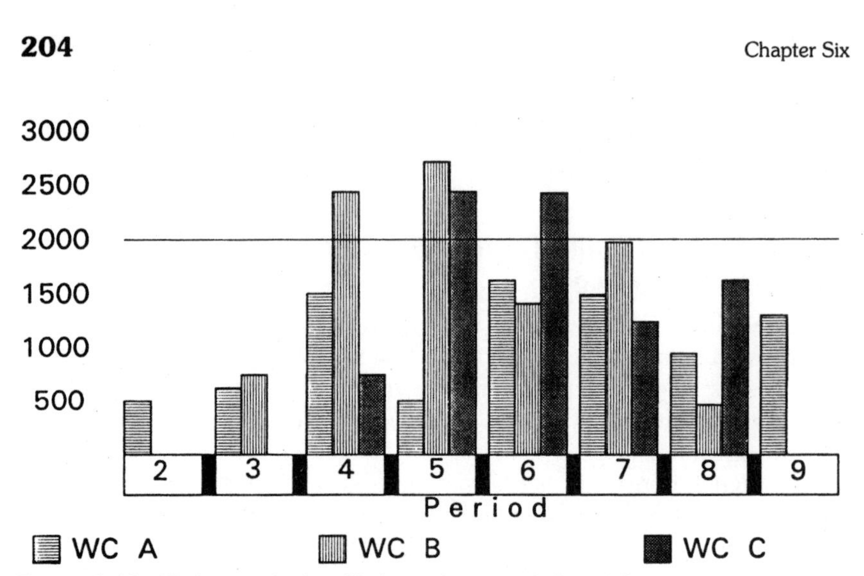

Figure 6-18. Work-center load profile for work centers A, B, and C.

ods 5 and 6. These overloads are large enough (about 38 percent) that they would probably be difficult to handle through overtime alone.

Extension of CRP Example

In the previous example, to make the manual tasks of backward scheduling and load distribution much simpler, we ignored all activities of the manufacturing cycle except the load, and arbitrarily used a lead time of one period for each operation. However, this is not the way that CRP operates in practice. CRP considers the five activities, queue, setup, run, wait, and move, defined in Figure 6-7, in determining exactly when a specific load will be imposed on a specific work center. CRP assumes that all setup is internal—that it requires the machine.

Figure 6-19*a* presents the full routing information on items *P, Q, R,* and *S.* We have combined setup time and run time into a load time and converted it to hours to make the chart of the schedule in Figure 6-19*c* more readable. In Figure 6-19*a,* the activities of the manufacturing cycle are indicated as follows: *Q* = Queue, *L* = Load, *W* = Wait, and *M* = Move.

To avoid excessive detail in this extended example, we are reducing the Planned Order Schedule to include only the last orders of *P, R,* and *S* indicated in Figure 6-19*b.* The figure shows the order quantities, with the load in hours (converted from minutes) above the quantity, and the work center, where the operation will be performed, below the order quantity.

In Figure 6-19*c,* we backward scheduled items *P, R,* and *S,* using the symbols for the activities, *Q, L, W,* and *M,* preceded by the symbol for the items *P, R,* and *S,* to indicate which operation we are scheduling. As before, we schedule from right to left and from bottom to top, starting with the last

Item	Operation and WC	WC A Q	WC A L	WC A W	WC A M	WC B Q	WC B L	WC B W	WC B M	WC C Q	WC C L	WC C W	WC C M
		Operation Times (Hours)											
P	1-A	16	21	8	4								
Q	1-A	16	14	8	4								
	2-B					24	7	8	4				
R	1-B					24	27	8	4				
	2-C									10	27	8	4
S	1-A	16	14	8	4								
	2-B					24	27	8	4				
	3-C									12	27	8	4

(a)

Period

2	3	4	5	6	7	8	9
							(21)
						(27)	400P1
					(27)	800R2	A
				(27)	800R1	C	
			(27)	800S3	B		
		(14)	800S2	C			
		800S1	B				
		A					

(b)

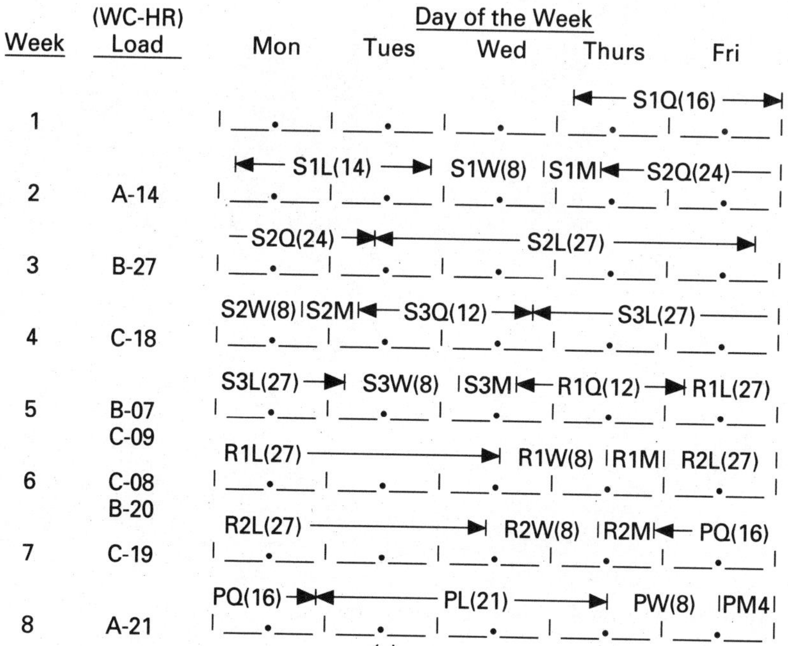

Week	(WC-HR) Load	Day of the Week: Mon · Tues · Wed · Thurs · Fri
1		(Thurs–Fri) ⟵ S1Q(16) ⟶
2	A-14	⟵ S1L(14) ⟶ S1W(8) | S1M | ⟵ S2Q(24) —
3	B-27	— S2Q(24) ⟶ ⟵ S2L(27) ⟶
4	C-18	S2W(8) | S2M | ⟵ S3Q(12) ⟶ ⟵ S3L(27) —
5	B-07 C-09	S3L(27) ⟶ S3W(8) | S3M | ⟵ R1Q(12) ⟶ R1L(27)
6	C-08 B-20	R1L(27) ⟶ R1W(8) | R1M | R2L(27)
7	C-19	R2L(27) ⟶ R2W(8) | R2M | ⟵ PQ(16)
8	A-21	PQ(16) ⟶ ⟵ PL(21) ⟶ PW(8) | PM4 |

(c)

Figure 6-19. (*a*) Routing information for items *P, Q, R,* and *S*; (*b*) Schedule of Operations for one lot of *P, R,* and *S*; (*c*) Backward Schedule, using full manufacturing cycle.

operation. Thus, we start with product *P* in week 9, followed by *R2* and *R1* in weeks 8 and 7. We then schedule operations *S3*, *S2*, and *S1* in weeks 6, 5, and 4.

By comparing Figures 6-19*a* and *b*, we can see that the distribution of load is considerably different, using the more detailed method of Figure 6-19*c*, as compared to the less detailed method of Figures 6-17*b* and *c* as summarized in Figure 6-19*b*. For instance, in Figure 6-19*b*, the loads for *S1* and *S2* were imposed on Work Centers A and B, respectively, in periods 4 and 5. In Figure 6-19*c*, these two loads were imposed one period earlier, that is, in periods 3 and 4.

CRP routinely calculates the loads of all operations of all products and distributes them to the work centers, as we have done for items *P*, *R*, and *S*, in Figure 6-19*c*. However, we will not do it by hand; the process becomes very detailed and very tedious, and that is why we use computers.

CRP Output

CRP's primary output is a Capacity Requirements Planning report, showing the load and capacity profile similar to Figure 6-20. The top half of the report plots the load for each *period* versus the capacity. The bottom half plots

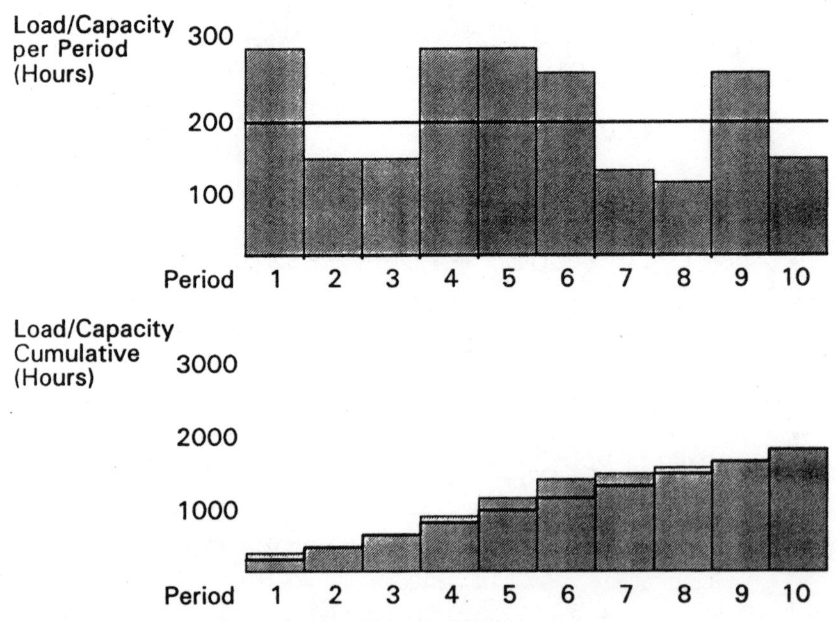

Figure 6-20. A typical work-center load profile report.

the *cumulative* load and capacity, that is, accumulating over the entire planning horizon of ten periods. The bottom half of the chart for period 1 is identical to the top half of the chart. In period 2, the bottom half equals the sum of periods 1 and 2 for the top half, for both capacity and load. In period 3, the bottom half equals the sum of periods 1 through 3 on the top half.

Ideally the load and capacity should be well matched on both a period basis and a cumulative basis. If the load and capacity are matched on a cumulative basis, but not on a period basis, the problem can usually be corrected by redistributing the load among the periods. If the load and capacity are mismatched on a cumulative basis, management must take action to balance them.

Using CRP Output Reports

Now that we have reviewed the format and the process for generating the work-center load profiles, let's look at how we use them. Capacity Requirements Planning enables us to balance, or match, load to capacity. The following are five basic actions that we can take in response to a CRP load report, if there is a disparity between available capacity and required load. These actions can be executed singly, or in various combinations.

- Redistribute Load

  Use alternate work centers
  Use alternate routings
  Adjust operation start date forward or backward
  Hold some work in Production Control for later
  Revise the MPS

- Increase Capacity

  Schedule overtime during week, or work weekends
  Add extra shifts
  Add equipment and/or personnel
  Subcontract one or more shop orders

- Reduce Capacity

  Temporarily reassign personnel (JIT would suggest using this time to invest in worker education, or to maintain equipment and facilities)
  Reduce length of shifts or eliminate shifts

- Increase Load

  Release orders early
  Increase Lot Sizes
  Make items that are normally buy items
  Increase the MPS

- Decrease Load

  Buy some items that are normally make items

  Revise the MPS

Now let's look at some sample Capacity Requirements Planning reports, and determine the action we should take. Figure 6-20 shows a CRP load report that is out of balance in some periods, but is balanced in the cumulative report. This is because some of the periods are overloaded, while others are underloaded. Because the total load and capacity are balanced, all we have to do is redistribute the load from one period to the next, using one of the methods suggested in the previous section.

Figure 6-21 illustrates a situation where the individual periods are relatively well balanced, but the cumulative load is considerably higher than the cumulative capacity, because each period's load is higher than capacity. This situation requires either an increase in capacity or a decrease in load, or a combination of the two. When a company uses Master Scheduling and Rough-Cut Resources Planning, these situations should only occur occasionally.

Figure 6-22 shows a situation where the period loads and capacities are not well-matched, and the cumulative load is considerably less than the cumulative capacity over the ten periods. This will require redistribution of

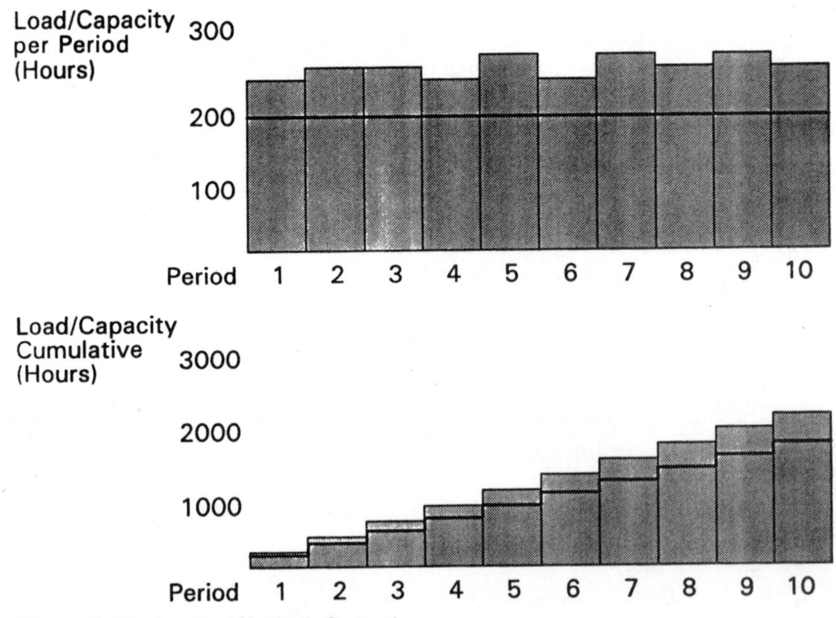

Figure 6-21. Load profile Work Center A.

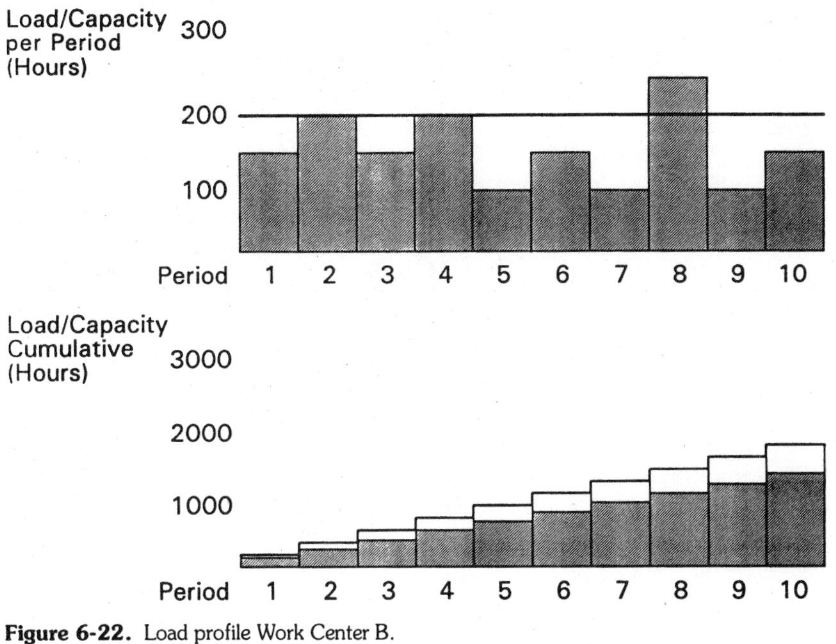

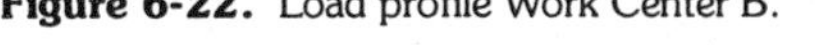
Figure 6-22. Load profile Work Center B.

the load, and either an increase in the load or decrease in the capacity or both.

Summary

Capacity Requirements Planning compares the projected load, by work center and period, against projected capacity, printing a report or screen that enables planners to take appropriate action to correct the imbalances. We discussed the following topics.

- Two loading options, finite and infinite loading. CRP operates in an infinite loading mode, which means that it will show the entire load on a work center, even though that load may far exceed the work center's capacity.
- Two scheduling options, forward and backward. CRP uses backward scheduling as a primary method, reverting to forward scheduling only when backward scheduling suggests that an order start before today.
- Major inputs to CRP, including work center, load (both Scheduled Receipts and Planned Order), and routing information.

- How CRP works, both with words then with a detailed example with simple data.
- Three possible scenarios, showing different capacity planning reports, suggesting the proper action to take for each one.

CRP can consume considerable computer resources to run (on the same magnitude as a full regenerative MRP run), and can create considerable quantities of reports. However, we suggest that CRP is a vital and necessary tool for companies that run MRP. CRP alone has the power to verify, at the detailed level, MRP's assumption of infinite capacity. Without CRP, the MPS and MRP must assume that the required capacity will indeed be available. If that assumption is wrong, customer shipments will be late, and costs will increase.

We are aware that most companies do not run CRP, because they can use Rough-Cut Resource Planning in the MPS, or schedule the load based on experience. This does not diminish the power and potential importance of this tool. On the contrary, one of the best ways to gain a competitive edge is to dare to do what others fail to do.

Selected Bibliography

Andreas, Lloyd: "40 Days to Due Date" by Willy Makeit: The Role of Capacity Management in MRP II," *MCRP Reprints,* APICS, Falls Church, VA, 1991, pp. 1–7.

Berry, William L., Thomas G. Schmitt, and Thomas E. Vollmann: "Capacity Planning Techniques for Manufacturing Control Systems: Information Requirements and Operational Features," *Journal of Operations Management,* vol 3, no 1, November 1982, pp. 13–25.

Blackstone, John H., Jr.: *Capacity Management,* South-Western, Cincinnati, OH, 1989, Chap 1 and 4.

Clark, James T.: "Capacity Management," *MCRP Reprints,* APICS, Falls Church, VA, 1991, pp. 6–9.

———: "Capacity Management—Part Two," *MCRP Reprints,* APICS, Falls Church, VA, 1991, pp. 10–16.

Deis, Paul: *Production and Inventory Management in the Technological Age,* Prentice-Hall, Englewood Cliffs, NJ, 1983, Chap 5.

Fogarty, Donald W., J. H. Blackstone, and T. R. Hoffmann: *Production and Inventory Management,* South-Western, Cincinnati, OH, 1991, Chap 13.

Donovan, R. Michael: "Production Scheduling and Capacity Management: The State of the Art," *APICS 1991 Conference Proceedings,* APICS, Falls Church, VA, 1991.

Lunn, Terry with Susan A. Neff: *MRP: Integrating Material Requirements Planning and Modern Business,* Irwin, Homewood, IL, 1992, Chap 9.

Plossl, George W.: *Production and Inventory Control: Principles and Techniques,* 2d ed., Prentice-Hall, Englewood Cliffs, NJ, 1985, Chap 9.

———, and Oliver W. Wight: "Capacity Planning and Control," *MCRP Reprints*, APICS, Falls Church, VA, 1991, pp. 60–96.

Schultz, Terry: *Business Requirements Planning: The Journey to Excellence,* The Forum Ltd, 1984, Milwaukee, WI, Chap 9.

Smith, Spencer B.: *Computer Based Production and Inventory Control,* Prentice-Hall, Englewood Cliffs, NJ, 1989, Chap 10.

Vollmann, Thomas E., W. L. Berry, and D. C. Whybark: *Manufacturing Planning and Control Systems,* 3d ed., Irwin, Homewood, IL, 1992, Chap 9.

Wemmerlöv, Urban: *Capacity Management Techniques for Manufacturing Companies with MRP Systems,* APICS, Falls Church, VA, 1984.

———: *Case Studies in Capacity Management,* APICS, Falls Church, VA, 1984.

7
Using M&CRP in MRP II

Understanding the Interactions

Introduction

M&CRP is just one portion, although perhaps the most significant portion, of an MRP II system. M&CRP works best when teamed with the other functions of MRP II, including Sales and Marketing, Engineering, Purchasing, Accounting and Finance, and so on. MRP II is the communication center of all activities, integrated and interrelated for the success of the company.

This chapter discusses the interactions between the various activities in the MRP II system, with emphasis on those activities that interact directly with the Material and Capacity Requirements Planning system. The activities of the MRP II system are shown schematically in Figure 7-1. Interactions between activities are indicated by lines connecting the activities, with the primary direction of interaction indicated by an arrow.

In the first part of the chapter, we present the overall operation of the MRP II system, stressing the interaction between activities, and emphasizing those activities that interact directly with the M&CRP module. Then we describe the operation of the MRP II system for four Demand Response Strategies that were discussed in Chapter 1.

Make-to-Stock

Assemble-to-Order

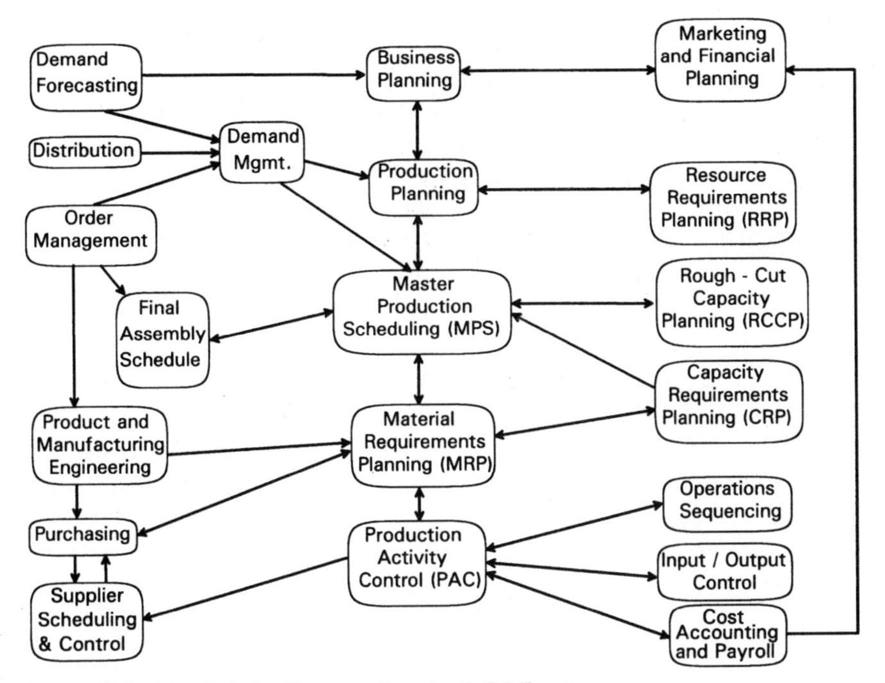

Figure 7-1. Manufacturing Resources Planning (MRP II) system.

Make-to-Order

Engineer-to-Order

In the latter part of this chapter, we describe the interactions between various activities, in more detail, again emphasizing interactions with M&CRP. The most important interactions of the M&CRP system are with: (1) the Master Production Schedule; (2) Production Activity Control; and (3) Purchasing modules. The MPS drives the M&CRP system. It provides the planned build schedule of end products that is exploded in the MRP system. Using the output from the MRP system, the CRP system determines the capacity required to execute the MPS.

M&CRP provides make and buy requirements for the Production Activity Control and Purchasing systems, respectively. In many cases, we can provide advance procurement information from the M&CRP system directly to the supplier to help them to ship on schedule. The M&CRP system drives the Production Activity Control and Purchasing systems in a manner similar to the way the MPS drives the M&CRP system.

MRP II Overview

MRP II is an explicit and formal manufacturing information system that integrates Sales and Marketing, Accounting and Finance, Engineering, and Production. It encompasses all aspects of a manufacturing company, from business planning at the executive level, through detailed planning and control at the tactical and operational levels, through execution in the shop and purchasing, with feedback from each level to the levels above.

Business Planning

Manufacturing plans and controls start with the business plan. The business plan is a statement of the major objectives of the company and the strategies for achieving these objectives in the next two to ten years. It is based on long-range forecasts, and includes inputs from Marketing, Engineering, Production, and Finance. In turn, the business plan provides direction to and coordination among Marketing, Engineering, Production, and Finance.

Marketing is responsible for analyzing the market and determining the firm's response: markets to be served, products provided, desired levels of customer service, pricing, and promotion. Engineering is responsible for quickly developing quality products based on Marketing input. Finance must determine the source of funds, and compute projected profits, budgets, and cash flows. Production must determine how to best satisfy market demands by setting production rates, inventory levels, and utilizing planned resources most efficiently.

Demand Management

After completion of the business plan, the next step in the MRP II process is Demand Management, which aggregates demand from all sources. The consolidated demand forms the basis for the production plan and the Master Production Schedule. As shown in Figure 7-2, Demand Management consists of three major activities: demand forecasting, distribution, and order management. Distribution can be additionally divided into two categories: traditional distribution and Distribution Requirements Planning (DRP). Order management can be subdivided into four areas: order consolidation, order entry, order promising, and order servicing.

Marketing and manufacturing jointly decide which of the company's products will be Make-to-Stock, Assemble-to-Order, Make-to-Order, Design-to-Order, and Make-to-Demand. This decision determines how customer demand will be satisfied and what the lead time of the products will be. This

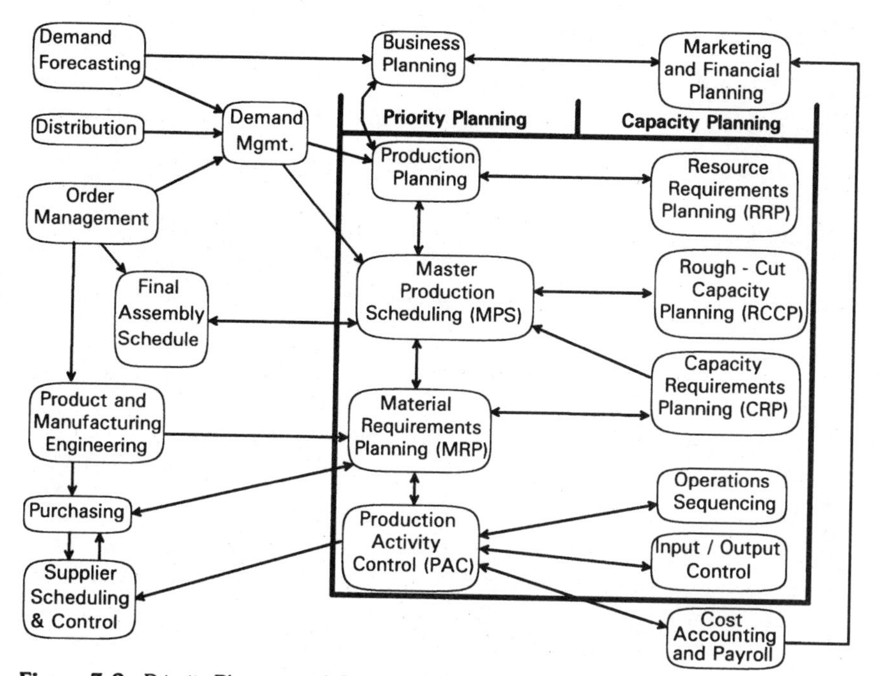

Figure 7-2. Priority Planning and Capacity Planning in an MRP II system.

is based on an understanding of the process and product factors discussed in Chapter 1.

Priority Planning and Capacity Planning

With Business Planning and Demand Management completed, we are ready to start on the hierarchy of priority planning and capacity planning, shown in Figure 7-2, which comprises the major part of MRP II. Priority planning determines what material is needed and when it is needed. Capacity planning determines the resources, or capacity, such as, machine hours, labor hours, and financial resources needed by manufacturing to produce the desired output.

For every priority planning activity there is, or should be, a comparable capacity planning activity. The manufacturing system cannot fulfill the desired priorities unless the necessary capacity is available. As indicated in the box in Figure 7-2, the four levels in the priority and capacity planning hierarchy are:

1. Production Planning and Resource Requirements Planning
2. Master Production Scheduling and Rough-Cut Capacity Planning
3. Material Requirements Planning and Capacity Requirements Planning
4. Production Activity Control and Input/Output Control

Both the priority planning activities and capacity planning activities operate over progressively shorter planning horizons, and produce reports in progressively greater degrees of detail, as we move from the top of the planning hierarchy to the bottom.

Production Planning and Resource Requirements Planning

Production planning, also called Sales and Operations Planning, is the process of setting the overall level of manufacturing output to best satisfy the planned level of sales and inventory. The production plan, which is the output of the process, states the manufacturing rate for each product group based on forecast, and calculates the resulting inventory. The production plan must be consistent with the business plan.

Production planners make plans that will satisfy market demand within the resources available to the company. Resource Requirements Planning evaluates the long-range resources, such as land, facilities, machinery, and work force, in order to determine if they are available or can be made available as needed, to satisfy the production plan. If the necessary resources are available, the production plan can be executed as is. If they are not equal, management must change the production plan, or acquire more resources. When the required and planned available resources are equal, the production plan can then be passed down to the next level for execution.

Master Production Scheduling and Rough-Cut Capacity Planning

The MPS breaks down the production plan to show, for each time period, the quantity of end items to be made. The production plan shows a production rate for a *group* of products, whereas the Master Production Schedule schedules quantities of *specific* end items in specific time periods.

Rough-Cut Capacity Planning estimates whether or not the planned resources are sufficient to carry out the Master Production Schedule. If the required and available resources are equal, the Master Production Schedule is passed down to the next level for execution.

Material Requirements Planning and Capacity Requirements Planning

Using Bills of Material, inventory, and order data in addition to the Master Production Schedule as inputs, the Material Requirements Planning module develops Planned Order Releases for the material needed to satisfy the Master Production Schedule. It also manages priorities by recommending that open orders be rescheduled when existing due dates differ from the actual need dates.

Capacity Requirements Planning determines if the projected capacity is sufficient to accomplish both the open orders and the Planned Orders generated by the MRP system. CRP uses routing files and work center information to calculate the scheduled load at work centers, assuming an infinite capacity. Using the CRP report, which compares the loads to demonstrated capacity for each work center, planners can act to correct the imbalances by reducing or rescheduling the orders, using alternate routings, reassigning, hiring, or laying off workers.

Production Activity Control, Input/Output Control, and Operations Sequencing

Production Activity Control represents the execution and control of manufacturing plans that have been developed in the previous levels. This is the level where the work is actually accomplished. This level also generates valuable feedback, which is used by the higher levels to improve their planning.

Production Activity Control develops detailed short-range schedules using shop order due dates from MRP and detailed routings. PAC releases orders to the factory, and controls these orders on the shop floor. Input/output control monitors the quantity of work arriving at, and leaving, a work center. After comparing actual arrivals and completions to planned amounts, planners can act to correct imbalances by scheduling overtime, transferring workers between work centers, transferring load to other work centers, or splitting and overlapping operations.

Purchasing and Supplier Scheduling

Purchasing and Supplier Scheduling performs the same function for external manufacturers (suppliers) as Production Activity Control performs for internal manufacturing. Purchasing often sends the MRP schedule directly to the supplier, so that the supplier can develop shop schedules that directly support it.

MRP Variations Based on Demand Response Strategy

Make-to-Stock

The Make-to-Stock mode minimizes customer order fulfillment lead time, because customer orders can be filled directly from stock on the shelf, usually within 24 hours, and often on the same day as order receipt. However, Make-to-Stock increases the total inventory investment, unless replenishment orders also have very short lead times. In the Make-to-Stock mode, very little in operations is keyed to actual customer orders; rather, the focus is on replenishment of inventory based upon forecasted usage. Thus, an accurate forecast is the most important output of Demand Management. With the rare exception of back orders, actual customer orders are not identified or tracked in the production process.

In a Make-to-Stock operation, the cycle begins with the producer, rather than the customer, specifying the product. The customer takes the product from stock if the product and price are acceptable, and the inventory is available. The finished goods inventory buffers the production process from actual customer orders. At any particular time, there may be little correlation between actual orders being received and what is being produced. The production system builds stock levels for future orders, not current ones. Current orders are filled from available stock.

In Demand Management, we explicitly define service levels and resultant Safety Stocks. We also set the requisite degree of flexibility for responding to mix or engineering design changes here, by defining buffer stocks and time fences. The Master Scheduler then maintains the required level of buffer stocks.

The Final Assembly Schedule and the Master Production Schedule are one and the same schedule, except when a Planning Bill is used for creating the Master Production Schedule. When a Super Planning Bill is used, the pseudo product of the Super Bill is scheduled in the top level of a two-level MPS, with the actual end product scheduled in the second level of the MPS and the FAS. See Chapter 3 for a more detailed description of planning with Planning Bills.

Engineering becomes involved in the Make-to-Stock mode only for Engineering Change Orders (ECO) and new products. When an ECO is introduced, product engineers and manufacturing engineers design the new product and process, as described in detail below, for the Engineer-to-

Order mode, consulting with Production Control to establish the effectivity date, or lot, for shifting from the old to the new product.

The objective of a Make-to-Stock operation is to meet the desired level of customer service at minimum cost. Thus, the key performance measures are customer service, utilization of production assets (inventory and capacity), and profit margin. The detailed performance measures might include inventory turnover, capacity utilization, overtime utilization, and percentage of orders filled from stock. We discuss additional performance measures in Chapter 12.

Assemble-to-Order

In the Assemble-to-Order mode, all subassemblies or modules are available in inventory. When a customer orders, the manufacturer assembles the specific options and base modules to fill that particular order. This Demand Response Strategy is used by companies having multiple end products that are made from common modules. The demand for the modules can be forecasted more accurately than can the demand for the final product. Thus, these companies can respond to customer demand by forecasting and stocking the modules, and then assembling the final product only upon receipt of the customer order.

In scheduling the MPS, the company can either forecast and schedule each module by using a Modular Planning Bill, or (better yet), they can create a Super Modular Planning Bill to drive a two-level MPS. In the Super Modular Planning Bill, the parent is an unbuildable collection of the modules, whereas the children are the actual modules exploded by percentages based on their historical usage. Some companies will inflate either the forecast or the usage percentages, or both, to ensure a high service level. When received, the customer orders are sent directly to the Final Assembly Schedule, and the lead time is equal to the time for assembly if the modules are in stock. Many Assemble-to-Order companies can ship within 24 hours of order receipt.

In the Assemble-to-Order mode, the key performance measures are customer order shipment time and product cost. Other critical measures include the accuracy of the forecast and the accuracy of the usage percentages for the modules, because these determine how quickly we can satisfy customer demand.

Make-to-Order

In the Make-to-Order Demand Response Strategy, we do not inventory many specific subassemblies; instead, we inventory the product designs, raw materials, and possibly some common components. We have made the products before. In Make-to-Order, we key our processing activities to individual customer orders. The order cycle begins when the customer specifies the desired product and requests a quote of cost and delivery time. On the basis of the customer's request, we will quote a price and delivery time. We can create the quotation immediately, if the order is for a standard product; for custom orders, the quotation may take longer. If the customer accepts the quotation, we assemble the product from components and/or build completely to customer specifications, then ship the product to the customer.

Demand Management converts specific day-to-day customer orders into MPS actions, consuming the planned materials and capacities. The key operations performance measures are the delivery time, the percentage of orders delivered on time, and profit per order.

Design-to-Order

In the Design-to-Order Demand Response Strategy, we keep *nothing* in inventory, not even the product design. Customers will request quotes for design and manufacture of unique items. The quotation process can be long and complex, requiring some initial design work. When a customer places an order, we first develop the detailed design, in consultation with the customer, knowing there is a distinct probability of additional changes. As we are designing, we start ordering the raw material in order to minimize the total time needed to complete the product. We then fabricate the components, assemble the product, and ship it to the customer. This Demand Response Strategy is useful for products that are new, unique, or heavily customized.

If we choose to include the engineering effort as a component of the end product, MRP II can track the progress of the design by using routings. We perform the design process, as outlined in Chapter 3, with the Engineering function working concurrently with Purchasing, Manufacturing, and Materials Control, as well as with the customer and/or Marketing.

The Master Scheduler includes the product in the Master Production Schedule, in conjunction with all the other products ordered by other customers. With the Master Production Schedule and the Manufacturing Bill of Materials as inputs, the M&CRP system explodes the product to determine the required quantity and timing for the manufacture and/or purchase of the subassemblies, parts, and raw materials needed to build the product on time.

The key performance measures of Design-to-Order operations are design and product quality, on-time delivery, and cost.

Detailed Interaction of MRP II Activities

In the following sections, we briefly describe how the various MRP II activities operate, and then describe, in detail, how each activity interacts with other activities, and in particular with M&CRP. We discuss M&CRP interactions with:

Final Assembly Schedule, Master Production Schedule, and Rough-Cut Capacity Planning

Production Activity Control and Short-Range Capacity Planning

Just-in-Time

Purchasing and Supplier Scheduling

Engineering

Total Quality Management

Demand Management

Accounting and Finance

FAS, MPS, and RCCP

Brief Description of FAS, MPS, and RCCP

The *Final Assembly Schedule* (*FAS*) is a schedule of end items to be assembled for specific customer or stock requirements. It is constrained by the availability of material, as provided by the MPS, and by assembly capacity. The Final Assembly Schedule defines what we are actually going to produce and when. When the MPS uses the end item as the item to be scheduled, the MPS and the FAS are one and the same schedule. The situations in which this occurs are discussed in Chapter 3.

Master Production Scheduling encompasses the variety of activities involved in the preparation and maintenance of the Master Production Schedule. The MPS is the anticipated build schedule for a company's products, stated in specific product configurations, for the required planning horizon. The MPS represents a statement of production, not a forecast of product demand. It represents manufacturing's disaggregation of the production plan into individual product items. The disaggregation should always summarize back to the production plan. Chapter 3 contains a more detailed description of the design and operation of the MPS system.

Rough-Cut Capacity Planning (*RCCP*), discussed in Chapter 2, determines the capacity required to produce the MPS. The planning horizon is the same as the MPS, usually one to three years. It converts the Master Production Schedule into capacity needs for key resources: workers, machinery, warehouse space, suppliers' capabilities, and, in some cases, investment funds and energy. Rough-Cut Capacity Planning utilizes one of three techniques:

1. *Capacity planning factors,* which specify the quantity of some resource required to produce the output. We derive capacity planning factors indirectly, from aggregate historical data, as opposed to directly from a detailed analysis of resources required to produce each component of the product. When these planning factors are applied to the MPS, RCCP can estimate overall labor and/or machine-hour capacity requirements. RCCP then allocates this overall estimate to an individual work center on the basis of historical data on shop workloads.
2. *Bills of capacity,* also called *bills of resources,* list the required capacity and key resources needed to manufacture one unit of a selected item or family. Rough-Cut Capacity Planning uses these bills to calculate the needed capacity for the MPS, so as to ensure that the production plans are attainable in the short-term future.
3. *Time-phased bills of capacity,* often called *product profiles,* list the capacity and key resources required for the production plan. The resource requirements are further defined by a lead-time offset in order to predict the impact of the item or product family on the load of key resources for a given time period. These are often used to predict the impact of the scheduled items and the overall impact of the scheduled load on key resources for the immediate production week.

Master Production Scheduling represents the culmination of all master planning activities. Demand Management supplies all the demand requirements; production planning provides the operating constraints within which the MPS must be constructed and maintained, and Rough-Cut Capacity Planning indicates which resources are potential bottlenecks. In turn, the MPS provides the principal input to M&CRP. The MPS forms the foundation for making customer delivery promises, utilizing plant capacity effectively, attaining the firm's strategic objectives, as reflected in the production plan, and resolving trade-offs between manufacturing and marketing.

M&CRP Interactions with FAS, MPS, and RCCP

The FAS, MPS, and RCCP are all one level above M&CRP, as indicated in Figure 7-1. Thus, they are in position to provide guidance and constraints to M&CRP, and to receive feedback from M&CRP. In general, M&CRP receives direct inputs only from the MPS, and not the FAS.

The selection of items to schedule in the MPS, discussed in Chapter 3, also determines the BOMs that M&CRP will use and explode. The planning horizon used for the M&CRP is usually shorter than the one for MPS, although sometimes it is the same length. It should not be longer, because there is no data beyond the MPS horizon. Management's selection and enforcement of proper time fences in the MPS will considerably reduce the disruption in the M&CRP caused by changes. It is especially important that no changes be made inside the Demand Time Fence unless they are truly emergencies, and that only product-mix changes be made inside the Planning Time Fence.

When the CRP function of the M&CRP indicates that the current MPS is infeasible from a capacity viewpoint, one of the options is to go back to the MPS to determine if the MPS can be changed in some way to achieve a feasible schedule. The revised MPS should be processed through MRP, then CRP, to determine if it is feasible, rather than RCCP, because CRP is a more accurate capacity planning procedure.

RCCP does not provide an accurate check of capacity needed to determine whether or not a revised MPS is feasible. CRP considers a number of things that RCCP, even when using Product Profiles, does not consider. RCCP does not consider the on-hand and in-process inventories of components. Neither does it consider the effects that Lot Sizing of components will have in determining when loads will occur, and the amount of setup time required.

PAC and Short-Range Capacity Control

Brief Description of PAC and Short-Range Capacity Control

Production Activity Control (*PAC*) executes the detailed manufacturing plans on the shop floor. It plans and releases orders to the factory, and controls these jobs on the shop floor. PAC schedules, controls, measures, and evaluates the effectiveness of production operations. Specifically, PAC is composed of the following production activities: order release, operation scheduling, operation sequencing (dispatching), data collection, production reporting, corrective action, and order close-out. These production activities may occur in a wide variety of production environments including job shops, process plants, and high-volume production facilities.

PAC controls short-term capacity by monitoring and adjusting the relationship between supply and demand. Capacity Control includes determining resource requirements, establishing resource availabilities, and continually balancing both requirements and availabilities to accomplish the plan in the short term.

Operations Sequencing is a simulation technique that models and predicts the sequence in which jobs will actually flow through the work centers, based on available capacity. It includes only Scheduled Receipts. It starts with the current date and simulates jobs moving forward (forward scheduling). It loads the highest-priority job in each work center first. If another high-priority job arrives in the work center's queue (as the program is simulating the future), that new high-priority job will take precedence over a job that has been in the queue longer. Thus, Operation Sequencing simulates the actual function of a queue, rather than using the fixed queue times from the work center records.

Operations Sequencing can be used as a more accurate prediction of required job completion than CRP or Input/Output Control. Operations Sequencing prioritizes the jobs in each queue, so that the workers know which job to start as an existing job completes. Operations Sequencing is basically a finite scheduler. As such, it is rather computer-intensive. Practitioners normally run it for only a few days out, and rerun it daily so that it will provide accurate and timely priorities.

Input/Output Control is an integrated process that includes planning the acceptable input and output ranges per time period in each work center; measuring and reporting actual inputs and outputs; and correcting out-of-control situations. In addition to controlling capacity, Input/Output Control is an effective technique for controlling queues, work-in-process, and manufacturing lead time (the time from the release of an order to its completion). It enables the planner to determine what action is necessary to achieve the desired output, work-in-process, and manufacturing lead time objectives.

M&CRP Interactions with PAC and Short-Range Capacity Control

Production Activity Control and Short-Range Capacity Control are one level below M&CRP, as indicated in Figure 7-1. They are in position to receive guidance and constraints from M&CRP, and to provide feedback to M&CRP. PAC provides feedback to the M&CRP system about how the actual manufacturing orders are being executed, and the status of work-in-process. Many of the PAC techniques are designed to execute the detailed material plans produced by the M&CRP system. The PAC system also closes the loop, as illustrated in Figure 7-1, by measuring actual performance and comparing it to the plan. Thus, PAC is an essential component of a closed-loop MRP II system. Kanban and other JIT techniques can replace most PAC functions.

Input/Output Control provides a method for monitoring the actual consumption of capacity during execution of detailed manufacturing plans produced by the M&CRP system. It can indicate the need to update capacity plans as actual shop performance deviates from current plans, as well as the need to modify planning factors used in CRP.

PAC Contributions to M&CRP

Material Requirements Planning depends upon Production Activity Control for several inputs. We describe each of them in the following paragraphs.

Inventory Accuracy. Even when stockroom inventory accuracy exceeds 95 percent, MRP may still malfunction because of inventory inaccuracy. Stockroom balances are only one part of the MRP calculation of projected availability. Two other inventory balances are also critical: (1) the material allocations against manufacturing orders, and (2) the manufacturing Scheduled Receipts (or open orders), including accurate projected completion dates. These two balances are part of the PAC system. A third critical inventory balance, purchased Scheduled Receipts, does not fall under the PAC.

Accurate Lead Times. Inaccurate lead times can kill an MRP system. Understated lead times cause excessive expediting. Overstated lead times, which are more prevalent owing to the tendency to be cautious, cause early release of orders, higher work-in-process, and associated control problems. The average process and interoperation times can be monitored by a PAC system that provides output reporting of these figures.

Accurate Bills of Material. An effective MRP system requires that Bills of Material be accurate and properly structured according to the way a product is really manufactured. An effective and inexpensive way to continuously validate the Bills of Material is to compare actual inventory issues with picking lists. The only activity that can truly validate the Bill of Materials is manufacturing, and the PAC is part of this feedback loop.

Valid Scheduled Receipt Data. CRP uses existing Scheduled Receipts, as well as Planned Orders, when it computes the load on the plant. PAC updates the Scheduled Receipts, so that CRP has valid status information on each order.

M&CRP Contributions to PAC

Just as M&CRP depends on PAC, so does PAC depend on M&CRP for three major inputs.

Material Availability. The philosophy of MRP is to ensure material availability at the time of order release; MRP will normally alert a planner to orders due for release with probable material shortages. However, if orders with material shortages have a high priority, PAC can check material availability at the time of order release, and then make appropriate adjustments to release quantities and/or revise due dates, to minimize the impact on the shop floor.

Capacity Planning. Without CRP the PAC system is exposed to the dangers of invalid due dates and production schedules that are overstated, when compared with the available capacity. The more volatile the demand, the more difficult this problem is.

Priority Planning. Basically, this is the function of determining what material is needed and when. Master Production Scheduling and Material Requirements Planning perform the basic planning process, but CRP validates the MRP schedules, and PAC maintains proper due dates and delivery dates on required materials.

Just-in-Time

Brief Description of Just-in-Time

Just-in-Time (*JIT*), also known as zero inventories and stockless production, is a philosophy of manufacturing. Its goal is to eliminate all waste in the entire manufacturing cycle, from designing the product to its delivery to the customer. Waste encompasses everything that does not directly add to the value of the product in the customer's eyes. Waste is defined as anything more than the minimum amount of materials, labor, machines, or tools necessary for production. Material movement, handling, counting, storage, and virtually all paperwork can be considered as waste. In particular, JIT's objectives are to produce products in the minimum possible lead time, and to provide just the right amount of material of the specified quality, just when it is needed at each stage throughout the production process.

JIT seeks to continuously improve a company's ability to respond economically to change. JIT deliberately exposes and prioritizes nonvalue-adding activities, with the intent to completely eliminate them. This process changes every functional area inside a company, as well as interfaces to suppliers and customers.

JIT employs a number of techniques that achieve the philosophy of waste elimination, such as minimizing inventories, reducing lot sizes, reducing setup time and cost, streamlining production processes, leveling production rates, implementing a pull system of moving goods, establishing high quality levels, emphasizing preventive maintenance, eliminating administrative overhead, and creating multifunctional workers. These techniques go beyond those traditionally considered part of manufacturing planning and control. JIT makes major changes in the actual practice of manufacturing, with the intent to simplify material planning, eliminate shop floor tracking, substantially reduce work-in-process inventories, and eliminate the transactions associated with shop floor and purchasing systems.

M&CRP Interactions with Just-in-Time

When changes take place on the shop floor, the M&CRP system must change in order to keep pace with these changes. This is particularly true when JIT is implemented in a company. The main factors that affect the M&CRP system when JIT is installed are: (1) the reduction in lot sizes, (2) the reduction in queue times and the resultant reduction in manufacturing lead times, (3) the potentially huge increase in the number of shop orders and inventory transactions reported, because of the combined effect of the first two factors.

One of the first requirements to support the installation of JIT in a company is to reduce the volume of inventory transactions. Cutting the number of times a lot has to be logged into and out of an inventory location, not only reduces transactions, but enables the material to move to the next operation more quickly. This clearly helps to increase the material flow velocity, as well as to reduce the lead times. When the volume and velocity of material flow reaches a very high level, the company should abolish detailed inventory transaction reporting and implement *post-deduct,* or backflushing, inventory movement. In the post-deduct system, the computer inventory of components is automatically reduced by the computer after completion of the end product, or when the product flows through a checkpoint. The reduction is based on the components that should have been used as specified in the Bill of Materials. Thus, instead of making many inventory transactions per product, we are making only one, by exploding the Bill of Materials and multiplying by the production count of the products assembled. If this approach results in a difference between the book record and what is physically in stock, we have an opportunity to improve our Bills of Material or our process. Given the minimum inventories carried in a true JIT system, the difference cannot be very large.

Whenever there is a combination of M&CRP and JIT in the same shop, information must flow between the two systems. A JIT cell in the middle of a plant under M&CRP control must communicate with the M&CRP system; there must at least be a hand-off from M&CRP to JIT at the start of the JIT process, and a transfer back to M&CRP at the end. One way of supporting this need is to create phantom bills of activities under JIT control. MRP can control raw material requirements, and JIT can control the factory (no shop orders or tracking). The phantom bill would ignore the creation of the detailed parts and assemblies performed under JIT scheduling, while the M&CRP system would pick up the completed part on assembly as a part number on the Bill of Materials at completion. Chapter 9 discusses the integration of JIT and M&CRP in more detail.

Purchasing and Supplier Scheduling

Brief Description of Purchasing

Purchasing includes all the activities involved in procuring materials, supplies, and services. The basic objective is to obtain the proper materials, supplies, and services, at the right time, in the right quantity, of the right quality, at the right price, and from the right source. Purchasing performs four major activities:

1. Selecting suppliers, qualifying their capabilities to be a long-term partner, negotiating the terms of purchase most advantageous to both parties, and issuing necessary purchase orders.
2. Coordinating delivery from suppliers to meet schedules, and negotiating any changes in purchase schedules dictated by various circumstances and customer demands. (As a company moves toward JIT, this function diminishes.)
3. Acting as liaison between suppliers and other company departments, including Engineering, Quality Control, Manufacturing, Production Control, and Finance, on all problems involving purchased materials. However, this does not mean shielding suppliers from others in the company. Purchasing can, instead, encourage frequent communications directly between suppliers and the appropriate levels within a company.
4. Looking for new products, materials, and suppliers that can contribute to a company's strategic objectives, acting as the company's "eyes and ears" to the outside world, and reporting on changes and trends in mar-

ket conditions, short- and long-term allocations of standard industry components, and other factors that can affect company operations.

Brief Description of Supplier Scheduling and Control (SSC)

A good *Supplier Scheduling and Control* system should extend well beyond the Purchasing department. The material acquisition process includes not only purchasing personnel, who select suppliers, negotiate prices, delivery, and payment terms, and who release written commitments of suppliers, but also requisitioners; materials planners, receiving dock, quality, and stockroom personnel; and finally the Accounting department. In fact, because the SSC system needs to track both internal and external activities, it often requires a more extensive planning and control system than that employed for internal shop orders.

Requisition Tracking. The SSC tracks a material requisition from the time it is initiated to the time it is covered by a released purchase order. For items used repetitively in production, the requisitioner need enter only the item number, due date, and quantity, and the SSC system provides all the basic information, such as item description, unit of measure, planning lead time, preferred supplier, normally assigned buyer, and so on. For first-time items, the requisitioner must provide all this information. (In most cases the requisitioner is a MRP planner, and the requisition is a planned order generated by the MRP system.)

The SSC system should also provide the requisitioner (MRP planner) with a list of all open requisitions in release date (due date minus planning lead time) sequence. This list can be used by the MRP planner/buyer or buyer/planner to follow up in order to assure that all open requisitions receive the appropriate attention.

Priority Dispatching of Open Requisitions. In a manner similar to a PAC dispatch report, the SSC system provides each buyer, on a daily basis, with a backlog report, in priority sequence, of all open requisitions not yet placed. The priority, like that of a shop order, should reflect the relative difference between the time left to achieve on-time delivery and the supplier's quoted lead time (the critical ratio).

Release-to-Dock-to-Stock Tracking. The progress of a purchase order can be tracked in much the same way as a Shop Order. To facilitate tracking, the SSC system must have the ability to assign a "purchase routing," which can vary by item. The operational steps to be tracked could include any of the following: (1) supplier acknowledgment of delivery date and quantity, (2) acknowledgment of receipt at dock (no count), (3) receipt count, (4) acceptance by quality and/or inspection, and (5) storage in a specific location. (JIT attempts to eliminate the need for inspection, counting, and material movement activities.)

The purchase routing can be maintained in the computer files, in the same manner as a manufacturing routing, and can be automatically assigned to each purchase order. Just as in the PAC system, individuals update the SSC system activity by entering the operation number and associated data at a terminal. The SSC system allows the status or location of any purchase order to be determined by on-line inquiry. It also accumulates labor hours by operation, and will thus allow direct costing of all dock-to-stock activities.

Matching Supplier Invoices. The final step in the SSC system is to match accounts payable entries to purchase order and receipts history to make sure that the quantity, price, terms, FOB point, and so on, match to the receipt and any subsequent dock-to-stock activity. This comparison provides automatic feedback to accounting indicating that it is "OK to pay," or listing discrepancies that must be resolved before payment can be made.

M&CRP Interactions with Purchasing and Supplier Scheduling

When a single supplier has been designated for an item, much of the purchase order release function can be completely automated. If MRP is working effectively and the Lot-Sizing rules are realistic, the SSC system can complete the purchase order to the point where the buyer/planner has to only approve the data before it is printed and released. This significantly reduces buyer/planner time and elapsed time from the requisition stage to order release.

The SSC system can also prepare a purchase order in a time-phased manner to coincide with the released and planned order logic of MRP. By informing the supplier (for their planning purposes) of changes to the material acquisition plan well in advance, late shipments can be significantly reduced. In a sense, revealing the probable rate of future purchase orders is like *buying* supplier *capacity*. It allows suppliers to adjust their production plan in the same fashion that Production Control adjusts to internal loads as calculated by CRP.

The practice of supplier scheduling greatly simplifies purchasing communications. Once a blanket order has been placed, the MRP schedule replaces purchase requisitions and order releases, and also eliminates the need for the buyer and the salesperson to communicate about schedules except on an exception basis.

Under manual purchasing systems, purchasing personnel spend most of their time doing clerical work and expediting purchase orders. M&CRP automates much of the clerical work and reduces the need for expediting because planning of requirements is more accurate. This means that buyers can spend more time in supplier evaluation and selection, negotiation, and value analysis. This reduction in administrative cost is one of the major benefits from implementing M&CRP systems. The planner/buyer keeps the production control supervisor, and the rest of the company, appraised of the progress of all purchase orders for components and raw materials by entering current data into the system.

Engineering

Brief Description of Engineering

Although the primary contacts with Engineering are with Product Design Engineering and Manufacturing Engineering, M&CRP also interacts with a number of other Engineering disciplines, namely: Industrial Engineering, Maintenance Engineering, and Plant Engineering.

Product Design Engineering is responsible for designing the company's products to satisfy the customer's needs, and for providing design information to all concerned. Manufacturing Engineering is responsible for designing and installing the manufacturing capabilities, and for determining how each product and component should be fabricated and assembled. Industrial Engineering is responsible for layout of the manufacturing facilities and for job design and operator training.

Maintenance Engineering is responsible for performing the necessary preventive and corrective maintenance to ensure reliable manufacturing operation. Plant Engineering is responsible for providing all plant services needed for the manufacturing system, such as electric, air, hydraulic power, heat, and light.

M&CRP Interactions with Engineering

Product Design Engineering creates the Engineering Bill of Materials and provides much of the information stored in the Item Master File that pertains to the identification and description of each item.

From the Engineering Bill of Materials, Manufacturing Engineering develops the Manufacturing Bill of Materials, which is one of the primary inputs to the MRP process. They also develop the assembly procedure for the product and the routing for the fabrications of each component, which are important inputs to the CRP process.

Industrial Engineering works with manufacturing to define work methods and time standards for the assembly of the product and the fabrication of the components. M&CRP uses these time standards for scheduling and for determining capacity requirements.

When engineering changes are made in products that are being manufactured, the implementation of these changes must be closely coordinated between Production Planning, Engineering, and the shop to ensure the customer receives the correct product. We discussed engineering change control in much more detail in Chapter 3.

When Maintenance Engineering is going to shut down particular machines and/or production lines for maintenance, they coordinate the shut-down schedule with Production Control. Also, Plant Engineering communicates any scheduled outages of power or other plant services to production control.

Total Quality Management

Brief Description of Total Quality Management (TQM)

Total quality includes the quality of goods, of services, of time, of place (high-quality buildings, plants, offices), of equipment and tools, of processes, of people, of environment and safety, of information, and of measurements. For each of these, we judge total quality in two ways. One is the average level of quality; the other is variability around that average.

The TQM view applies equally to goods and services, but we focus more on *processes* than the goods or services themselves. TQM emphasizes improvement, and better averages with less variability, rather than notions of goodness. When quality of the processes, run by people listening to their customers, improves, the goods and services do too.

TQM de-emphasizes the concepts of acceptable quality levels, fixation on quality of end products, service quality as a "fuzzy" art, and quality as a separate profession. Instead, Total Quality Management emphasizes the following concepts:

- Defining quality in terms of performance relative to customer's needs (and customers include both the final consumer of the product, plus the next person who uses or touches my work, plus all interested parties in between)
- Continued and rapid improvement as the primary goal

- Focus on quality of *processes* and resources, both human and physical, that support these processes
- Proper response to the final and next-in-process customer
- Quality as everyone's job

M&CRP Interactions with Total Quality Management

The interactions between M&CRP and TQM are many and varied. First, TQM has considerable effect on the schedules prepared by M&CRP. If the degree of quality of the manufacturing system is poor and/or unpredictable, it will be almost impossible to develop accurate schedules that will meet customer due dates. Consistent quality is a prerequisite for developing accurate and realistic schedules.

On the other hand, M&CRP must allow sufficient time in its schedules for such actions as preventive maintenance, stopping the line when a defect is found, and statistical process control measurements, if high quality is to be attained. The schedule must not be achieved at the expense of quality, or neither will be achieved in the end.

The change in the approach to quality, including eliminating time-consuming product quality inspections on the production line, will improve not only the efficiency of the production line, but also the ability to schedule and control the line. The same is true of purchased material. The reduction of incoming material inspection by improving the quality at the supplier improves the material procurement and handling process significantly.

By reducing lot sizes and lead times for items, M&CRP can streamline the flow of items through the shop. This streamlined flow provides an improvement in quality through faster feedback; defects are detected and can be corrected before many nonconforming items have been produced.

Demand Management

Demand Management is the gateway module for MRP II systems as illustrated in Figure 7-1. Demand Management provides coordination between all of manufacturing and the marketplace, which includes sister plants and warehouses, as well as conventional customers. Although Demand Management interacts with the M&CRP system primarily through MPS, rather than directly with M&CRP, the effect of this interaction on M&CRP is very great.

Brief Description of Demand Management

As shown in Figure 7-3, Demand Management consists of three major activities: demand forecasting, distribution, and order management. Distribution can be subdivided into two categories: traditional distribution and Distribution Resources Planning (DRP). Order management can best be understood by subdividing it into four areas: order consolidation, order entry, order promising, and order servicing. We define each of these briefly.

Demand Forecasting. Demand forecasting attempts to predict product sales so the company can have the right items in stock at the right time. Forecasts may be based upon an extrapolation of the past or insights gained from current events. For companies making Make-to-Stock products, demand forecasts are the basis of decisions for acquiring plant and equipment, planning the workforce, purchasing materials, and scheduling production. Even for companies making Make-to-Order products, plant and equipment decisions must still be based on demand forecasts, and competitive pressures often require that companies acquire materials before firm customer orders are in hand.

Traditional Distribution. Distribution refers to the activities associated with the movement of material, usually finished products and service parts, from the manufacturer to the customer. These activities encompass the functions of transportation, warehousing, inventory control, material handling, order administration, site and/or location analysis, industrial pack-

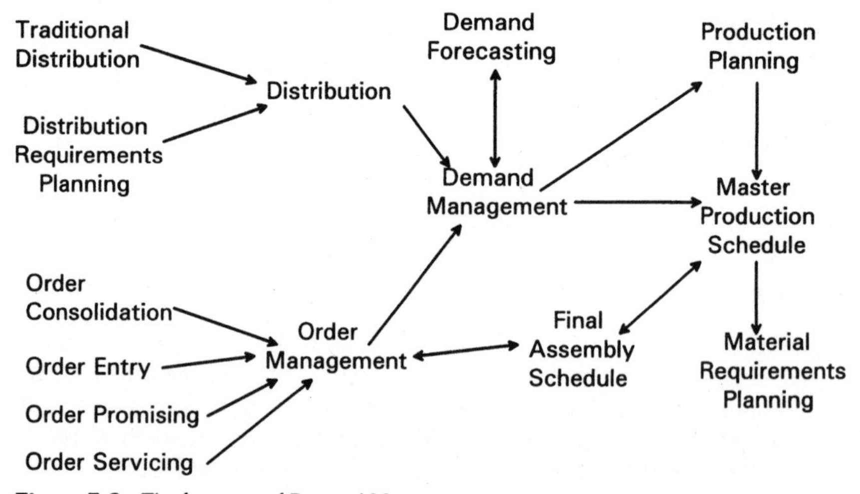

Figure 7-3. The functions of Demand Management.

aging, information processing, and the communications network necessary for effective management. In many cases, the materials and products move through one or more field-warehouse levels. In M&CRP, we are most interested in the order administration, information processing, and communications network aspects of distribution.

Distribution Requirements Planning (DRP). DRP primarily determines the need to replenish inventory at branch warehouses. A time-phased order-point approach creates planned replenishment orders at the branch warehouse level using MRP logic; these orders become the Gross Requirements on the supplying source. In the case of multilevel distribution networks, this ordering process can continue through the various levels of regional warehouse, master warehouse, factory warehouse, and so on, and become an input to the Master Production Schedule.

Distribution Requirements Planning is the logical extension of the use of time-phased materials planning techniques to the distribution-warehouse locations. The planned replenishment orders for DRP provide MPS with more specific Gross Requirements so that its plan is more realistic. In turn, the MPS requirement then allows M&CRP to provide a better plan of components and raw materials. Overall, the materials planning function has more time to react to changes in demand patterns at all company inventory locations.

Order Management. Order management is the overall function that encompasses receiving, consolidating, entering, promising, and servicing orders from customers, distribution centers, service parts, and all other sources, and then interacts with the materials organization to enter these orders into the Master Production Schedule. Order management is comprised of four functions: (1) order consolidation, (2) order entry, (3) order promising, and (4) order servicing.

1. *Order consolidation.* Order consolidation recognizes all potential sources of orders, such as forecast customer orders, branch warehouse requirements, interplant transfers, international requirements, and service parts demand, and consolidates them in a form suitable for input into the Master Production Schedule. The Master Scheduler then develops a Master Production Schedule that will satisfy the consolidated demand.
2. *Order entry.* Order entry is the process of accepting and translating what a customer wants into the producing company's terms, that is, into product numbers that are used in the MPS and M&CRP systems. This can be as simple as creating shipping documents for finished goods in a product line, or it might be a more complicated series of activities, including engineering effort for Make-to-Order products.

3. *Order promising.* Order promising is the process of making a firm delivery commitment based on availability of material and capacity. For Make-to-Stock products, this usually involves a check of uncommitted material; for all other types of products, this also requires verifying the capacity availability.
4. *Order servicing.* Order servicing is responsible for responding to customer inquiries about their orders. Order servicing and order entry are often performed by the same people.

M&CRP Interactions with Demand Management

Demand management interacts frequently with M&CRP, through MPS. Details vary significantly between Make-to-Stock, Assemble-to-Order, and Make-to-Order environments. In all instances, however, the underlying concept is that, over time, actual customer orders consume forecasts.

Demand Management explicitly defines service levels and resultant Safety Stocks. Demand Management also defines the requisite degree of flexibility for responding to mix or engineering design changes, by determining buffer stocks and timings. The master scheduler is then responsible for maintaining the required level of buffer stocks.

Demand Management converts specific day-to-day customer orders into MPS actions. Through the Demand Management function, actual demands consume the planned materials and capacities. Actual customer demand must be converted into production actions whether the firm manufactures Make-to-Stock, Make-to-Order, or Assemble-to-Order products.

In the following paragraphs we discuss the interactions of Demand Management with MPS-M&CRP in more detail for the Make-to-Stock, Assemble-to-Order, Make-to-Order, and Make-to-Demand environments by covering: (1) the general nature of the interaction, (2) the causes of uncertainty and how they can be countered, and (3) how day-to-day customer orders are handled.

Make-to-Stock. In the Make-to-Stock environment, very few actual customer orders directly enter the manufacturing process, because the demand is usually satisfied from inventory. Thus, the MPS task is one of providing inventory to meet forecasted future customer orders.

Uncertainty in the Make-to-Stock environment is largely in the demand variations around the forecast at each inventory location. Planners can set levels of Safety Stock in order to provide the service levels required. Sales and marketing professionals can develop close relationships with important customers to obtain the best possible future demand information.

In day-to-day operations, Make-to-Stock companies normally do not provide customer promise dates, because the material is in stock and the order is shipped immediately from inventory. If on-hand inventory is insufficient for an order, the order entry clerk must tell the customer when material will be available. Customer orders serve to trigger the resupply of inventory from which the order was filled.

Assemble-to-Order. For Assemble-to-Order firms, providing viable customer promise dates is vital. Many ATO manufacturers can accept an order, assemble it, and ship within one day. In other cases, customer orders will be booked for several periods into the future. The MPS scheduler uses the available-to-promise concept, discussed in Chapter 3, for each module or customer option in order to manage the conversion from forecasts to booked orders.

For an Assemble-to-Order firm, the uncertainty involves not only the quantity and timing of customer orders, but product mix as well. Safety Stocks can be used, and hedging is a valuable technique, as discussed in detail in Chapter 9.

For day-to-day operations in the Assemble-to-Order case, order entry sends customer orders to the Final Assembly Schedule.

Make-to-Order. A firm using Make-to-Order usually has a large backlog of customer orders. Some orders can be in progress, even though they aren't completely specified and engineered. This means the MPS scheduler is concerned with controlling these custom orders as they progress through all steps in the process. All of this has to be coordinated with customers as the orders become completely specified.

Uncertainty in a Make-to-Order firm is usually not the time or quantity of the customer order, but rather, how much of the company's resources will be required as the engineering is finally completed and exact requirements are determined.

For day-to-day operations, the primary activity is controlling customer orders, on the shop floor and in purchasing, to meet customer delivery dates. This must be related to the MPS to determine the impact of any engineering changes on the final customer requirement.

Linking with Sales and Customers through Demand Management

Demand Management, and more precisely order management, is M&CRP's gateway for contact with sales and the customer. To answer the question, "How do we position ourselves so our production will allow us to be competitive in terms of the lead time demanded by the marketplace?," sales and materials personnel jointly decide which of the company's products will be Make-to-Stock, Assemble-to-Order, Make-to-Order, and Make-to-Demand. This information needs to be available and understood by all sales personnel, so they can use lead time as a competitive weapon in their sales effort. The Available-to-Promise information provided by the MPS provides item by item information that can help sales personnel significantly in negotiating with their customers.

The sales and materials organizations must maintain open channels of communication. For example, whenever special sales programs are being promoted, the sales organization needs to alert the materials organization well in advance. Similarly, if an Engineering Change Order is being implemented that would affect delivery, materials needs to inform the sales organization about the ECO effective date and how it will affect deliveries.

Accounting and Finance

M&CRP creates and maintains the database that is necessary for the computerized operation of a manufacturing concern. By coupling this manufacturing database with accounting data files, a whole new generation of financial and management reports has evolved. The major M&CRP and accounting files are shown in Figure 7-4.

When integrated with the M&CRP system, financial and cmst-accounting systems provide the capability to evaluate cost performance, to cost inventory for financial statement purposes, and to project manufacturing and assembly costs for sales estimates and financial forecasts. The accounting systems also link when M&CRP ships an order to a customer, and when it receives products from a supplier. The following sections describe some of these integrated efforts.

Inventory Accounting

Perhaps the most useful accounting benefit of M&CRP is that it provides information that can be used to drive a cost-accounting system. In many manufacturing organizations, inventory control has an information system completely independent of the cost-accounting department. This dual information, of course, causes a double work load and a highly questionable balance sheet.

However, with a consolidated system available through the M&CRP database, the entire organization monitors inventories, and is ultimately responsible for inventory cost accuracy. In a standard-costing environment, once a manufacturing order is completed, the M&CRP program compares the accumulated actual cost against a standard cost (developed from engi-

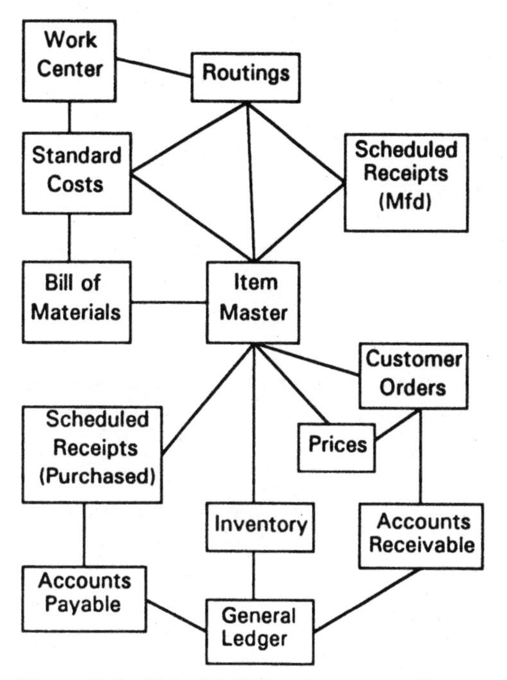

Figure 7-4. Major M&CRP and accounting files.

neering estimates), and records the manufacturing cost variance. M&CRP then transfers the completed items from work-in-process to finished goods that are valued at standard cost. Therefore, the work-in-process is carried on the balance sheet at actual cost, whereas finished parts, assemblies, and raw materials are valued at standard cost.

Inventory Management

M&CRP, properly implemented, can greatly reduce the burdens on management personnel responsible for ordering, stocking, and controlling inventory. Managers have access to information, going well beyond what they need for ordering, stocking, and controlling inventory. By combining the Bill of Materials and product cost files with the inventory file, we have the capability to provide inventory dollars in three ways: balance by descending dollar-value sequence, balance by stocking location, and inventory value by product line. The system can also provide information to determine past and projected inventory turns for each item and each product line, and offer the information in lists of descending dollar amounts. Thus, inventory managers can also see where the dollar value of the inventory is located and how these dollars relate to their various product lines. More detailed measurements of inventory are discussed in Chapter 12.

Financial Management

The M&CRP interface produces several significant reports that are unique to financial management. A forecast of sales and gross profit by product can be obtained by using the Bill of Materials, standard product cost, and selling price in conjunction with the sales forecast. Similar information can be obtained on current shipments.

A simplistic cash-flow projection can be obtained by using the standard product-cost file in conjunction with the sales forecast. This calculation includes a major part of the cash expenditures over the time span of the sales forecast. Future cash collections also can be forecast by substituting the selling-price file for the product-cost file. Both of these cash-flow calculations would require built-in assumptions for lead time on expenditures and receivables.

Accounts Receivable

When M&CRP ships an order to a customer, it feeds the shipment information to the Accounts Receivable module of MRP II. This triggers the billing process, in a timely and error-free manner. M&CRP also reduces stock-on-hand in the finished goods inventory, which reduces the value of the inventory for the general ledger.

Accounts Payable

Likewise, when the company receives materials from a supplier, the receiving transaction increases the inventory on hand, which increases the value of the inventory for the general ledger. The receiving transaction simultaneously informs the Accounts Payable module of MRP II that the company has received materials, fulfilling one prerequisite for paying the supplier's invoice.

Summary

This chapter began with a discussion of the overall operation of the MRP II system, stressing the interaction between activities, with particular emphasis on those activities that interact directly with the M&CRP module. We outlined the MRP II process, and then described the adaptation of MRP II for four Demand Response Strategies: (1) Make-to-Stock, (2) Assemble-to-Order, (3) Make-to-Order, and (4) Design-to-Order.

In the latter part of the chapter we described the interactions between various activities in more detail, with emphasis on interactions with M&CRP. The most important interactions of the M&CRP system are with the Master Production Schedule, Production Activity Control, and Purchasing activities. The MPS drives the M&CRP system. It provides the build schedule of end products that is exploded in the MRP system. Using the output from the MRP system, CRP determines the capacity required to execute the MPS.

We also described the many interactions between M&CRP and JIT, TQM, the Engineering disciplines, the Demand Management function, and Accounting and Finance functions.

Although it almost seems self-evident or trite, we strongly recommend that you intelligently integrate these functions and activities. This will, in fact, provide a competitive edge, because achieving this integration is much more difficult to accomplish on a day-to-day basis than it might appear.

Selected Bibliography

Blackstone, John H., Jr.: *Capacity Management,* South-Western, Cincinnati, OH, 1989, Chap 5, 6, and 7.

Browne, Jimmie, J. Harhen, and J. Shivnan: *Production Management Systems: A CIM Perspective,* Addison-Wesley, Reading, MA, 1988, Chap 7.

Campbell, Robert J. and Thomas M. Porcano: "The Contributions of Material Requirements Planning (MRP) to Budgeting and Cost Control," *MCRP Reprints,* APICS, Falls Church, VA, 1991, pp. 216–218.

Carter, Phillip L. and Chrwan-jyh Ho: "Vendor Capacity Planning: An Approach to Vendor Scheduling," *MCRP Reprints,* APICS, Falls Church, VA, 1991, pp. 219–229.

Deis, Paul: *Production and Inventory Management in the Technological Age,* Prentice-Hall, Englewood Cliffs, NJ, 1983, Chap 8 and 12.

Fogarty, Donald W., J. H. Blackstone, and T. R. Hoffmann: *Production and Inventory Management,* South-Western, Cincinnati, OH, 1991, Chap 3, 14, 15, 17, and 18.

———, T. R. Hoffmann, and P. W. Stonebraker: *Production and Operations Management,* South-Western, Cincinnati, OH, 1989, Chap 3, 14, 16, 18, and 20.

Ford, Quentin: "Distribution Requirements Planning—and MRP," *MCRP Reprints,* APICS, Falls Church, VA, 1991, pp. 190–193.

Harding, Michael and Marylu Harding: *Purchasing,* Barron's, 1991.

Ling, Richard and Walt Goddard: *Orchestrating Success: Improve Control of the Business with Sales and Operations Planning,* Wight Publications, 1988.

Makridakis, S. and S. Wheelwright: *Forecasting Methods for Management,* 5th ed., Wiley, New York, 1989.

Martin, Andre: *DRP: Distribution Resource Planning,* Oliver Wight, Essex Junction, VT, 1983.

May, Neville P.: "Which First: MRP or Production Activity Control?" *MCRP Reprints,* APICS, Falls Church, VA, 1991, pp. 194–198.

Melnyk, Steven A. and Philip L. Carter: *Production Activity Control,* Irwin, Homewood, IL, 1987.

Plossl, George W.: *Production and Inventory Control: Principles and Techniques,* 2d ed., Prentice-Hall, Englewood Cliffs, NJ, 1985, Chap 4, 10, 11, and 12.

Schorr, John E. and Thomas E. Wallace: *High Performance Purchasing,* Oliver Wight, Essex Junction, VT, 1986.

Schultz, Terry: *Business Requirements Planning: The Journey to Excellence,* The Forum Ltd, Milwaukee, WI, 1984, Chap 10, 11, and 12.

Smith, Bernard T.: *Focus Forecasting: Computer Techniques for Inventory Control,* APICS, Falls Church, VA, 1984.

Smith, Spencer B.: *Computer Based Production and Inventory Control,* Prentice-Hall, Englewood Cliffs, NJ, 1989, Chap 4, 8, 12, 13, and 14.

Stickler, Michael J.: "Purchasing—The New Frontier, *MCRP Reprints,* APICS, Falls Church, VA, 1991, pp. 239–243.

Umble and Srikanth: *Synchronous Manufacturing,* South-Western, Cincinnati, OH, 1990.

Vollmann, Thomas E., W. L. Berry, and D. C. Whybark: *Manufacturing Planning and Control Systems,* 3d ed., Irwin, Homewood, IL, 1992, Chap 3, 5, and 8.

Wassweiler, William R.: "The Impact of MRP on Shop Floor Control," *MCRP Reprints,* APICS, Falls Church, VA, 1991, pp. 199–200.

Wight, Oliver W.: *MRP II: Unlocking America's Productivity Potential,* CBI Publishing, Plano, TX, 1981, Chap 9, 10, 11, 12, 13, and 14.

8

The Technical Aspects of an M&CRP System

Introduction

Drivers who depend on their cars for their living, such as race drivers and taxi drivers, must learn about the automobile's mechanical structure. Likewise, a materials practitioner, who depends on an M&CRP system for his or her livelihood, must understand the basics of the internal structure of that system. A practitioner who understands a little about the technical aspects of an M&CRP system, will understand better what it can and cannot do, and, therefore, will know how to best use it (and how *not* to use it, which can be equally important). In this chapter, we explain, in as nontechnical terms as possible, the technical underpinnings of an M&CRP data system, including:

Basic concepts

Database and Database Management Systems options

Database contents

Data transaction processing

M&CRP system sources

Hardware options

Centralized vs. distributed systems

Data accuracy

Basic Concepts

The basic unit of information in business data is the *character,* which may be a letter, a number, or a punctuation mark. Characters are grouped together into a *field* that represents a name, description, quantity on hand, and so on. Groups of related fields are combined into one *record.* A user typically handles a record as a unit. For example, a record might contain all the planning information and other static descriptions of a single item. A logical grouping of records, such as grouping the item records for all items together, is known as a *file.* A relational database replaces the term field with *entity,* record with *row,* and file with *table.* In some respects, this is easier to understand; think of a spreadsheet, such as Lotus™ or Excel™. An entity, or field, would be a cell, where one row and one column meet; a record, or row, would be a single row, and a file, or table, would be a table, with both rows and columns.

Data files (tables) are the basic building blocks of a data system. A *Database Management System* (*DBMS*) uses data files as its foundation, adding the logical links between the data in the various files. Figure 8-1 illustrates the three types of data files used in M&CRP:

1. *Sequential or flat files,* which contain only data files and no indexes. This is like a filing cabinet with file folders in it. To locate part number 12345 in a flat file, the computer would read each record in the file, starting with record number 1, until it found the one for part number 12345. In a large file (such as one with 100,000 records), this process can take considerable time. If a program requires reading many records from a large

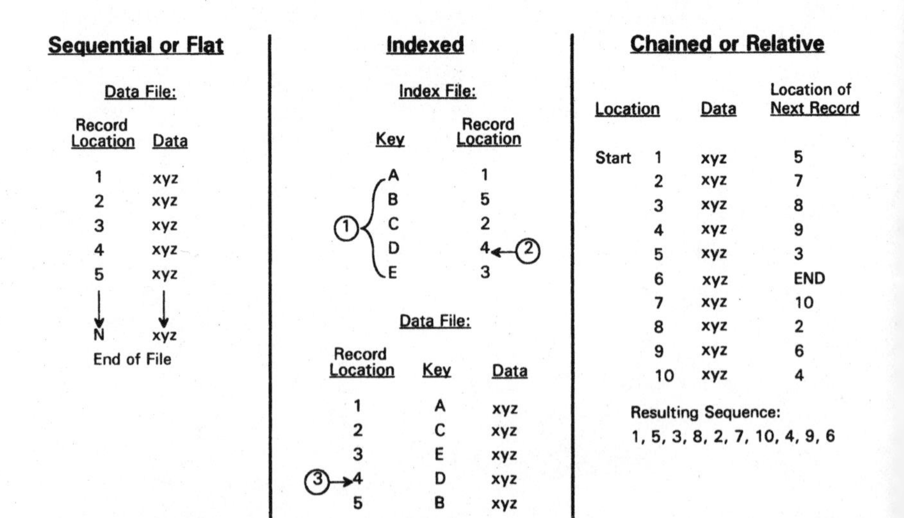

Figure 8-1. Sequential, indexed, and chained file organizations.

unindexed file, the program execution time will probably be unacceptably slow.

2. *Indexed files,* which use two types of files: data and index. Each data file has one or more "key fields" (for example, the most obvious key in an Item Master File is the item number; the keys in a customer file might be the customer number, the territory, and the sales rep). Each key field is organized in a separate index for more rapid access. The computer uses the keys to quickly find the desired record. For example, to reach the Item Master for part number 12345, the computer would rapidly read the part number index file, which is sorted by part number, to find the exact location of the record for part number 12345 in the Item Master File. The computer would then use that location information to directly read the part record from the Item Master File. Although the method is beyond the scope of this chapter, computers access indexed file data in a fraction of the time required to access the same file data in sequential fashion.
3. *Chained or relative files,* which create a chain of records through a large file or files. In this method, the location of the next record in the chain is maintained in physical data record itself. A good example is the chaining of records between the Bill or Materials file and the Inventory Records file for the Bill of Materials explosion process. Because of the different nature of the data contained in these two files, they cannot be consolidated or organized similarly. However, by appropriately chaining the related records in each, they can be permanently linked for the Bill of Materials explosion. Chained files may also be indexed, providing the same speed of execution as indexed files.

Database and Database Management Systems

As we saw in the last section, we can store data in sequential, indexed, and chained files. In this section we describe a database and a Database Management System. The difference between a database and a file is analogous to the difference between a thoroughly cross-referenced set of files in several cabinets, and a single file in one cabinet that is not cross-referenced to any other file. A database can identify and handle relationships among various data sets; a file cannot.

A Database Management System is software that performs all file maintenance tasks, including indexes, chains, record definitions, and so on, separately from any particular application program. Thus, how the data is to be accessed is maintained separately from the actual data. The DBMS can construct any number of different access paths through the same collection of actual data. We outline below some of the more widely recognized approaches to databases and DBMSs.

Commercially Available Databases

Database software suppliers sell and support DBMSs. The suppliers might also sell computer hardware, such as IBM, DEC or HP, or software only, such as Oracle, Sybase, Cincom, or Borland (Paradox and dBASE). All other things being equal, a commercially available database is the best platform for an M&CRP system, because it will probably continue to be enhanced and supported.

Commercially available databases also include report-writing and query (screen inquiry) tools that can be used by persons who are not computer programmers. These tools considerably increase the usefulness of the package, because they allow end users to customize their searches so they can find answers for themselves, without waiting for MIS to staff a project. There are, currently, three basic types, or architectures, of databases, with a fourth, *object-oriented,* starting to appear on the horizon:

1. *Hierarchical databases* strongly resemble a Bill of Materials; each data record (which is analogous to a file folder in a filing cabinet) is related to at least one other data record hierarchically above or below it. For example, a customer order would be related to the customer (above it). Each line item on the order would be related to the its individual part master (above it). This type of database is rather complex, because the only way to establish relationships is by going up and down the relationship chains. Additionally, expanding the size of data records (so that additional data, such as shipping dimensions, can be stored) can be relatively difficult; expanding the size of a field in the record, such as the part number field, can be equally difficult in some of these databases.
2. *Network databases* resemble a single-level Bill of Materials; the top level contains only indexes, and the bottom level contains all the data. A given set of data may have several indexes. These databases generally provide greater flexibility and speed than hierarchical databases; however, they can be as difficult to change as hierarchical databases.
3. *Relational databases,* the most recent of the three types currently in use, have by far the greatest flexibility. Relational databases have a sound mathematical foundation. They allow the expansion of field and record sizes with relative ease. They have a reputation of being slower than other database types, but their speed has been improving. We anticipate that relational databases will become the database architecture of choice, as suppliers continue to address the speed problem.

Internal Databases

An internal database is not available separately from the M&CRP package in which it is used; it is "hard-coded" directly into the M&CRP package. Internal databases were very common 10 to 20 years ago, before the advent of fast, reliable, commercially available databases. Internal databases are now rapidly waning in importance and use.

Database Contents

In the previous sections, we discussed the types of file organizations and types of database organizations that could be used in a M&CRP system. In this section we describe the conceptual contents of several of the major files, or tables, in the M&CRP data system. The files support one or more of the functional modules of the M&CRP-MRP II system, shown in Figures 1-18 and 7-1. In the following sections, we present a logical, simple organization of the M&CRP database. Some M&CRP suppliers have chosen substantially different ways to implement the actual database design, for reasons of performance and maintainability.

In the following tables, we list typical contents of each of the major files to enable you to better understand the types of questions the system can easily answer. Sometimes a file is known by more than one name; we have included most of the commonly used terms.

Item Master (Part Master)

This is the backbone of any inventory system, whether M&CRP or Order Point. The Item Master data set contains one record, or entry, for each item. Each record contains many pieces of data for that one item, including the ones shown in Figure 8-2.

Sometimes, particularly in systems for large companies, the Item Master File is divided into two files, with the static information for an item staying in the Item Master File, and the dynamic information being placed in a separate file variously called the Inventory File or Subordinate Item Master File, which we describe in the next section.

Inventory

The Inventory file contains quantity records of each location, batch, and/or lot of inventory. Having a separate file means that there are virtually no limits as to the number of locations, batches, and so on, that the M&CRP system can track. This file usually contains the types of data shown in Figure 8-3.

Part Number	123	456	789
Description	Washer	Flange	Widget
Revision Level	C		F
Stock On Hand	5,462	25	394
Last Transaction Date	11/15	11/19	12/3
Source	P	M	M
Product Line		Household	Industrial
Planner Code	KCL	NBW	TRO
Commodity Code	Fasteners	Metal Parts	Finished
Phantom?	N	N	N
Lot Control?	N	N	Y
Serial Control	N	N	N
Stocking Unit of Measure	EA	PR	EA
Buying Unit of Measure	GR	PR	EA
Selling Unit of Measure	EA	PR	DOZ
Cost Method	Std	Std	Std
Cost	$.0035	$4.372	$52.418
Last Cost Update Date	12/31	12/31	12/31
Order Method	ROP	MRP	MRP
Order Policy	Period	LFL	Fixed
Order Quantity	10 days		120
Order Quantity Multiple	1,440	1	1
Scrap %	2	5	1
Safety Stock	1,000	6	12
Lead Time (work days)	3	2	4
YTD usage	54,791	5,912	11,374
Current Selling Price		$8.75	$98.65
New Selling Price Date		12/31	
New Selling Price		$8.95	
ABC Class	C	B	A
Chart of Accounts	04-01-1202	04-01-1301	04-01-1405
Obsolete Date			

Figure 8-2. Item Master record.

Item Number	123	456	789
Warehouse	Portland	Portland	Chicago
Stockroom	Main	Main	
Location	24-5-3	29-32-4	A-1-3
Lot			BCA3214
Lot Date			9/24/93
Serial Number		F6513841	
Date Last Counted	12/15/93	11/13/93	
Batch	PM3931551		
Use By Date	4/15/94		
Last Transaction Date	2/11/94	1/22/94	12/17/93

Figure 8-3. Inventory record.

Bill of Materials (Parts List, Product Structure)

The Bill of Materials file connects multiple components to a given parent, in a single-level relationship. M&CRP systems store data in the Bill of Materials file on a single-level basis, that is, only information linking each parent and its immediate children. By successively linking the part numbers, M&CRP can create a full Bill of Materials for each item, either downward (a standard Bill of Materials), or upward (a Where-Used Report). This file also supports cost roll-ups. Finally, M&CRP uses the effective dates and scrap factors in this file. The BOM file contains many fields, including those shown in Figure 8-4.

Parent Part Number	456	456	789
Parent Part Rev Number		A	F
Component Part Number	123	124	123
Effective Start Date	9/15/89	1/16/93	
Effective Stop Date	1/15/93		
Quantity Per One Parent	5	5	8
Scrap %	4	3	3
Lead Time Offset			2
Operation Used In	30	30	65
Backflush?	Y	Y	N

Figure 8-4. Bill of Materials record.

Figure 8-4 shows that part 124 replaces part 123 as a component of 456, effective 1/16/93. Engineering has changed the revision number of 456, from a blank to an *A* to show this change. (There might be other concurrent changes to 456, all part of revision *A*.

Work Center

The work center records define the work centers, which are the basic building blocks of CRP and Production Activity Control. Each routing record defines a single operation for a specific item that must be performed in a work center. CRP computes projected loads by work center, and compares them to the actual time available, by work center. The Operations Sequencing and Input/Output Control modules operate on a work-center basis. Cost accountants use work centers, singly or in groups, to assign overhead. Work center data usually includes the fields shown in Figure 8-5.

Figure 8-5 shows a first shift and a second shift in the same work center,

Work Center ID	Assem-1	Assem-2	Weld
Work Center Description	Assembly 1st Shift	Assembly 2nd Shift	Welding
Cost Center/Department ID	Assembly	Assembly	Fab
Shift	1	2	1
Effective Start Date	10/15	8/07	5/22
Effective Stop Date	2/28	2/28	
Labor Capacity - Hours/shift	120	80	24
Crew Size	15	8	3
Labor Efficiency Factor	0.95	0.85	0.9
Number of Machines	8	8	4
Machine Efficiency Factor	1	0.9	0.9
Utilization	0.9	0.75	0.8
Default Setup Crew Time (hrs)	0.25	0.3	0.5
Default Setup Machine Time	0.25	0.3	1
Default Queue Time (hrs)	0.5	0.5	5.7
Overhead Factor	Labor	Labor	Mach
Default Labor Rate	9.75	10.25	13.86
Actual Efficiency - current year	0.94	0.87	0.83
Down Time - current year (mach hrs)	58	65.5	22
Average Setup Time - current year (hrs)	0.32	0.35	0.48

Figure 8-5. Work center record.

each with its own work-center record. Their staffing levels, workday length, and efficiencies are different.

Routing

Each routing record contains data for one routing operation for one part. Routing records are used to compute actual operational lead times, to track the status of a manufactured Scheduled Receipt, and by PAC as the basis for labor reporting. After a Scheduled Receipt is completed, the routing records retain their data for future analysis. This is where PAC gets its information about past labor requirements for a given job. Figure 8-6 includes the important data elements that routing records typically contain.

Requirements (Demand)

The Requirements file contains one record for each future requisition from stock, whether the part will be shipped to a customer (in which case the requirement is for a customer order), or used internally as a component of a higher-level assembly (in which case the requirement is an internal requirement). After the requirement has been filled, the record serves as a history of what we really did. Each Requirements record usually contains the elements shown in Figure 8-7.

The concept of allocation is discussed in Chapter 4. The first column shows a demand that has actually taken place. The M&CRP system retains the information in its files for future reference (e.g., when a Cycle Counter asks for all transactions for part 123, because the inventory is no longer in balance). The second column depicts a demand that is still a future demand. The third column shows the same information for a line item from a customer order.

Planned Order and Scheduled Receipt

Planned Orders and Scheduled Receipts are supply orders because they will replenish stock-on-hand in the stockroom. (Customer orders decrease stock-on-hand in our stockroom, so they are future requisitions.) Supply orders basically fall into two categories: manufactured (also called work orders or shop orders), and purchased. The major data elements of supply orders are shown in Figure 8-8.

This group of three supply records provides a snapshot of a company as it has received the purchase order for the washers, is currently welding the flanges, and is about to start assembling the widgets. Note that the part description is not carried in these records; when the computer needs to dis-

Part Number	456	789	789
Routing Operation	30	10	20
Operation Description	Weld end to base	Assemble washer to bolt; insert bolt	Tighten nut onto bolt
Alternate Routing Op Number		10A	20A
Effective Start Date	7/8/88	5/21/91	5/21/91
Effective Stop Date		10/19/93	10/19/93
Work Center	Weld	Assembly	Assembly
Setup Time—Machine	0.5	0.1	0.05
Setup Time—Labor	1	0.1	0.05
Setup Labor Grade	7	4	4
Setup Crew Size	1	1	1
Setup Quantity	1	0	0
Run Quantity	47.5	3.5	4.2
Run Unit of Measure	Pcs/Hr	100 Pcs/Hr	100 Pcs/Hr
Run Crew Size	1	1	1
Run Labor Grade	6	4	4
Run Machine Quantity	1	1	1
Last Time Study Date	8/16	4/22	4/22
Outside Processing Lead Time			
Overlap %	50	95	95
Overlap Quantity	10	50	50
Move Time—Hrs	1	0.01	0.01
Drawing Number	D 1022	D 945	D 945
Tool Number 1	F 2894	F 372	F 372
NC Tape Number			
Shrink Percent	6	0.5	0.5
Alternate Operation	31		

Figure 8-6. Routing record.

play the part description information, it retrieves the description from the Part Master record, thereby eliminating redundant data.

Transaction Processing

A transaction is a logical single interaction between the computer system and a sender or receiver of data. For example, receiving material into the

Part Number	123	123	789
Revision Number	C	C	F
Quantity	1,200	1,200	24
Date Required	10/11	10/14	10/15
Parent Part Number	456	789	
Customer Order # and Line #			4215-3
Allocated?	Y	N	Y
Work Order Number	11,582		
Date Issued	10/11		
Quantity Issued	1,200		
Stock Clerk ID	LJB		
Quantity Returned			

Figure 8-7. Requirements record.

stockroom requires a transaction, which increases the quantity on hand, and updates the outstanding order. Likewise, inquiring about the balance on hand of a specific part requires a transaction. Transactions are defined by two important characteristics: function and processing interval.

Transaction Function

Transaction function describes what the transaction does to the data in the system. There are two basic types:

1. Update transactions modify the data in the system. These can be further defined as:

   Add transactions, which add new data to the system (for example, adding a new part number or a customer order).

   Change transactions, which alter existing data, such as the balance on hand for an existing part (for example, a stockroom issue decreases the on-hand balance for an item).

   Delete transactions, which remove data from the system (for example, deleting an existing Item Master or a customer order).

2. Inquiry transactions report on data in the system without modifying it. These transactions can present data on a screen or in a report.

Processing Interval

The *processing interval* indicates how quickly a transaction is processed, and can be either on-line or batch.

Part Number	123	456	789
Revision Number	C		F
Required Quantity	2,880	122	12
Start Quantity	2,880	129	122
Required Date	10/10	10/14	10/20
Start Date	10/7	10/10	10/14
Parent Part/Customer Order	456	789	SO 3214
Quantity Actually Started		130	
Quantity Received into Stock	2,736		
Quantity Scrapped		3	
Quantity awaiting MRB	144	2	
PURCHASE ORDERS			
Supplier	Acme		
PO Number	5,462		
PO Placed Date	10/7		
PO Revised Date	10/8		
Promised Receipt Date	10/10		
SHOP ORDERS			
Work Order Number		6,892	
Date Started		10/10	
Last Operation		30	
Last Quantity Reported		125	

Figure 8-8. Planned Order and Schedule Receipt record.

On-line in an M&CRP system is characterized by having the person who entered the transaction wait for the result. The computer processes the single transaction immediately. Response time for on-line transactions can be measured in seconds, from the time a person presses the "Enter" key on the keyboard, to the arrival of the first response character on the screen.

Response time of less than a second is excellent, although 1 to 3 seconds is usually quite acceptable. Some research suggests that subsecond response time improves productivity far in excess of the modest time saved, by keeping a person's attention fully focused on the task at hand. Response times exceeding a second apparently invite a person's attention to wander, requiring them to remember what they intended to do next after the computer responds. This, in turn, significantly increases the error rate, as well as the amount of time required for the human to start performing the next task.

Response time acceptability is relative. When driving a car, a 1 to 3 second delay between touching the brake pedal and having the brakes engage is not acceptable. Likewise, for process control, such as controlling the functions of a steel mill or a computer-controlled lathe, 1 to 3 seconds is not acceptable. However, for a stock receipt transaction, 1 to 3 seconds is usually acceptable. For updating a general ledger, 1 to 3 *days* is usually acceptable, except during a monthly closing cycle.

Batch processing in an M&CRP system is characterized by holding transactions until they can all be processed together in a batch. One classic example of batch processing is the monthly general ledger. Another is MRP itself, which (in a regenerative mode) processes all part numbers, all open and planned replenishment orders, and all open and planned requirements. For a large database, a regenerative MRP run can require 10 or more hours. In fact, the length of MRP run time is the major reason why software suppliers originally developed net change MRP.

Perspectives on Processing Intervals

Initially, most M&CRP systems did not include on-line processing; the software was too difficult to write and too unreliable. In the late 1970s and early 1980s, many companies switched to on-line processing as the preferred alternative. Processing frequency (batch versus on-line) is almost a cultural issue in a company, with two different viewpoints:

1. Many accounting-oriented individuals see the world from an analytical viewpoint; they determine trends and variances from historical data. They edit transactions in batches until the batch balances; then they approve the batch for processing. The earliest inventory systems were, in fact, designed as inventory accounting systems.
2. Operations-oriented individuals see the world in an entirely different light; they use information to predict and manage the future as, for example, in the making and shipping of a product to specification, and on time. They need to know what has to be done in order to achieve these goals, based on the latest changes. They expend considerable energy reorganizing their departmental resources to adjust to the constant changes in customer demand, and the availability of material and personnel. They do not appreciate being told this Thursday that they were producing nonstandard parts on some unspecified day last week. They want to know what it will take to prevent nonstandard parts from being produced in the first place, or, at the very least, to be alerted as soon as a

nonstandard part has been produced, so they can minimize the waste. In this environment, batch processing is too unresponsive to be useful or relevant.

Figure 8-9 lists some typical transactions of an MRP II system, and indicates the suggested processing frequency for each. Some transactions can be effectively processed either by batch or on-line. However, even batch processing should include initial on-line editing, so that any errors in the starting parameters can be corrected prior to processing.

A multiuser business computer routinely processes more than one transaction in one second. For example, it can accept data from one user, while it accesses the disk for data for another, performs calculations for a third, and prints a report for a batch program. The operating system keeps each activity separate from the others, so that each user has the impression that he or she is the only person using the computer.

When the transaction volume at a particular time is greater than the computer system can handle, the result is similar to the traffic on an interstate highway in a major city at five o'clock on a rainy Friday night: slow throughput and frustrated people impatiently waiting to achieve their objectives. Curing this type of situation does not always require more or faster hardware; sometimes the operating system parameters can be adjusted, or applications programs revised. A new Master Scheduling program once required more than 60 hours to execute for 1000 master scheduled items. The customer notified the software supplier, who redesigned it for greater speed and reduced the run time by 90 percent!

TRANSACTION	ON-LINE	BATCH
Stockroom issue	X	
Order entry	X	
MRP regeneration		X
MPS "what-if"	X	
On-hand inquiry	X	
BOM change	X	
Item master maintenance		X
Accounts payable check	X	X
Receive cash	X	X

Figure 8-9. Typical transaction processing—on line versus batch.

M&CRP System Sources

M&CRP systems can come from several sources. In this section we outline the four basic sources for M&CRP systems in the order of increasing cost.

Standard Packages

Standard packages can come from M&CRP software suppliers, database software suppliers, and hardware suppliers. The more flexible packages utilize a relational database, and have been created by using a fourth generation language (4GL). The larger suppliers of these packages maintain and enhance their packages, and offer consulting and education to their customers. There are over 200 M&CRP packages on the market, running on everything from single PCs to mega-mainframes, and ranging in price from a few hundred dollars per module to over a million dollars for the entire package.

Customized Packages

These are standard packages that have been extended or modified for the customer. Customization can be performed by:

The supplier

The customer's in-house MIS (Management Information Systems) staff

A third-party customization house

Although customization is sometimes required, the true total costs of customization frequently far exceed the initial perceived cost. Whenever a new system revision comes out, the person who is responsible for the customization must review each line of customization, to determine whether the new version eliminates the need for the customization, ignores it, or conflicts with it. Customization can also cause substantial difficulty in working with the original supplier, if the customer thinks they have found a "bug," or unintended result, in the program.

Our experience indicates that over half the changes that users "must have" before implementing an M&CRP system become unnecessary within three months of implementation, if the project team can successfully defer installing these changes before implementation. We strongly encourage the project team and the steering committee to insist on absolutely minimal changes to the "vanilla-flavored package, straight out of the box," until at least three months (and perhaps six months) after implementation. Then, the committee can prioritize the change and enhancement request list.

Many companies that use PCs are creating additional functionality for their purchased packages. Some of the more common extensions include forecasting, additional inventory management reports, EDI, and data collection. These add-on capabilities do not compromise the integrity of the original package code; however, they still must be maintained.

Custom Development In-House

A custom development house will create software to meet a specific company's specifications. It is analogous to an architect and building contractor who design and build a house that is totally unique. However, this custom software will probably cost several times as much as a standard package. Most major software development houses measure their investment in their packages in the millions of dollars, and perhaps hundreds of staff-years. Even with the currently available tools, including CASE (Computer-Assisted Software Engineering), 4GL (Fourth-Generation Languages), RDB (Relational Data Bases), and ADD (Active Data Dictionaries), we strongly encourage any company to seriously look at existing packages before developing its own system. Additionally, custom developed packages normally lack one or more critical attributes:

Formal, professional education

On-going maintenance and enhancements

Decent documentation (although some packaged software documentation is also notoriously weak)

The largest risk is in the design itself. Many practitioners, designers, and developers, especially those who have spent many years at a single company, lack sufficient understanding of the various design alternatives to design a flexible, well-founded system. The following examples, all taken from real life, illustrate the dangers of custom modification and development.

The most important aspect of a completed application is its flexibility—its ability to support a changing environment. One manufacturing company designed its own customer order entry and customer service system, because "they were unique." They created the entire design around the sales territory number of each customer; each customer had only one sales territory. Unfortunately, the designers also assumed (because "they had always done it this way" in their manual system) that the sales territory number would be the salesperson's initials. After they invested six staff-years, they discovered the inherent flaws in their logic:

- Salespeople change: one changed her name when she married; another moved and was replaced.

- Territories change: a major customer moved to a new territory mid-year.
- Sales responsibilities change: the company had to realign territories mid-year for strategic reasons.
- Customers change: some customers (generally the largest and best) have many locations in many territories; the newly designed system was incapable of associating Division A of XYZ with Division B of XYZ.

Finally, even modifications to existing code carry considerable risk. We have personally watched a manufacturing company purchase a relatively full-featured M&CRP package with the intent to make only the "bare minimum critical" changes. A year later, this company requested a new, updated version of the code; their changes had become so complex that they scrapped their entire change effort and wanted to start with a fresh copy. A year later, they did the same thing! Thus, after two years, they had achieved absolutely no operational benefits from M&CRP, but had incurred extensive costs in development efforts.

However, custom development can yield some stunning benefits. If a company chooses to use information systems as a strategic tool, to push the leading edge, it will probably be forced to create its own custom system for two reasons:

1. There will be few, if any, existing packages that perform the function (that is the definition of leading edge), and
2. The company does not want the competition to have access to the developed system.

American Hospital Supply (AHS) provides one example of creating a strategic advantage with an information system. In the late 1970s and early 1980s, AHS realized that their customers would strongly prefer to receive their orders on the next day, rather than waiting several days to a week, as was the custom in the hospital supply industry at that time. So AHS placed on-line terminals in the supplies purchasing areas of its customers, with the guarantee that 95 percent of the items that a customer ordered would be delivered the next day. AHS rose from being the sixth largest hospital supplies provider to become the largest in the United States in large part because of the strategic information system application.

Don't change *anything* until you truly understand:

What you want it to do

What it currently does

How to make it do what you want

How your change will affect every other program in the entire system

Systems Integrators

Systems integrators are consulting and systems development houses that specialize in integrating two or more stand-alone packages (for example, a classic M&CRP system for order entry, material planning, purchasing, and financials, with a specialized system for the control of a shop floor of a printing company, with scheduling based on light-to-dark colors and printed page size). They leave the original modules as unaffected as possible. They can also create a custom package to specification. Systems integration projects are subject to the same risks as custom development projects, because the stand-alone packages were not necessarily designed to work with each other. In fact, in this example, both the M&CRP system and the specialized printing system had a shop order file. The integrator must ensure that each package can use the integrated shop order file.

Additionally, systems integration includes the same risks as package modification. Every upgrade for the base package must be carefully scrutinized by the integrator, to ensure that it will function acceptably with the integrated packages.

Hardware Options

The basic computer hardware options discussed in this section are actually defined more by their respective operating systems than by their processing capabilities. Large minicomputers can now perform work that was done exclusively by mainframes 5 to 10 years ago. Even more interesting, high-powered microcomputers now easily perform work that required minicomputers 5 to 10 years ago, and which required mainframes 15 to 20 years ago.

Mainframe

Mainframes are the oldest class of computers and, traditionally, the largest. Mainframe operating systems architectures were developed in the 1960s. IBM, Unisys (formerly Burroughs and Sperry), NCR, Control Data, Bull HN (formerly Honeywell), Fujitsu, Hitachi, and ICL are representative mainframe suppliers. Each mainframe has a proprietary operating system. This class of computers is characterized by major investment, sometimes in the millions of dollars for the hardware alone, although mainframe suppliers are also bringing smaller and smaller machines to market to compete with minicomputers and microcomputers. Application software tends to be relatively robust, full-featured, and expensive. Mainframes are designed to process many transactions per second, and/or to run large batch jobs very quickly. In their earliest architectures, these systems were designed as batch (not on-line) machines.

Minicomputer

Minicomputers, or "minis," are the second oldest class of computers. Minicomputers appeared in the early 1970s, from manufacturers such as DEC, Hewlett-Packard (HP), Data General (DG), Prime, and Wang. Most mainframe suppliers also design and market minicomputers. The most widely used minis are the DEC VAX, the Hewlett-Packard HP 3000, and the IBM System 3X series, including the System/32, System/34, System/36, and System/38, and the AS/400 series. Minis originally had a proprietary operating system; some are now using more standard operating systems, such as UNIX, or one of its derivatives. Applications software tends to be relatively robust and full-featured, but substantially less expensive than mainframe software. These systems are usually designed to support multiple users by processing many on-line transactions per second, with secondary emphasis on batch processing. Most of them can be used in a distributed fashion, in which multiple minis share data while they each perform their own tasks. Minis are now also starting to be used as very robust file servers in Local Area Networks, which we discuss later in this chapter.

Microcomputer

Microcomputers are the newest architecture. Microcomputers, or PCs (Personal Computers) originally appeared as single-user novelty and hobby devices in the mid 1970s. They became useful to business in the late 1970s when an MBA student persuaded a friend to write a program to calculate spreadsheets on an Apple PC. The vast majority of microcomputers for single users in North America utilize one of two operating systems: MS-DOS (or PC-DOS, for IBM PCs), or the Macintosh operating system. In Europe, UNIX is more popular. Whereas the DOS machines are based on a standard created by IBM, literally hundreds of companies assemble, sell, and service IBM-compatibles, or "clones." The major supplier of Macintosh class machines is Apple Computer. Competition is extremely intense in the DOS arena; in 1990, price-performance was improving by a factor of approximately 2 every 18 months, and the improvement cycle is becoming faster. A large PC can serve as the only computer in a small company, replacing a mini. In this instance, the PC generally utilizes the UNIX operating system, or one of its derivatives.

Work stations are a distinct class of microcomputers. Although they were originally intended to be used by an individual to perform intensive calculations, such as 3-D engineering design, they can also be used as the main computer for a small to medium-sized company. Most work stations now utilize UNIX, or a derivative, as their operating system. They are extremely powerful and deliver impressive price-performance. Major work station suppliers include IBM, Sun, Hewlett-Packard/Apollo, and DEC. In the early 1990s, work station performance started to eclipse the performance of the largest mainframes twenty years earlier.

The use of *PC networks,* usually *Local Area Networks* (*LANs*) continues to increase dramatically. PC networks require a PC rather than a "dumb terminal" for each user, because the processing is performed on the user's computer. The architecture requires one or more computers (which can be microcomputers, minicomputers, or even mainframes) with very large disk drives to act as "file servers" (to store and transmit data to whichever PC requests it). The major suppliers of these network operating systems in a DOS environment are Novell, Microsoft, IBM, and Banyan. This architecture is often called "*client-server,*" because the "client," or user, performs the processing on the local PC, getting data from the central file server. Client-server systems provide exceptional value, not only with respect to price and performance when originally purchased, but also owing to their much lower maintenance costs and inherent flexibility.

Hardware and Software Costs

Although it may be difficult to imagine, in the early 1900s companies carefully rationed telephones to those who really needed them, because phones were expensive. One phone might serve an entire department. Personal computers are following the same acceptance and utilization curve as telephones, though much more rapidly. Figure 8-10 illustrates the relative price- performance and power of the three basic hardware classes.

Additionally, there have been large differences in the software costs for the three hardware types. Traditionally, mainframe-based software has been priced two or more times as high as mini-based software with the same functionality. Likewise, mini-based software has traditionally been substantially

	1981	1985	1989	1993
Mainframe	$420,000	$200,000	$100,000	$54,000
Mini	$170,000	$85,000	$20,000	$6,000
PC/Workstation	$15,000	$6,000	$900	$100

Figure 8-10. Relative price-performance of computers. [*Sources 1981–1989:* Business Week, *March 6, 1986 (data: Gartner Group); 1993: Langenwalter & Associates estimate.*]

more expensive than PC-based software with the same functionality. However, mini and mainframe software and hardware prices have started to drop to reflect PC-based competition.

In the information systems world, "rightsizing" (or "downsizing") refers to replacing a larger computer (such as a mainframe or a mini), with one or more smaller computers (such as minis or PCs). Rightsizing can also mean putting smaller computers into departments or divisions, allowing the smaller computers to access the corporate computer. In the last few years, rightsizing has become increasingly feasible owing to the price-performance advantages of networked smaller computers. Although rightsizing appears to be extremely economically attractive (even compelling, at times), it also carries some less obvious costs:

- Retraining the organization on the new applications
- Fewer or less robust features in the new operating system
- Potential lack of supplier support resources (owing to the rapid growth of the smaller machine classes)
- Potential scarcity of applications software designed for larger businesses
- Integrating the rightsized remote division with the rest of the organization

Centralized vs. Distributed Systems

Originally all M&CRP-MRP II systems were centralized, owing to the enormous cost of early computers. Because of the drastically lower cost of computers and the increased capability of small computers, many of the early advantages of centralized systems no longer apply. Distributed systems, consisting of a group of computers linked in a network that permits communication and interrogation between machines, often provide a better capability at lower costs. However, distributed systems are not necessarily the best choice for a given company. Some of the advantages and disadvantages of centralized and distributed systems follow.

Centralized Systems

- Advantages:

  All data are in one place, facilitating sharing by many users on an efficient basis.

  Big computers can sometimes run complex jobs faster than small ones.

Better security, backup, and control of database owing to a more professional operation.

Can support a very large number of terminals on-line at the same time.

- Disadvantages:

Single system failure may shut down the entire business, unless there is backup equipment.

Much of a centralized system's resources are spent resolving internal contention for machine resources, which leads to very complex operating systems, databases, and operating environment.

If the centralized computer is a mainframe, the system software normally does not perform truly interactive processing.

Distributed Systems

- Advantages:

The total system work load is spread throughout the system.

The whole system never goes down; only one part of it will fail at a time.

Places local files at the local site, not off in some physically remote location.

Enables small computer responsiveness for many more users in a large system, because only a few users are logged onto any one machine.

Eliminated much complexity and much of the contention for resources.

No real practical limit on the expansion of the total system, particularly if all computers do not have to be linked directly together.

- Disadvantages:

The network can become inefficient, if more than a small percentage of data must be passed between computers.

Inter-computer communication is complex and may require considerable overhead in each machine.

Users may overload small computers by running large jobs on the small machines.

Hardware and software compatibility may become a problem; everything must match for communications to work properly.

Hardware, and some software maintenance, must be performed at all locations instead of one place.

Data Accuracy

The accuracy of data in the database is absolutely critical for M&CRP system success. Decisions based on inaccurate data can be extremely expensive. Errors in the files in manual systems are frequently overridden by knowledgeable planners; computer-based systems are less knowledgeable and less forgiving. For example, the MRP system will rapidly explode an error in the quantity of a component per assembly in a BOM record through lower level components and materials, to generate shop and purchase orders for the wrong quantities. As one comic stated, looking at a computer printout, "It would have taken a hundred clerks a hundred years to mess things up this thoroughly!" We will now look at some approaches that can increase data accuracy.

Input Cross-Checking

The sooner a data entry error is discovered, the easier it is to correct. Computerized cross-checking, or validation, can discover many errors at the time of entry. For example, if an employee wants to report receiving 100 pieces of item 57604 into the stockroom, from completing shop order 31492, the automatic cross-checking might include the following:

Is the employee authorized to enter this information?

Is there a production order 31492 for item 57604?

Has the previous operation of order 31492 been reported completed?

Was 100 units reported completed at the last operation?

Is the unit of measure used in reporting the receipt, the same as the unit of measure in the Item Master File?

In an on-line system, any inconsistencies outside pre-specified tolerances immediately show up on the terminal for correction. An analyst should reconcile any inconsistency that cannot be corrected immediately.

The addition of a check digit can be valuable in catching most transcription errors for any pure numeric entry, such as an item number. When the number is originally created, the computer calculates the check digit as a function of the preceding digits; the check digit becomes the last digit in the number. Then, when the numeric data is being entered, the computer recalculates the check digit and compares it to the check digit that has just been entered. If they are different, the computer flags the data for attention. A check digit will catch about 95 percent of data entry errors. Bar codes and EDI are now replacing manual data entry, reducing the need for check digits.

Shop Order Closeout

When a manufactured Scheduled Receipt is reported to be complete, and the items have been duly received into the stockroom or shipped, the computer closes the Scheduled Receipt and updates the data base. This includes the marking of the Scheduled Receipt as closed, and adjusting the on-hand quantity. Before closing the order, the computer program can validate:

Item number received versus the part number on the shop order

Quantity received versus the quantity reported

Reviewing closed shop orders can frequently point up errors in the master files. Unplanned material transactions, such as parts returns or supplementary requisitions for parts for the assembly department, may suggest errors in the BOM file. A tool listed on the routing for an order, but not withdrawn from the tool crib, can indicate an error in the routing file.

Purchase Order Closeout

The Receiving department is typically the place where purchased material is first received. Most companies do not perform a detailed count of the parts at this point. Rather, the receiver matches the number of cartons, total shipment quantity, and so on, against the purchase order. The stockroom often performs a detailed count prior to putting the items away, although some companies have their receiving inspectors count the received materials.

File Consistency Checks

Maintenance runs can determine if the files are consistent. Every item designated "make" in the Item Master File, should have a routing and a BOM. (New items that are just being added might not have routings or BOMs yet.)

One way of checking the accuracy of the BOM files for new products is through the cost roll up capability. Suppose product *A* has been in production for some time, and product *B* is being introduced. *A* and *B* differ only in some minor components. The total standard cost of each product should differ only by the difference in standard cost of the parts that differ. If this is not the case, the BOM file data for product *B* probably contains an error. The size of the discrepancy may suggest the location of the error.

Bar Codes

A bar-code system can be used for identifying parts anywhere, from receiving through the shop floor, and checking orders prior to shipping. Bar codes can also assist in counting, using the following procedure:

- Prepare the bar-code labels. This involves weighing the empty container, weighing a counted sample of parts, and entering the part number and sample size on the keyboard. Prompts for these steps are provided on the screen.
- Print a label that gives the empty weight of the container, the weight per piece, and part number. Attach this label to the container.
- To count an item, the operator reads the bar-code label with the scanner, and weighs the loaded container on the scale.
- The computer displays the number of parts on the screen and records them in the files, either locally or on the main computer.
- If the count is part of a physical inventory, the operator may also scan a bar code label attached to the rack or shelf, giving the location.

This system is a more efficient and more accurate method of counting large numbers of small parts than the normal method of counting by weighing.

Data Element Dictionary

The data element dictionary defines the security and accessibility for each element in the database. The dictionary usually contains the following information:

Name of the data element

Size of the field (number of characters or digits)

Relationships to other data

Default value

An active data dictionary can prevent an Item Master from being deleted while it still has stock-on-hand, outstanding Scheduled Receipts, or other transaction activity. It can prevent an unauthorized person from either changing or viewing the data. It forms the basis of the newer database management systems.

Cycle Counting

The on-hand balances in data records must exactly match the physical on-hand balances. The only way to be sure of this match is to compare the two numbers by physical counting. This is the purpose of the annual physical inventory. But the annual inventory count is designed primarily for financial purposes and relates to the overall imbalance, which includes compensating errors (e.g., if we find about as many as we lose, the dollar value of the company is unchanged). For M&CRP purposes, this is not good enough. A shortage of one part and an excess of another are two errors, both of which may seriously impact our ability to deliver products to customers, and to run our shop smoothly. For example, if a shoe manufacturer has 1000 left shoes but no right shoes, when the records said that 500 of each were in stock, the dollar value of the inventory is accurate, but the manufacturer cannot ship to customers.

Cycle counting, or continuously monitoring on-hand balances, is one alternative to the annual physical inventory. One way to cycle count is to have personnel whose entire job is to cycle count, checking some item on-hand balances every day. Another way is to have stockroom personnel levels set to readily handle peak loads, with slack times used for cycle counting. A third alternative is to cycle count for an hour before the stockroom opens each day.

Another issue in cycle counting is how to select the items to be counted. There are several approaches used by companies to select items to cycle count. These approaches can be, and usually are, combined.

- *ABC Classifications.* *A* items could be counted monthly (with different *A* items counted each day of the month), *B* items quarterly, and *C* items annually. The count list for a given day would include 1/20 of the *A* items (because there are 20 work days in the month, and we want to count all *A*s in each month), 1/60 of the *B* items, and 1/240 of the *C* items.
- *Zero Balances.* When the inventory record shows a zero balance, we can check the bin. If the bin is indeed empty, this count is obtained at minimum cost.
- *Negative Balances.* A negative balance indicates that an error has occurred and the inventory should be checked.
- *Randomized Lists.* List of items to count may be selected at random, perhaps with some limits such as *ABC* categories.
- *Each* X *Transactions.* This approach assumes that items tend to become inaccurate as the number of transactions increases. For example, if we counted an item in a location three months ago, and there have been no transactions, the probability of that item still being correct is relatively high. On the other hand, if there have been three or four transactions per day, the probability of that item still being correct is relatively low. Thus, we can define the number of transactions that will occur for a given item, before we want to count it again.

Performance Measures

A company can establish goals for database accuracy, and the procedures for measuring performance to those goals. Chapter 12 covers this topic more fully.

Summary

The major facets of the technical foundation of an M&CRP system, and our recommendations, are follows:

- Data files, the building blocks of the M&CRP data system, can be of three types: sequential, indexed, or chained; we prefer indexed or chained for speed.
- Database and DBMS options, in which we recommend using a commercially available database, and prefer the newer relational database structure for flexibility and ease of use.
- Database contents, in which we present many of the more common data fields typically used by an M&CRP system.
- Data transaction processing, in which we strongly prefer on-line to batch processing for responsiveness and accuracy.
- M&CRP system sources, in which we strongly recommend using packaged software if at all possible, because of the costs and risks of custom development.
- Package modification, in which we strongly suggest that you minimize modifications. We do not discourage people from adding to, or extending, their packages, although this practice also creates a maintenance burden at some point.
- Hardware options, in which we suggest that PC-based systems and networks provide the best price-performance and flexibility.
- The advantages and disadvantages of a centralized system versus a distributed system, in which we have no recommendation.
- Some steps to ensure data accuracy.

Selected Bibliography

Browne, Jimmie, J. Harhen, and J. Shivnan: *Production Management Systems: A CIM Perspective,* Addison-Wesley, Reading, MA, 1988, Chap 8.

Caruso, David: "Making Sense of New Information Technologies," *The Performance Advantage,* APICS, October 1991, pp. 32–35.

Deis, Paul: *Production and Inventory Management in the Technological Age,* Prentice-Hall, Englewood Cliffs, NJ, 1983, Chap 12.

Orlicky, Joseph: *Material Requirements Planning,* McGraw-Hill, New York, 1975, Chap 9.

Smith, Spencer B.: *Computer Based Production and Inventory Control,* Prentice-Hall, Englewood Cliffs, NJ, 1989, Chap 3.

Vollmann, Thomas E., W. L. Berry, and D. C. Whybark: *Manufacturing Planning and Control Systems,* 3d ed., Irwin, Homewood, IL, 1992, Chap 2.

9

Adapting M&CRP-MRP II to Nontraditional Manufacturing Environments

"So near, and yet so far . . . " ANONYMOUS

Introduction

What happens when you adapt an excellent tool to a function for which it was not originally designed? The traditional Material and Capacity Requirements Planning system was designed to work in the *traditional* manufacturing environment. This environment includes the fabrication and assembly industries that manufacture discrete products. Many companies use a job shop to fabricate the parts, and a small batch-flow line to assemble the final products. Traditional M&CRP uses both Make-to-Order and Make-to-Stock Demand Response Strategies, depending upon the nature of the product and the competition.

Software developers are extending the usefulness of MRP II to satisfy the more specialized needs of various markets. We describe a number of the more significant adaptations of M&CRP-MRP II in this chapter, including:

Repetitive Manufacturing

Just-in-Time

Process Manufacturing

Flexible Manufacturing Systems and Computer Integrated Manufacturing

Agile Manufacturing Systems

Job Shop

Tool Requirements Planning

Maintenance Planning

Program Management

Space limitations preclude full coverage of all the adaptations of M&CRP-MRP II. We briefly summarize some of the other applications at the end of the chapter.

Note that we are no longer restricting our discussion to M&CRP alone. These topics involve the entire functionality of MRP II, which is based on M&CRP theory, and uses the MRP and CRP modules as its heart.

Repetitive Manufacturing (RM)

Introduction

In *Repetitive Manufacturing,* various items with similar routings are made across the same process whenever production occurs. Products may be made in separate batches or continuously. Production in a repetitive environment is *not* a function of speed or volume.[1] The products involved are discrete in nature, and differ from fluids, powders, and processes involving chemical change.

Three factors differentiate Repetitive Manufacturing from traditional job shop manufacturing:

1. The type of product
2. The production processes requiring specific equipment and their arrangement

[1] *APICS Dictionary,* Seventh Edition, APICS, Falls Church, VA, 1992.

3. Manufacturing may occur on lines of work stations set in assembly sequence, or at a single bench assembly station

Figure 9-1 compares the characteristics of a job shop and a repetitive shop.

It is what happens on the shop floor, not in the planning process, that makes the Repetitive Manufacturing environment a unique application of MRP. Although the traditional MRP procedures are theoretically valid, they are too costly from an administrative standpoint. The system must function in an environment where material lot integrity is not always maintainable, where status of materials and work schedules on the shop floor is not precisely known, where certain inventory record balances are less accurate, and where schedule changes are frequent and rapid.

Let's examine some of the characteristics of Repetitive Manufacturing in more detail:

- *Production tends to be continuous, rather than in discrete lots,* making it difficult to identify specific manufacturing orders. Where lot identity does exist, contiguous lots of similar items are often run on the same production line.
- *Items flow through the production process on a relatively fixed routing.* Fixed production lines, where people and machines are grouped according to the product being produced rather than the production function being performed, are common. This contributes to the difficulty in maintaining manufacturing order integrity, and reduces the usefulness of machine center performance data for capacity planning purposes.
- *Production may be to a schedule rather than to specific manufacturing orders.* Schedules will normally be prepared for each production line or logical schedule group. Planners control production by volume produced over time (generally by production line or schedule group), rather than by a specific quantity of a specific item due in a specific time period.

Attribute	Job Shop	Repetitive Shop
Planning	Shop orders	Blanket schedules
Run Length	Short	Long, continuous
Product	Custom or nonstandard	Standardized
Material Issues	Specific to jobs (orders)	Nondiscrete flow issues
Job Tracking	Discrete started/completed jobs	Cumulative production
Material Accountability	Tight	Loose, accumulative

Figure 9-1. Comparison of job shop and repetitive shop.

- *Internal lead times tend to be relatively short,* once the basic production line has been set up. The RM environment can provide a high degree of flexibility within a very short period of time, assuming the required production rate is being met, and the required changes are within the parameters of the production line or schedule group.
- *In-line processing is common.* Fabrication and subassembly production is often performed simultaneously with the final assembly operation in which the parts are consumed. This means that the assembly schedules must be released before all the components are available.
- *Operations, as described on the process sheet, are often combined or split* as staffing on production lines is changed to meet changing production requirements. This makes reporting by operations almost impossible, and limits meaningful reporting at the work-center level.
- *Shop floor progress can be reported at paypoints only.* A line could have dozens of operations, but only 2 to 5 paypoints. Paypoints can also trigger component backflushing, or post-deduct, in which the computer automatically computes and relieves the inventories that have been consumed.
- *Work-in-Process tends to be smaller* when compared to annual sales rates. Because products tend to be completed quickly after they are first placed into a Repetitive Manufacturing environment, work-in-process is a smaller percentage of total inventories, and also of annual sales.

These parameters require a different approach to Manufacturing Planning and Control to suit the RM environment. The issues are very basic. We need to monitor performance, without depending on discrete manufacturing orders, and to minimize the cost of obtaining meaningful data.

The adaptations of M&CRP-MRP II for Repetitive Manufacturing are the same as those for JIT, discussed in the following section.

Just-in-Time (JIT)

In the 1980s, Just-In-Time started being used in the United States. JIT is not a Manufacturing Planning and Control system. It is a philosophy of manufacturing within which MRP and other techniques can not only exist, but even thrive. JIT had evolved during more than two decades in Japan, where it was based on United States supermarkets and Henry Ford's assembly line. In JIT, operations (processing, movement of material, and so on) occur only when they are needed or demanded, and not before. JIT deliberately attempts to minimize inventories at all levels, which is why it is also known as Zero Inventories. JIT emphasizes continual effort to remove waste and inefficiency from the production process by emphasizing small Lot Sizes, high quality, and teamwork. It includes process simplification, quick setups, and synchronization through uniform plant load.

Rather than following the approach of M&CRP of trying to control the complexities of manufacturing by developing a complex control system, JIT actively seeks to reduce these complexities. This dramatic reduction in complexity permits and even encourages, much simpler controls. In fact, simple manual controls are feasible in many cases. Let's look at how JIT resolves many of the problems of implementing M&CRP in a manufacturing shop.

- Material flows rapidly through the shop; the amount of material at any place and at any point in time is so small that elaborate tracking systems are both impractical and unnecessary. With operations linked closely together in space and time, the product is always visible, which greatly simplifies control. Because material is *pulled,* formal material tracking is no longer necessary. The problem of collecting labor data for costing is resolved by using *process costing,* rather than job order costing.
- The Production Planning function may be the only capacity planning function required (eliminating the need for Production Activity Control), especially if planning is by major families. The entire materials continuum is synchronized to a Final Assembly Scheduling function. In JIT, the Production Planning process balances the production plan to the market forecast through varying cell staffing levels. By tracking the variability of customer ordered options, the JIT Production Planning process can include an option planning capability.
- Material planning by part number through the use of Planned Orders does not occur. The *pull* system both authorizes and prioritizes work, thus eliminating the need for the order release function of MRP II.

The characteristics outlined above strikingly resemble the Repetitive Manufacturing characteristics presented earlier in this chapter. From a logical viewpoint, JIT converts a manufacturing floor into a repetitive production line, in which the production rate of the final product paces all other operations, including the fabrication shop and suppliers.

Integrating MRP II and JIT

Many manufacturing companies use both MRP II and JIT to obtain the advantages of both systems. MRP II provides a stable Master Production Schedule, visibility into future material requirements, excellent internal and external communications, and Capacity Requirements Planning. JIT provides a pull system for executing production and material plans that respond to changing conditions on the shop floor. Integrating MRP II and JIT requires software that can support both systems in an integrated manner. The software for MRP II remains the same, but must be extended and/or modified to support JIT. Discenza and McFadden[2] list three major areas where MRP II software packages require extensions or changes: manufacturing specifications, materials planning, and manufacturing control.

Manufacturing Specifications. The manufacturing specifications area consists of two capabilities: Bills of Material, and stock areas and deduct lists.

- *Bills of Material* contains a flat, or single-level, BOM for each product to be manufactured, in contrast to the multilevel BOM normally used in MRP. To convert MRP-oriented bills to JIT, you can perform actual BOM maintenance to flatten your bills so they reflect how you now make your product, or you can change the intermediate levels to "phantoms."
- *Stock Areas and Deduct Lists* module defines the manufacturing process. A stock area is a production line, or work center, where parts are consumed. A deduct list is a list of parts and quantities that are consumed within a given stock area.

In JIT, inventory accounting is normally performed when a unit of the finished product is completed. When the product is completed, all components on the single-level Bill of Material are deducted from inventory. This procedure is referred to as post-deduct, or backflushing. This is in marked contrast to classic MRP, where inventory activity is triggered by shop orders, and inventory accounting for components is performed before production begins. However, many MRP systems now support post-deduct and point-of-use in the Bill of Materials module. To gain greater visibility of work-in-process, you can establish deduct points at intermediate work stations, as well as at the end of the production line. A deduct list defines a list of components consumed at each deduct point, or paypoint.

Materials Planning. The materials planning component consists of two capabilities: rate-based Master Production Scheduling and JIT-oriented Material Requirements Planning.

- *Rate-based Master Production Scheduling* is a management planning tool appropriate to JIT production. With this tool, monthly production of each

[2]Richard Discenza and Fred R. McFadden, "The Integration of MRP II and JIT through Software Unification," *Production and Inventory Management Journal*, APICS, Falls Church, VA, 4th Qtr, 1988.

end item in the Master Production Schedule is expressed as a *daily* rate. This module assists the analyst in scheduling production to meet shipping schedules and inventory objectives. For example, the MPS for product *X* for the next four months might appear as shown in Figure 9-2.

- *JIT-oriented Material Requirements Planning* determines the timing and quantities of parts to support the Master Production Schedule. Unlike conventional MRP, JIT MRP does not compute Planned Order Releases. Instead, it determines, on a daily basis, the availability of each part and the additional parts required to support the MPS.

Manufacturing Control. The manufacturing control area has three capabilities: production reporting and post deduct, inventory accounting, and material cost reporting.

- *Production Reporting and Post Deduct* reports actual production completed, and relieves stock areas of inventory consumed in production. As each parent item is completed and passes a deduct point or paypoint, this module deducts the components on the deduct list from the appropriate stock area (the parent item is then credited to the following stock area). The module thus provides continuous inventory tracking and the ability to report actual versus planned production.
- *Inventory Accounting* maintains the current inventory status of each stock area, whether in a production area or in a storeroom. This module also supports cycle counting and other inventory management functions.
- *Material Cost Reporting* summarizes accounting information on materials consumed during each accounting period. It provides exception reports and summarized material cost reports for the period. This module also provides an interface to the general ledger.

MONTH	JAN	FEB	MAR	APR
No. working days	20	20	22	21
Prod. Rate (units/day)	20	25	20	20
Backlog orders	50	100	0	0
Orders forecast	400	375	500	400
Total Orders	450	475	500	400
Prod. (units/month)	440	500	440	420
Beginning Inventory	50	40	65	5
Ending Inventory	40	65	5	25

Figure 9-2. JIT Master Production Schedule for product X.

Integrating M&CRP and JIT

JIT software can function as a stand-alone system, or it can be integrated with MRP II software to form a complete system. Many MRP software vendors are integrating their MRP II software with JIT software. An overview of this integration is shown in Figure 9-3, where JIT and MRP II share certain software. This integration is consistent with the practice, in many repetitive manufacturing companies, of using MRP II as a planning system while using JIT as the execution system.

Shared Software. MRP II and JIT share several database modules, plus the bar code application module, as follows:

- *Item Master.* The Item Master File contains one record for each item in inventory, and is common to JIT and MRP.
- *Bill of Materials.* MRP includes a multi-level BOM, whereas the BOM for JIT is single-level. However, MRP can process single-level bills for JIT, so that both JIT and MRP bills can be maintained in just one system.
- *Work Center.* A work center may be a stockroom, an assembly line (or a portion of a line), or some other workplace. Definitions of stock areas are common to both systems.

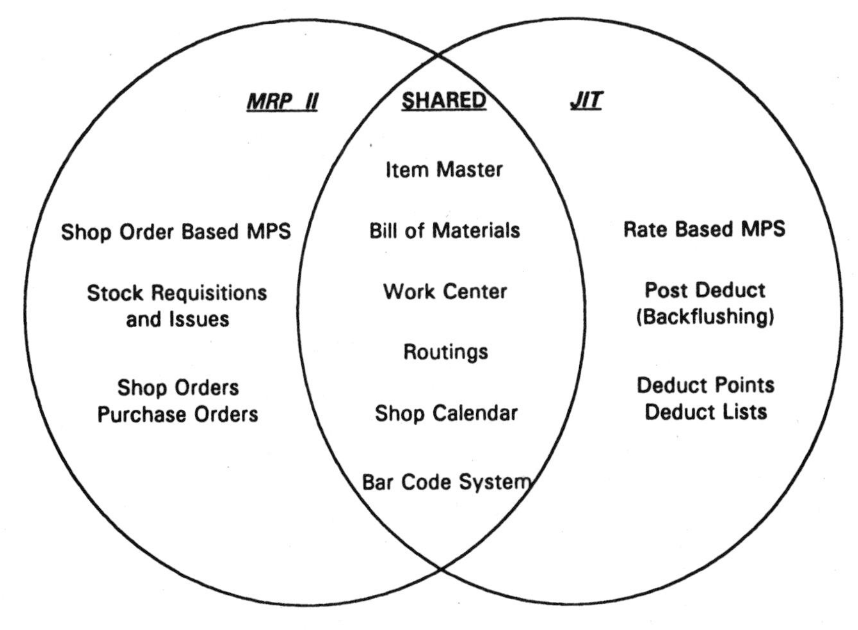

Figure 9-3. Integration of MRP and JIT.

- *Routings.* The routings file defines the routing for each parent item whether in MRP II or JIT (although a routing in JIT tends to be fixed over a period of time).
- *Shop Calendar.* The shop calendar facilitates scheduling by identifying each working day. Both systems use the same shop calendar.
- *Bar-Code System.* This is an application module that can be used to capture data for either system.

Stand-Alone Software. The following modules are unique to MRP II or JIT, although there may be some linking.

- *Master Production Scheduling.* In MRP, the MPS is typically based on weekly time buckets, whereas in JIT the MPS is based on a daily rate.
- *Inventory Accounting.* In MRP, inventory accounting is based on stock requisitions and issues before production starts, whereas in JIT this accounting is based on post deduct. However, more MRP systems offer post-deduct capabilities.
- *Push versus Pull System.* MRP uses work orders and purchase orders that are pushed through the system. JIT pulls items through the system and uses deduct points and deduct lists to track materials and account for such movement. JIT uses purchase orders as blanket orders, for a long-term supply of a commodity or group of commodities.
- *Cost Accounting.* JIT supports Activity-Based Costing, with many overhead pools allocated to the activities that cause the particular overhead expenses. MRP generally supports standard costing, with a single, large overhead pool.

Process Industries

Introduction

The process industries tend to be associated with products at the high volume, high standardization, or commodity end of the spectrum, as shown in Figure 1-11. As a result, the process industries tend more toward the continuous line flow or flow shop end of the spectrum, as indicated by Figure 1-11. On the other hand, the fabrication and assembly industries tend more toward low volume, low-standardization products that are typically made in jumbled flow or job shops. A flow shop is a manufacturing facility in which materials flow through the plant with a fixed routing. A job shop is a manufacturing facility in which materials flow through the shop with routings being dependent on each job. Figure 9-4 provides a more complete comparison of job shops and flow shops.

In addition to the difference between a flow shop and a job shop, process industries differ from the fabrication and assembly industries in a number of other aspects, including:

Variability in raw material quality

Variability in Bills of Material

Product yields may vary widely

May have large demands for intermediate products

Co-product demands must be balanced

Product or raw materials may have a shelf life

The process industries often obtain their raw materials from mining or agricultural industries. These raw materials have natural variations in quality, which leads to variations in Bills of Material. For example, variations in the acidity, concentration, and so on, of the raw material may cause variations in the proportion of the ingredients or components required to make a finished product.

A company's production and inventory planning system should be con-

ATTRIBUTE	JOB SHOP	FLOW SHOP
Routings	Variable	Fixed
Material Handling Equipment	Variable Path	Fixed Path
Layout	Process	Product
Equipment	Flexible	Specialized
Volume	Low	High
Lead Time to Increase Capacity	Shorter	Longer
Capacity Definition	Difficult	Easy
Skill Level of Equipment Operators	Craftsmen	Operators
Operator Task	Build the Product	Operate Equipment
WIP Inventories	High	Low
WIP Warehoused?	Often	Seldom
Job Overlapping	Seldom	Usually
Equipment Failure Affects	Machine	Plant
Late Purchased Part Receipt Affects	Customer Order	Plant

Figure 9-4. Comparison of job shop and flow shop.

sistent with its position on the product-process matrix. Firms producing commodities in a flow shop environment require a significantly different planning system than firms producing custom products in a job shop environment. We will discuss these differences under the headings of strategic, tactical, and operational planning.

Strategic Planning

Process industries develop extensive long-range resource requirement plans. Because the process industries are more capital intensive than the fabrication and assembly industries, capital budgeting decisions have a greater impact on financial performance in the process industries. Two important strategic decisions are plant capacity and plant location. Long-range plans for materials, manpower, energy, and waste disposal are also important in the process industries. Fabrication and assembly industries have less emphasis on long-range resource planning and more emphasis on product design. A flexible manufacturing facility that can quickly produce new product designs is more important for these firms.

In many process industries, products are made on a regular cycle. The sequence of products is often dictated by product or process technology. Determining the target sequences and cycle length is an important production planning problem in the process industries.

Tactical Planning

With respect to tactical planning, the major difference between process industries and fabrication and assembly industries is the approach to Master Production Scheduling. The motto of fabrication and assembly industries is, "Get me the parts and I'll make the product." In order to minimize investment in materials, MRP plans the need dates for components and purchased parts by using backward scheduling from the finished product due date. Once the materials requirements are scheduled, the closed loop M&CRP system examines the feasibility of the schedule against equipment and manpower capacities. This approach is the material-driven Master Production Scheduling method, because materials are scheduled first and then capacities are checked. Virtually all M&CRP systems operate in this manner.

Because of the high capital investment, process industries concentrate on achieving high equipment utilization, and usually have a good estimate of the capacity of their line or plant. Thus, process industries tend to first schedule capacity and then materials. This is called capacity-driven Master Production Scheduling. A capacity-driven procedure schedules production runs of various products on each production line, so that capacity is utilized at the rate specified in the production plan. Having determined the production schedule, the capacity-driven Master Schedule then calculates raw material requirements.

A scheduling system must plan both material and capacity. But which should be done first? (Virtually all M&CRP software still performs these tasks sequentially, forcing a company to make this choice. In Chapter 13, we predict that the next generation of M&CRP systems will integrate these two constraints into a single scheduling algorithm.) Minimizing investment in materials requires a material-driven procedure, such as a closed-loop MRP system. Utilizing equipment efficiently requires a capacity-driven technique. For a job shop producing custom products, a material-driven scheduling method is more appropriate. For a flow shop producing commodity products, a capacity-driven scheduling method is more appropriate. Firms in between these two extremes must either select a material-driven or capacity-driven procedure, and adapt it to fit their special situation.

Besides the emphasis on capacity-driven Master Production Scheduling, major characteristics of process industry scheduling systems include:

- More emphasis on distribution requirements planning and inter-plant transfer planning
- Closer coupling of Master Production Schedule with forecasts
- Master Production Schedule is less compliant to customer requests
- Lot sizes may be dictated by
  Facilities design
  Manufacturing practices for ensuring product quality
- The Master Production Schedule is the authority to produce
- Production plan or Master Production Schedule determines sequencing
- Schedules generally have smaller time intervals

Operational Planning

After the Master Production Scheduling stage, material and capacity-driven scheduling techniques diverge. *Material-driven planning* concentrates on material availability to meet product demand schedules. Material-driven planning controls priorities by satisfying the demand within the constraint of the Master Schedule. The MRP explosion process establishes all dependent material requirements, based on the released Master Schedule. The Capacity Requirements Plan identifies the manufacturing resources (people, work centers, tools, and equipment) required to support the material plan.

Capacity-driven planning does not work this way. Master Scheduling requires the finite management of key facility capacity. Materials planning is relatively simple; product plans can be exploded directly into purchasing contracts. Master Schedules primarily control the production activity on specific equipment or production lines. Capacity-driven planning controls priorities by controlling the demand within the constraint of available capacity. The Master Schedule is maintained in terms of load against key facilities, rather than products. The Master Schedule also determines the materials required to support the capacity plan. In summary, the *planning sequences* for the two methods are as follows:

Sequence	Material-driven	Capacity-driven
1	Product	Product
2	Material	Capacity
3	Capacity	Material

Material-driven planning calculates manufacturing priorities without regard for capacity. In the correct manufacturing environment, this type of planning can be effective, for example, in discrete products companies that operate one or two shifts per day, and that have the flexibility to run overtime. Capacity can be treated as a flexible resource that permits planners to concentrate on resolving the conflicts over material availability.

However, when operations run 24 hours per day, 7 days a week, as they do in most process industries, capacity is extremely finite. Management must resolve scheduling conflicts within the confines of available capacity. In fact, in some companies utilization of available capacity is the primary focus. As soon as capacity is planned finitely, the Master Scheduler becomes responsible for managing the sequence of the scheduled activities.

A Master Scheduler in a material-driven planning company must coordinate the flow of numerous parts to final assembly, whereas the Master Scheduler in a capacity-driven planning company must manage the sequence of activities across fully loaded equipment.

Process Flow Scheduling (PFS)

Process Flow Scheduling, a scheduling approach for process firms, relies on a diagnosis of the process flow to guide the scheduling steps. In this approach, described by Taylor and Bolander,[3] the process flow is organized into a hierarchy of process trains, stages, and units.

[3]Sam G. Taylor and Steven F. Bolander, "Process Flow Scheduling," *American Production and Inventory Control Society, 1989 Annual International Conference Proceedings,* APICS, Falls Church, VA, 1989, pp. 98–99.

- A process train is a sequential series of processing equipment in which a family of products is produced with a fixed routing. No material is transferred from one process train to another; each process train is a totally complete system to manufacture finished parts. A process train could be for one or more similar products; it might include an assembly line with some fabrication. It could be one or more cells.
- Process stages are the areas in a process train that are relatively self-contained. A work cell is one type of a process stage; an assembly line is another. Process stages must be decoupled by inventory buffers that allow each stage to be scheduled as a separate entity. These inventories permit the scheduling of products through process stages in different sequences, or in different Lot Sizes (somewhat like scheduling jobs through a job shop).
- At the lowest level of the process is the process unit, which transforms the flowing material in some way. A single machine is an example of a process unit.

Each stage in the process structure may be scheduled by either the Material-Driven Scheduling (MDS) or Capacity-Driven Scheduling (CDS). The selection of MDS or CDS for a given stage depends on the particular scheduling environment. In general, MDS should be used when: (1) materials are relatively expensive, (2) there is excess capacity, and (3) setup costs are negligible. Conversely, CDS should be used when: (1) capacity is relatively expensive, (2) the stage is a bottleneck, or (3) setups are expensive. CDS is often used when products must be scheduled in a natural sequence, such going from light to dark colors. Multiple process stages can be scheduled by combining MDS and CDS for each stage in a process structure. Scheduling logic can be combined in many possible ways.

The most significant difference between PFS and M&CRP is the order in which materials and capacity are planned. M&CRP is based on *product* structure, including Bills of Material for material planning, and product routings for capacity planning. In contrast, PFS is based on the *process* structure, and uses material lists for passing requirements between stages.

This fundamental difference in scheduling logic leads to three other differences.

- MRP first schedules materials at all levels in the Bill of Materials and then checks capacity for all processes. In contrast, PFS usually checks materials and capacity at each stage, before moving on to the next stage.
- MRP allows use of the same equipment at different levels in the Bill of Materials. Thus, a drill press could be used to perform similar operations at different levels in the Bill of Materials. On the other hand, PFS re-

stricts equipment usage to a single stage. Because of these differences in equipment usage restrictions, MRP is well suited for job shops, whereas PFS is better suited for flow shops.

- MRP systems do not perform detailed shop floor planning; a shop floor control module normally develops detailed schedules. In contrast, PFS schedules are so sufficiently detailed that they do not need another program, such as shop floor control, to break them down to shop execution level.

Because of the difference in detail, *MRP and PFS can be used in the same planning system.* MRP can be used to develop raw material plans by using a two-level Bill of Materials that contains finished products and raw materials. However, the PFS system would be used for managing work-in-process inventories and for detailed scheduling.

Summary Comparison of Discrete and Process Manufacturing

Process and discrete manufacturing companies plan their processes in very different ways, regardless of which scheduling method is used. Bruun and Frank[4] have outlined the fundamental differences, which are included in Figure 9-5.

[4]Richard J. Bruun and Donald N. Frank, "A Changing View of Capacity Management," *APICS: The Performance Advantage,* APICS, Falls Church, VA, in press.

FUNCTION	DISCRETE	PROCESS
Master Schedule	Unit Driven Production	Process Driven Family
Constraint	Material Availability	Capacity Utilization
Planning Issues	Schedule Attainment	Process Continuity and Yields
MRP	Primary Tool for Detailed Planning	Secondary Tool, or not used for Detailed Planning
CRP	Secondary Tool for Detailed Planning	Primary Tool for Detailed Planning
BOMs	Deep and Complex	Shallow and Simple
Routings	Many with Alternatives	Fixed by Process
Work Centers	Discrete and Flexible	Fixed and Dedicated
Planners	Commodity and Part Focused	Process Focused

Figure 9-5. Comparison of discrete and process planning.

Flexible Manufacturing System (FMS)

MRP II systems are traditionally weak in the execution and feedback of the plans once they have been put in place. A Flexible Manufacturing System provides one approach to tighter integration.

A FMS consists of a factory area that is regulated by a computerized control system that is usually distinct from the MRP II system, and is often even on a different computer. The FMS may contain numerical control machines, manual work centers, assembly areas, quality assurance areas, a tool preset area, an input/output area, and an Automated Storage and Retrieval System (AS/RS). The FMS area often starts out as only one section of the factory, with additional machines and work centers being added later. Thus, any attempt at integrating MRP II and FMS must recognize that the FMS may include only a select number of work centers. The MRP II system must be able to work with the FMS work centers, as well as with the remainder of the work centers in the factory.

The Flexible Control System contains the interactive applications and software to direct the material handling, manufacturing, and tool control for the work centers and machines within the FMS. The FCS needs to have access to much of the same information that MRP II uses to schedule shop orders:

Sequence of operations required to make the part

The work center and machines required

What tools, fixtures, Numeric Control programs, and material must be available before the order can be scheduled

Scheduling Methods

Both MRP II and FMS systems perform a scheduling function, however, they use entirely different methods.

M&CRP-MRP II Scheduling. In MRP II, the Master Production Schedule schedules products that are needed to satisfy both customer orders and forecasts of future orders, roughly testing the schedule against capacity. MRP explodes the product using the Bill of Materials, and then uses inventory netting and lead time offsetting to obtain the quantities and due dates for all shop orders, yielding requirements for parts and raw materials. CRP again tests the MRP plan against capacity constraints.

The Dispatch List in the Shop Floor Control (SFC) module utilizes the quantities and due dates of the shop orders to schedule a planned start and due date for each operation of each order. In many packages, the Dispatch

List utilizes average work center queue times, specific operation setup and run times, operation quantities, average move times, work center efficiencies, and operation overlap factors to compute dates for each operation. In most cases, end and start dates will be determined by backward scheduling. Some systems use both backward and forward scheduling. In either case, the MRP II system can develop Dispatch Lists for each work center, which provide the work center supervisor with information on each job order (due date, priority, quantity, and so on) to assist in scheduling. Because this SFC scheduling method arbitrarily adds average queue time, without regard to what is really happening on the shop floor, the SFC Dispatch List often requires manual intervention to make it workable.

FMS Scheduling. The scheduling approach for a Flexible Manufacturing System is similar to that required by a dynamic job shop. The FMS processes a variety of parts simultaneously through the machines. The FMS can choose standard or alternative machine sequences for a given operation, basing the decision on the status of the machines and the mix of parts available for processing at that time. Because the machines involved in a FMS are very expensive, a basic objective of a FMS is to maximize machine utilization. To provide excellent customer service, however, most companies have the dual objectives of maximizing utilization and minimizing job tardiness.

Achieving these dual objectives requires the FMS to make several complex decisions in a very short time. These include:

- Dividing overall production targets into batches of parts
- Assigning production resources in a manner that maximizes resource utilization
- Responding to changes in upper level production plans or material availability
- Work order scheduling and dispatching (which part to introduce next into the FMS, and when)
- Movement of workpieces and the material handling system (which machine to send this workpiece to the next, which cart to transport it, and so on)
- Tool management (selecting tools for the job, keeping track of tools and tool wear, reacting to tool breakage, and so on)
- Fixtures management (selecting fixtures for the job, moving fixtures to the work area, and so on)
- NC program management (selecting NC programs, downloading and up-loading at appropriate times, and so on)

- Reacting to disruptions (failures of machines, tools, and material handling system, or sudden changes in production requirements)
- System monitoring and diagnostics

All these decisions must be made correctly and made on time, if the FMS is to operate effectively and efficiently. Because of their complexity, these decisions can *not* be made manually in the requisite time frame. Besides, the normal operating mode of a FMS does not have any people readily available to make these decisions.

The FMS scheduler (a computer program) works very differently from the SFC-MRP II scheduler. The FMS scheduler must know the planned start and due date of the shop order, so that it can schedule the operations to meet the due date. However, the FMS scheduler is also aware of the entire work load that must be processed within a given work center, or machine, over a given time period. Using this information, the FMS scheduler may decide to *not* schedule shop orders. Instead, it may decide to schedule *production* orders, which consist of several shop orders that can use the same basic machine setup. The FMS scheduler can examine the entire work load scheduled for a work center, or machine, and group together the operations on shop orders that best utilize existing tools, fixtures, and materials, and which can be produced to meet the due date of each order. This process provides an opportunity to reduce setup time and queue times.

The FMS scheduler can check tool, fixture, material, and NC program availability, because with the FMS, the scheduler is responsible for developing the sequence of jobs in a work center. This schedule is then used to drive other FMS activities. The tool and fixture requirements for each production job can be prepared when the individual NC programs for each operation within the production job are available. Thus, unlike the MRP II operation scheduler, which only suggests dates, the FMS scheduler is really the driver of activities performed within the FMS.

The MRP II Scheduling Program in the FMS Environment. MRP II and FMS systems have some common functions. Both can schedule the operations for a work center, prepare Dispatch Lists, and develop tool kits. However, because the FMS scheduler is more precise, and more in tune with current activities on the shop floor, its output is much more meaningful as a driver for day-to-day activities. This does not mean that the MRP II scheduler has no function. On the contrary, the FMS still needs to know what manufactured orders are required to meet the Master Schedule, and when these orders need to be complete. The MRP II scheduler provides this function by computing the start and due dates for each shop order. It can also provide the FMS with an initial operation schedule that can be used as a starting point for the FMS scheduler. In addition, MRP II still schedules all those other work centers that are not part of the FMS area. It is also not unusual for a combination of FMS and non-FMS work centers to appear on the same manufacturing routing.

Many MRP II systems have very sophisticated tools for looking at both current and future planned work loads. There is often little need to duplicate such capability with the FMS. However the short-term load schedule developed by the FMS is much more precise than that suggested by the MRP II system. Thus, when FMS capacity is a problem, it may be desirable in some situations to pass the FMS revised schedule dates to the MRP II system. A planner would then be able to see short-term capacity figures, using the tools provided by the MRP II system.

The short-range approach to solving the gap between MRP II and FMS entails the internal development and implementation of either:

A software interface to a commercially available finite scheduling package, or

A scheduling package and interface that is developed completely in-house.

Regardless of how the near-term interface is developed, the development process will be long and expensive, and will probably provide less than satisfactory results.

Data Files and Their Use

Because MRP II and FMS systems have certain common functions, they require much the same type of information. Both need basic information on resources and routing to perform their respective functions. Several types of entities can be considered as resources, such as work centers, machines, tools, fixtures, and NC programs.

Work Centers and Machines. In order for MRP II to suggest a realistic initial schedule for the FMS, it needs basic capacity information for each FMS work center. An MRP II planner could then review and adjust the planned load for the FMS. Otherwise, an MRP planner should assume some available capacity (probably infinite capacity) for the FMS as a whole, and let the FMS scheduler operate on all potential orders.

The basic capacity information includes the number of machines or people with a work center, the number of hours per day the machines and people work, and the anticipated schedule of preventive maintenance, or planned vacations. Because all non-FMS work centers within the MRP II system provide this information, FMS work centers can provide it as well.

Routings and Shop Orders. The MRP II system uses routings to produce shop orders for the shop floor. The shop order can be used to prepare a list of tools required, to generate a detailed operation schedule for each work center, and to develop load profiles and Dispatch Lists for each work center. This information is the foundation of the Shop Floor Control module of the MRP II system. However, because the FMS may decide to schedule a production job that consists of several shop orders, detailed operation schedules, load profiles, and Dispatch Lists generated by the MRP II system will probably be unusable in the FMS environment. The FMS generates its own list of tool requirements, operations schedules, and work center Dispatch Lists.

Tools, Fixtures, and NC Programs. The tools and fixtures required for each operation of a shop order can be obtained from several places, including the MRP II routing, the FMS, and NC programs. The MRP II routing can indicate the tools and fixtures for manual work centers and machines, but it can indicate only the NC program number for numerical control machines. For FMS work centers, the tool and fixture requirements are still required on the NC program. Only the FMS control system can develop the complete tool and fixture requirements for a FMS production job.

Information Flow between M&CRP-MRP II and FMS-CIM

MRP II and FMS should be able to pass data to each other, no matter which computer(s) they reside on. In this section we discuss the information that flows between the two systems.

From MRP II to FMS. The basic work center and machine capacity data on the MRP II system can be downloaded to the FMS. This ensures that both systems use similar work center capacity details with their respective scheduling algorithms. The MRP II system sends released shop order details to the FMS, including new Scheduled Receipts and changes to existing Scheduled Receipts. MRP II can also optionally include tool and fixture requirements, if they are not included as part of the NC program requirements.

Feedback from FMS to MRP II. The FMS control system is responsible not only for scheduling the FMS work areas but also for reporting the status of the operations completed within the FMS. Because people generally utilize the MRP II system to see the status of an order within the plant, the FMS system must pass back to MRP the status of FMS milestone operations. This status feedback includes actual operation start date, completion date, and changes to quantity that could result from an inspection reject. The MRP II system would then have available the correct operation information necessary to schedule the remaining downstream operations of an order, which may or may not include additional FMS work centers.

When FMS short-term capacity may be inadequate, the FMS can send its more precise schedule dates to the MRP II system, so that a planner can utilize basic Capacity Requirements Planning tools to adjust the MRP II plan. Such information is not required, if the FMS has sufficient capacity to meet the MRP II plan.

Agile Manufacturing Systems (AMS)

We foresee that a hybrid manufacturing system, as defined in Chapter 1, which combines the best aspects of MRP II, JIT, and FMS, will become the cornerstone of manufacturing company operations in the 1990s. This system will be required for companies to effectively compete with short response times and high flexibility. We have chosen to call this system an Agile Manufacturing System.

An AMS is a manufacturing system that is designed to optimally serve its customers in the most efficient manner. First, it can shift manufacturing processes rapidly and efficiently, in order to provide the products that the customers desire. Second, it can provide these products in the time desired by the customer. Third, it operates with the utmost efficiency, using all the productivity techniques learned from JIT. An AMS provides the same efficiency as a line-flow operation, with the flexibility and responsiveness of a job shop.

The core of an AMS is its ability to communicate, both internally and externally, thus requiring the communications capabilities that are the hallmark of an effective MRP II system. The front end of an AMS, that part which interfaces to the customers, will vary from industry to industry. However, it will allow companies that are producing to stock now (severely limiting their ability to meet specific customer requirements quickly) to consider an Assemble-to-Order environment. We now look at how an AMS can work for the customer service and order entry areas, as well as the manufacturing and material acquisition areas, for each of the major Demand Response Strategies.

Make-to-Stock

For Make-to-Stock products, an AMS uses a Master Production Schedule that is driven by forecasts, and compared frequently to actual sales. Customer service relies heavily on EDI, minimizing customer order transit time and potential errors. Theoretically, as customers and suppliers continue to forge tighter relationships, the AMS tracks on-hand inventories at the customer locations, so that the customer does not have to place an actual order or release. This is the electronic equivalent of the bread or potato chip jobber who checks the store shelves daily, replenishing quantities to a preset amount.

The manufacturing area will probably be mixed-model JIT, with simple Kanbans that regulate the flow of materials throughout the organization. Suppliers can also use either manual or electronic Kanbans (supplied by the AMS). MRP plans the flow of materials through the pipeline, possibly with the same modifications suggested earlier for JIT implementation.

Assemble-to-Order

For Assemble-to-Order products, an AMS utilizes a configuration order entry Bill of Materials to edit and accept each customer order, then prints the order in the final assembly area for immediate assembly and shipment. Again, the manufacturer could be tied directly to the customer with EDI, so that the customer's computer would order legal configurations. There is no technical reason why the manufacturer's AMS could not track on-hand inventories at the customer locations (as in Make-to Stock), predicting when stock-on-hand would be running low.

The final assembly area might very well be JIT, selecting the proper assemblies and components from floor stock, and replenishing the floor stock through Kanbans. As in Make-to-Stock, suppliers can also use manual or electronic Kanbans. Alternatively, the final assembly area can be a sophisticated FMS. In this case, an order can actually be accepted and produced without any humans touching a keyboard. For example, the order could be:

- Electronically transmitted directly into the manufacturer's computer from the customer's computer
- Processed through the features-and-options order entry configurator in MRP II
- Translated to a manufacturing order by MRP
- Transmitted to the FMS from MRP
- Produced on the shop floor by the FMS directly controlling the appropriate machines
- Put in a box ready for shipment

Although this might seem to be impossible for some companies, others are already doing it. For example, with the advent of high-speed, high-quality, digitally-based copiers, a printer of technical manuals can now provide this quick-turn, highly accurate service to customers. The printing company can produce one copy of a given manual, and put it on the delivery truck within four hours after the customer's computer discovers the need, error-free and cost-effectively, without human intervention. Consider the impact of this technology on the competitors, who still require days to set up conventional printing presses, causing an order of less than 500 copies to be very costly.

Make-to-Order

A standard Make-to-Order environment provides considerable flexibility for the customer, at the expense of timeliness. An AMS attacks the timing issue by fully integrating the entire communications system. The AMS provides access to various components, utilizing a configuration order entry Bill of Materials, perhaps with expert systems rules. It uses MRP to create and print the orders for the subassemblies and other components, then sends the resulting items for final assembly and shipment. Additionally, it uses CRP to ensure that sufficient capacity exists. Ideally, the AMS includes a fully integrated MRP and finite CRP, in order to provide much better estimates of shop order lead times for material release, and to avoid overloading the plant in the planning phase. The AMS will also provide direct access to the engineering design applications, such as CAD. The ultimate AMS would link the manufacturer's CAD system to the customer's CAD system, or equivalent.

On the manufacturing and shop floor side, a FMS would probably make more sense than JIT and Kanban, because each product is increasingly different. JIT would still be appropriate for common components and subassemblies; FMS would be more appropriate for the "specials." The FMS could use CAM, CNC, or any of a host of methodologies for directly controlling machines. As in all these environments, the suppliers would be directly linked to the manufacturer.

Design-to-Order

If Make-to-Order is both flexible and relatively slow, Design-to-Order is even more so. Thus, an AMS that fully integrates the communications and planning capabilities for a Design-to-Order manufacturer will provide a highly competitive edge. Here, the AMS emphasizes the design tools, interfacing them directly to the customer's design tools. These design tools also directly link to the shop floor. The AMS uses MRP heavily, to quote customer orders and to plan the materials for each order. As in Make-to-Order, MRP should include an integrated finite CRP to provide realistic schedules, both from a lead time standpoint, and a capacity standpoint.

Given the uniqueness of each design, JIT and Kanban probably have no direct application as far as material management. However, the JIT principles of minimal waste and maximum responsiveness still apply. The ties to the suppliers will be similar to those for Make-to-Order, except that we might also need to tie our design tools directly to our suppliers' design systems. The AMS will tie directly to the FMS for shop floor control, in order to reduce communications time to an absolute minimum.

Job Shop

Job shops occur in two distinct varieties: "independent," those that sell to outside customers, and "captive," those that are within a larger manufacturing company, whose customer is the company.

Independent job shops differ from the Make-to-Stock environment of an M&CRP shop in three important ways, which stem from the ability of the job shop to produce whatever a customer wants in a reasonably short time span, rather than to produce standard products. These three important differences are:

- Job shops have much greater difficulty scheduling and forecasting, because they make so many varied products.
- Job shops often create estimates, or quotations, for customers before receiving the actual purchase order from a customer.
- Job shops track customer orders directly on the floor. In most job shops, the work order number and the sales order number are identical, and the work order directly identifies the end customer.

Except for scheduling, these issues do not significantly affect dependent captive job shops.

Scheduling

Scheduling a job shop, using MRP II, requires a Master Production Schedule that drives Material Requirements Planning. In turn, the Planned Orders from MRP drive Production Activity Control to develop the detailed schedules for the shop floor. MRP II implicitly assumes that the components being fabricated in the job shop will be assembled into end items, which will be scheduled by the Master Production Schedule to meet customer demand.

However, if the job shop is responding to independent demand created by external customers, instead of by an internal dependent demand created by the production schedule for final products, developing a Master Production Schedule to drive the rest of the scheduling process is more difficult.

One approach is to develop Super Family Bills for families of products having similar demand patterns. More accurate forecasts can be developed at the family level than at the individual product level. The Master Scheduler and marketing personnel review and derive the percentages for the actual products in the Super Bill from the product mix sold during the past six months. The Master Scheduler develops a two-level Master Production Schedule, using the family forecast and the Super Family Bill to determine the schedules for the actual products that comprise the second level. The schedules for the actual products in the second level actually drive MRP and PAC in the normal manner, as described in more detail later in the chapter.

A job shop produces a large number of products on a Make-to-Order basis. The low volume of each product, and the differing design of the various products, dictates that a job shop configuration is the best process (a repetitive flow line just does not work in such a situation). Because of the many possible product configurations, accurately forecasting the demand for each individual end product is virtually impossible. Likewise, each product may be relatively small and simple, but the total number of products produced makes the scheduling problem difficult.

Forming Families and Forecasting Family Demand

To facilitate demand forecasting, Marketing first segregates the products into families, based on the similarity of demand pattern. This similarity can be determined from past sales figures, by knowing that the product is used by a particular industry that has a distinctive demand pattern, or by other factors. The fewer the families the better, provided all members of the family have the same demand pattern.

Marketing and Sales create the sales forecast for the family. This family forecast should be more accurate than a forecast for any individual product in the family, for two reasons: (1) the errors in the forecasts for each product tend to cancel each other when aggregated into a family forecast, and (2) more sophisticated forecasting techniques can be used, because fewer forecasts are required.

Developing Super Family Bills

Marketing and Sales can determine the percentages in the Super Family Bill, by comparing the sales of each product in the family with the total family sales for some period in the recent past (the last six periods, for example). We discussed these techniques in some detail in Chapter 3; now we can put them to good use. For example, let's assume that family *A*, with products *A*1, *A*2, and *A*3, had sales during the last six periods as shown in Figure 9-6.

From the data in Figure 9-6, we can determine the percentages for the Super Family Bill by dividing the total sales for each product by the total sales for the family as follows:

$$A1 = 120/600 = .2 = 20\%$$
$$A2 = 180/600 = .3 = 30\%$$
$$A3 = 300/600 = .5 = 50\%$$

The Super Bill of Materials for family *A* would then appear as shown in Figure 9-7. Let us assume that the forecast for family *A* for the next six periods is as shown in the top line of Figure 9-8. To more simply show the concepts, we will assume that the Master Schedule is revised on a periodic basis, that there is no Safety Stock, and that we have chosen lot-for-lot lot sizing.

The Master Schedule for family *A* explodes, using the percentage shown in the Super Bill to obtain the forecasts for products *A*1, *A*2, and *A*3, as shown in Figure 9-8. Once the Master Schedules for the products are complete, they can be used to drive the M&CRP and PAC.

If there is a lack of confidence in the family forecast, or in the forecast for one of the products, we can use *hedging*, rather than Safety Stock, to prevent stock-outs, if the forecast is low. To illustrate, let's assume that we lack confidence in the forecast for product *A*3, and we wish to employ hedging for safety. The actual manufacturing lead time for product *A*3 is two periods, but for competitive reasons, we desire to quote a response time of one period to all customers. Thus, the first period of manufacture of *A*3 will be based on forecast, rather than orders. (This should encourage us to cut

Product/Period	1	2	3	4	5	6	Total
A1	15	17	18	20	24	26	120
A2	24	28	31	29	32	36	180
A3	41	45	47	51	57	59	300
Total	80	90	96	100	113	121	600

Figure 9-6. Family *A* sales data for last six periods.

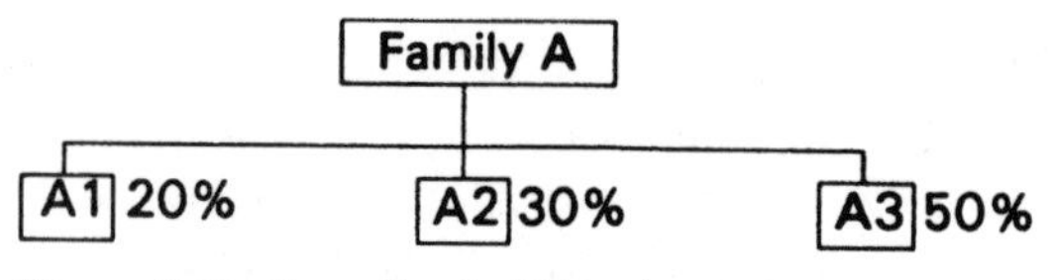

Figure 9-7. Super Family Bill for family *A*.

FAMILY "A"		1	2	3	4	5	6
Forecast		120	110	100	110	120	130
Customer Orders		101	59	20	5	0	0
Available	0	0	0	0	0	0	0
Avail to Promise		19	51	80	105	120	130
MPS		120	110	100	110	120	130

x.2

A1		1	2	3	4	5	6
Forecast		24	22	20	22	24	26
Customer Orders		20	14	5	0	0	0
Available	0	0	0	0	0	0	0
Avail to Promise		4	8	15	22	24	26
MPS		24	22	20	22	24	26

x.3

A2		1	2	3	4	5	6
Forecast		36	33	30	33	36	39
Customer Orders		30	15	5	0	0	0
Available	0	0	0	0	0	0	0
Avail to Promise		6	18	25	33	36	39
MPS		36	33	30	33	36	39

x.5

A3		1	2	3	4	5	6
Forecast		60	55	50	55	60	65
Customer Orders		51	30	10	5	0	0
Available	0	0	0	0	0	0	0
Avail to Promise		9	25	40	50	60	65
MPS		60	55	50	55	60	65

Figure 9-8. Two-level Master Schedule.

A3	1	2	3	4	5	6
Forecast	60	55	50	55	60	65
Customer Orders	60	30	10	5	0	0
Available 0	0	0	0	0	0	0
Avail to Promise	0	45	40	50	60	65
MPS	60	75	50	55	60	65

A3	2	3	4	5	6	7
Forecast	55	50	55	60	65	65
Customer Orders	55	40	25	10	0	0
Available 0	0	0	0	0	0	0
Avail to Promise	0	30	30	50	65	65
MPS	55	70	55	60	65	65

Figure 9-9. Using hedging in lieu of Safety Stock.

manufacturing lead times to less than one period, so that we do not have to forecast at all.)

When we use hedging, the Master Scheduler does not roll the hedge quantity across the Demand Time Fence into the current period for final production, unless we get firm customer demand. Thus, these hedge parts could actually be used wherever required, rather than specifically for this particular customer. To ensure that we start enough *A3*s, we will use a hedge quantity of twenty, that is, we will start 75 units in period two, rather than 55, as shown in the top record of Figure 9-9. If the hedge quantity, or any part of it, was not required for period two, the hedge quantity of 20 would be rolled over to period three and the process repeated, as indicated in the lower record of Figure 9-9.

In essence, we are carrying a Safety Stock of 20 partially built units for product *A3*. The fact that the products are only about one-fourth finished means that we are getting the benefit of Safety Stock with only one-fourth the investment.

Estimating-Quoting

Most independent job shops and custom manufacturers must estimate, or quote, on a specific job, before the customer awards them the work. Job shops traditionally only receive 25 to 35 percent of the work that they quote, and (in most cases) the customer is unwilling to pay for the quotation. This presents three opportunities for job shops to:

- Quote accurately, with minimal resources required by the quoting process
- Monitor outstanding quotes and manage the sales cycle with potential customers, so that they can win the highest possible percentage
- Use the quote as the basis for the actual work for each job that is won, so they can know quickly if they are going over budget

Quoting accurately can use the same basic technique as forecasting. We can define templates of typical products, using typical configurations, and modify those easily for specific customer requests. This customer-specific request then becomes a quote. When we win the job, we can report our actual labor, materials, and other expenses against this specific job. If the actual experience differs considerably from our estimate, we can either change the way we work the next time, or revise our future estimates. As we perform more jobs of a given type, our experience base continues to improve our estimates.

Most companies who use computers to assist with estimating use spreadsheets of varying complexity. Although this can reduce the effort required for creating the estimate, these spreadsheet quotes can be difficult to interface to the project accounting system, and must be manually revised each time the template changes. Finally, they have no inherent ability to compare the estimates to the actuals. Some MRP II systems now include quoting and estimating modules for just these situations. Other MRP-like packages started in the job shop community, using quotes and estimates as their design center. In addition to tracking estimates, these packages allow us to include multiple items with no MRP part number as part of the quote, including special parts, subcontract labor, and various overhead charges.

Finally, with automated quotes, we can track and analyze the ones we win and lose with relative ease, and even track the probable reasons why we lose.

Tracking Specific Customer Orders on the Floor

In a traditional MRP environment, the shop makes assemblies from parts. Parts are interchangeable and not customer-specific. Job shops with independent demand, on the other hand, make items to customer order. They stock very few parts, and most items that they do stock will probably be raw materials. Job shops only fabricate and assemble when they get actual customer orders. Their standard manual paperwork uses the sales order number, which might also be the customer PO number, as their shop order number. They track all expenditures for labor and materials, at all levels, against this number, so they can determine whether or not the job was profitable. The people in the shop can easily coordinate the various pieces that comprise a single customer order, ensuring that they will all be available at the right time.

For an MRP II system to work effectively in a job shop, the system must allow the customer name and order number to be carried all the way down to the lowest component. Some packages can accomplish this by merely using the lot-for-lot Lot-Sizing technique; others require additional effort or modification.

Job Shop Summary

Job shops can forecast more accurately and easily by using family Bills of Material, than by forecasting each potential item by itself. The crucial part is identifying families of products that have similar demand patterns, and whose aggregate demand can be forecast fairly accurately. If the forecast is still questionable, we recommend the use of hedging to provide some safety at a minimum cost. A two-level Master Production Schedule that uses the family forecast and Super Family Bill drives M&CRP and PAC in the normal manner.

Job shops also require specialized estimating and quoting capabilities that are not part of the traditional MRP environment. The estimates must be easy to create and relatively accurate. Many companies use spreadsheets for estimating; some job-shop oriented MRP-like packages and estimating modules, within the MRP II system, can also assist with this.

Finally, no matter how the job is estimated, planned, and sold, job shops almost always track progress on the shop floor, in purchasing, and in accounting by customer job number.

Tool Requirements Planning (TRP)

A Tool Requirements Planning system plans and controls tool inventories in order to reduce those inventories, eliminate tooling shortages, and minimize the risk of uncontrolled tool obsolescence. A Tool Requirements Planning system functions the same as a Materials Requirements Planning system in determining which materials (tools) are required at which time.

The TRP has much simpler procedures, because using standard MRP stockroom procedures will create an uneconomically large number of transactions. Because control can be exercised at many points, the system designer must determine how to operate the system effectively, with the fewest possible control points. In a typical shop, tools can be:

Ordered

Received

Inspected

Released to inventory

Released to production

Returned to inspection when worn

Either reconditioned or discarded

To reduce the number of transactions, the TRP can be treated as consisting of two functions:[5]

1. The tool replenishment function
   Tool Requirements Planning system
   Purchasing process
   Shipping and receiving
   New tool inspection
2. The operations function
   New and used tool storage
   Issuing tools to production
   Used tools inspection
   Reconditioning

This system requires only two transactions: (1) receiving new tools into inventory after passing inspection, and (2) issuing tools to the working set, after an old tool is scrapped.

Maintenance Planning

Maintenance management is tantalizingly similar to manufacturing management; both deal with the application of labor and materials to accomplish tasks. Maintenance, however, differs from manufacturing in that many of its activities are not routine, and have very loose standards, if any at all. Maintenance includes three types of activities that we describe using automotive examples.

[5]Randall M. Savoie, "Tool Requirements Planning: A New Approach," *APICS 1988 Conference Proceedings*, APICS, Falls Church, VA, 1988.

- Planned (preventive) maintenance, such as the routine lubrication, oil change, and change of oil filter on a car.
- Predictive maintenance, such as the driver of a car noticing a vibration that increases with speed. The mechanic diagnoses a U-joint starting to fail, and replaces it at a mutually convenient time.
- Nonscheduled maintenance, such as the U-joint failing while the driver is en route to an important appointment, causing the car to stop without warning. This totally disrupts the plans of the driver.

Maintenance planning supports the preventive maintenance management cycle. A sophisticated maintenance planning system might try to forecast the predictive and nonscheduled maintenance activities as well. A maintenance planning system must answer three questions:

1. What maintenance work needs to be done?
2. What resources does it take to do the work?
3. When would we like to do the work?

To answer these questions, the system must review the backlog and/or forecasts of work (what work needs to be done?), the activity work plans and the related bills of resources (what resources does it take to do the work?), and the desired priority of the work and the availability of the operating equipment (when would we like to do the work?). In this way we derive the resource needs and timing, known as "time-phased resource requirements." A large maintenance organization, with considerable experience, *knows* that air conditioners tend to break in the summer, and that the big press generally goes down every four to five months. To assist with the predictive maintenance, experienced maintenance departments also train machine operators to detect changes in the way machines operate, sound, or look. In fact, the forecasting capabilities of MRP II systems might be useful in scheduling a maintenance department for predictive maintenance, complete with spare parts kits. Likewise, the Master Production Scheduling capabilities could be utilized to reserve productive capacity for nonscheduled maintenance, if nonscheduled maintenance occurs in a somewhat steady manner.

Once the time-phased resource requirements are aggregated, management must answer three additional questions:

1. What resources do we have?
2. What resources do we still have to get?
3. Can we get them in time to meet the desired need date?

As management answers these questions, they develop a feasible maintenance schedule.

Where possible, maintenance activities can use the Bill of Materials and routing databases of an MRP II system. Through time, the database will improve in accuracy and become an invaluable tool. However, repairing a machine that won't run can be as simple as replacing a worn electric motor or as complex as a total rebuild of the entire drive mechanism. This extreme variability makes standards much more difficult to create and monitor. Maintenance stockrooms and cribs can use the inventory control transactions to keep their perpetual inventories.

Using MRP II for Military Program Management

The use of MRP II for military or government program management requires developing a Bill of Resources.[6] The Work Breakdown Structure, or WBS, required by MIL-STD-881A in most government programs, specifies the deliverable product, which is any combination of hardware, software, documentation, and services. This is essentially a Bill of Resources. To use MRP for program management, you establish a Master Schedule for the program, with a quantity of one, due for "shipment" on the program completion date. This is perfectly consistent with using a Bill of Materials to produce a single assembly at the end item level. The project, or program, has a quantity of one for virtually any project, because it is based on a single budget authorization, regardless of the number of items.

Because the project Bill of Resources has a structure that contains item, quantity, and unit of measure, you can enter these structures into MRP II part master files, as if they were physical components required for assemblies. The computer BOM processor can process these part master records as indented bills.

Master scheduling a project involves creating a part master for each major subproject, such as engineering, documentation, and so on. Once this is done, an MRP run will explode the requirements for each resource by time period, creating Planned Orders for the resources needed to meet the schedule, based on the lead times for resource acquisition. CRP provides forward visibility of capacity constraints for all programs. This schedules each program, and alerts management of bottlenecked labor operations, whether the bottleneck is a contract officer or a machine center.

[6]Keith R. Plossl, "Using MRP II for Program Management: Theory and Practice," *APICS 1988 Conference Proceedings,* APICS, Falls Church, VA, 1988.

Additional M&CRP-MRP II Adaptations

MRP has been adapted in many other equally important areas. In fact, it is primarily limited by our own imaginations. In this section, we present brief descriptions of several other nontraditional areas that have used M&CRP-MRP II.

Apparel

Although apparel manufacture appears to fit traditional MRP, it requires additional information in the form of size and style. For fashion apparel especially, style is all-important. Last year's dress will probably not sell for even the material cost, let alone at a profit. This category also includes footwear, and such household goods as curtains, drapes, and tablecloths.

Foods and Pharmaceuticals

Producers of foods and pharmaceuticals are required by government agencies to track their products by lot. Special capabilities within many MRP II packages track lots from suppliers through products, to wholesale and retail locations. Additionally, food and pharmaceutical manufacture is generally capacity-constrained, rather than material-constrained. It, therefore, uses the materials planning side of M&CRP less than it does the production planning and resource planning capabilities of a capacity-oriented MRP II system.

Medical Equipment, Complex Equipment

Medical equipment and complex equipment manufacturers must frequently maintain serial number control of their products, not only through the manufacturing process, but throughout their entire life. The serial number control can be used for recalls, if a certain component batch starts to fail. It also serves as the basis for the field service that many manufacturers provide for complex equipment. When a unit goes into service in the field, the manufacturing and materials system should turn over to the field service support system the complete record of actual products used, by lot and serial number. Because field service is actually a type of maintenance, some field service organizations have adapted MRP II systems for their own use.

Distribution Requirements Planning

Companies that have a distribution network with multiple warehouses need DRP support. This capability requires logic that tracks actual on-hand quantities, and rolls up requirements at the warehouse level into consolidated demand in the Master Production Schedule.

Hospitals

Hospitals have used MRP to manage their very extensive inventories. However, they must also add some extra data elements for some types of inventories, such as medication. These additional data elements could include the doctor that prescribed a medication, the nurse or other person who administered the medication, and so on.

Rebuild and Remanufacture

Rebuilding and remanufacturing is very similar to initial manufacturing, with the addition of some of the uncertainties discussed earlier in the section on maintenance management. Rebuilding an item involves disassembling and inspecting the original product, determining which item(s) must be cleaned, which can be repaired, which must be replaced (and whether they should be replaced with new or rebuilt parts), then performing the work. In the automotive aftermarket industry, companies have been remanufacturing starters, alternators, transmissions, and the like for years. A newer industry involves remanufacturing the toner cartridges for laser printers.

Summary

One of the chief virtues of good M&CRP-MRP II systems is that they are extremely flexible. They offer many options that enable them to be tailored to the user's unique business environment. Additionally, they usually reflect good management practices.

MRP II systems are evolving into what can be called "adaptive" systems. An adaptive MRP II provides the features and functions necessary to support a company that starts as one type of manufacturer (e.g., as a traditional job shop), and evolves into a different type (e.g., a repetitive or process manufacturer).

We recommended that M&CRP-MRP II evolve to integrate more fully with JIT and repetitive environments, including rate-based Master Scheduling, post-deduct material processing, and generally minimizing transactions. It was also suggested that process-oriented manufacturers utilize Process Flow Scheduling, and we discussed how M&CRP and MRP II can and must integrate with FMS to achieve their potential benefits.

The strategy, tactics, and operations of process industries were outlined, and we showed the differences between material-driven scheduling and capacity-driven scheduling, noting that process industries, by their nature, use capacity-driven scheduling.

We then described how an Agile Manufacturing System can operate in each of the four traditional Demand Response Strategies: Make-to-Stock, Assemble-to-Order, Make-to-Order, and Design-to-Order. We recommended that companies move toward Agile Manufacturing, so that they can become as competitive as possible. The adaptation of M&CRP and MRP II to the two types of job shops, independent and captive, was discussed, and we outlined how MRP concepts apply to maintenance management.

In looking ahead, we predicted that MRP, CRP, and finite scheduling will be fully integrated in future software packages, thus enabling M&CRP to more directly meet the needs of capacity-driven organizations. Finally, we have shown how M&CRP-MRP II has been limited in its adaptation only by our imaginations. The power of M&CRP-MRP II to project the future and communicate throughout an organization is second to none.

Selected Bibliography

Banam, James W.: "How to Design Systems for Job Shops," *MCRP Reprints,* APICS, Falls Church, VA, 1991, pp. 154–157.

Fogarty, Donald W., J. H. Blackstone, and T. R. Hoffmann: *Production and Inventory Management,* South-Western, Cincinnati, OH, 1991, Chap 11.

Belt, Bill: "MRP and Kanban—A Possible Synergy?," *Production and Inventory Management,* First Quarter, APICS, Falls Church, VA, 1987, pp. 71–80.

Hall, Robert W.: "Stockless Production for the United States," *MCRP Reprints,* APICS, Falls Church, VA, 1991, pp. 149–153.

———: *Zero Inventories,* Dow Jones-Irwin, Homewood, IL, 1983.

———: *Attaining Manufacturing Excellence,* Irwin, Homewood, IL, 1987.

Hay, Edward T.: *The Just In Time Breakthrough,* Wiley, New York, 1988.

Gessner, Robert A.: *Repetitive Manufacturing Production Planning,* Wiley, New York, 1988.

Lewis, Earl R. and Paul G. Conroy: "The Synthesis of MRP, Group Technology, and CAD-CAM," *MCRP Reprints,* APICS, Falls Church, VA, 1991, pp. 200–202.

Lubben, R. T.: *Just-in-Time Manufacturing,* McGraw-Hill, New York, 1988.

Pizak, Stanley C. and William A. Thurwachter: "Multi-Plant MRP," *MCRP Reprints,* APICS, Falls Church, VA, 1991, pp. 97–100.

Proud, John F.: "The Factory in 1990, Will it be Dark and Empty?," *APICS 1986 Conference Proceedings,* APICS, Falls Church, VA, 1986, pp. 344–347.

Putnam, Arnold O.: "MRP for Repetitive Manufacturing Shops: Flexible Kanban System for America," *MCRP Reprints,* APICS, Falls Church, VA, 1991, pp. 161–169.

Sandras, William: *Just-in-Time: Making It Happen,* Oliver Wight, Essex Junction, VT, 1990.

Schonberger, Richard J.: *Japanese Manufacturing Techniques: Nine Hidden Lessons in Simplicity,* Free Press, New York, 1982.

———: *World Class Manufacturing,* Free Press, New York, 1986.

———: *Building a Chain of Customers,* Free Press, New York, 1990.

Suresh, Nallan C.: "Optimizing Intermittent Production Systems through Group Technology and an MRP System," *MCRP Reprints,* APICS, Falls Church, VA, 1991, pp. 207–210.

Taylor, Sam G., Samuel M. Seward, and Steven F. Bolander: "Why the Process Industries are Different," *MCRP Reprints,* APICS, Falls Church, VA, 1991, pp. 170–185.

Thurwachter, William A.: "Capacity Driven Planning," *APICS 1982 Conference Proceedings,* APICS, Falls Church, VA, 1982, pp. 384–388.

10
Selecting and Justifying an MRP II System

"If you don't know where you are going, you might take the wrong road" ANONYMOUS

Introduction

The ultimate embarrassment is to discover that a system does not fit, and cannot be made to fit, the company's business, after you are six months into implementing a $500,000 MRP II package that has been approved by the Board of Directors.

This chapter provides an overview of the process of selecting the proper MRP II system for a manufacturing company. The process of selecting an M&CRP system necessarily involves the entire manufacturing information system (normally an MRP II system). In any case, the selection methodology is module-independent; it works just as well in selecting a new accounts payable system, a CAD system, or a demand deposit system for a bank.

Although the approach and the activities outlined in this chapter appear to be easy, appearances are deceiving. The approach embodies the corporate equivalent of the maxim for individuals, "Know Thyself." The process usually takes from several weeks to several months, for most organizations; however, the results are well worth the effort. The approach includes the following steps:

Defining corporate mission, objectives, and strategy

Defining enabling tasks

Defining the information system objectives that support the corporate objectives

Defining the information system functions and features

Preparing the nontechnical portion of a Request for Proposal (RFP)

Sending the RFP to selected systems suppliers

Selecting 2 to 3 finalists

Selecting the winning proposal

Running the Pre-Selection Pilot

Creating an implementation plan

Justifying the new system

Preparing the final recommendation for board approval

Figure 10-1 is a Gantt chart that presents an overview of these steps. In some cases, one or more of these steps can be skipped, or combined with others, with no substantial increase in risk. Actual times for each of these steps may vary greatly from this projection.

Given the amount of time and resources required by this approach, many organizations short-cut the process by skimping on defining either the corporate objectives or the detailed functions and features. This can be very costly, because the selected system will form the backbone of the corporate information network for the next 5 to 10 years, and will dramatically affect the organization's ability to compete effectively.

We address the issues of MRP II justification, outlining both potential benefits and potential costs. We also caution each company to develop its own numbers for the categories, rather than using the percentage ranges that are provided as guidelines.

The chapter concludes with an outline of some of the many alternatives available in today's information systems.

Defining Corporate Objectives and Strategy

Two of the three major reasons that implementations of new technologies (such as MRP II) fail, according to the APICS Systems and Technologies Certification Review Course, are the limited understanding of strategic goals, and ignorance of how to combine techniques to fit our company's needs. The third, which we cover in detail in the next chapter, is the inability to manage the organizational dimension of implementation.

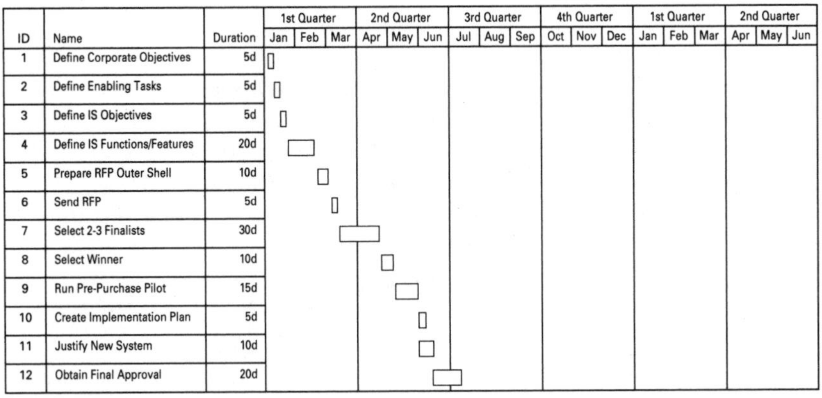

			1st Quarter			2nd Quarter			3rd Quarter			4th Quarter			1st Quarter			2nd Quarter		
ID	Name	Duration	Jan	Feb	Mar	Apr	May	Jun	Jul	Aug	Sep	Oct	Nov	Dec	Jan	Feb	Mar	Apr	May	Jun
1	Define Corporate Objectives	5d																		
2	Define Enabling Tasks	5d																		
3	Define IS Objectives	5d																		
4	Define IS Functions/Features	20d																		
5	Prepare RFP Outer Shell	10d																		
6	Send RFP	5d																		
7	Select 2-3 Finalists	30d																		
8	Select Winner	10d																		
9	Run Pre-Purchase Pilot	15d																		
10	Create Implementation Plan	5d																		
11	Justify New System	10d																		
12	Obtain Final Approval	20d																		

Figure 10-1. MRP II system selection steps.

If we do not understand our corporate strategic goals, we cannot design support systems, either human or electronic, with any assurance that they will fit our actual needs. Unfortunately, most organizations have never clearly defined their strategic objectives. And even if top management has agreed on the organization's strategy, most employees do not understand it, or do not know what their department's mission is to support those strategies. This gives rise to disagreements between departments concerning values and priorities.

Likewise, many companies install techniques in a "technique du jour" mentality; they have tried JIT, TQM, MRP, ISO 9000, Theory Z, EDI, CAD/CAM, and the entire panoply of latest approaches. Although each of these techniques can be very appropriate and cost-effective, they must be integrated into the company's operating style and culture, and with all other existing techniques. Thus, if we have MRP and are implementing JIT, we must determine what MRP techniques we need to change in order to accomodate JIT. Also, we need to ask what JIT techniques we need to modify to fit our current MRP operating style.

Why define corporate (or division, or subsidiary, or company) objectives and strategy? What does this have to do with information systems? Companies have two choices:

1. Try to interface existing systems that were never intended to work with each other (the bottom-up approach)
2. Start from the top to determine the actual corporate information requirements (the top-down approach).

Having seen the considerable inefficiencies and expense caused by the first approach, we strongly encourage companies to consider the second. This provides a framework into which systems must be fitted, and specifies how systems must work together.

The statement, "Information Systems should support corporate objectives" seems to be both completely obvious and unarguably true. However, most corporate information systems have evolved over the years to serve the needs of various functional departments. For example, the Inventory Control department requested inventory planning and control systems; the Accounting department requested general ledger, accounts payable, and accounts receivable systems. Unfortunately, many of these systems required information from other areas in order to work effectively. A purchasing system, for example, should be connected to the material planning system, so that projected purchase orders for production materials will automatically become available to the purchasing agent's system. Likewise, when purchase orders are placed, the company has created a future liability that will affect accounts payable. Thus, a purchasing system that is designed only for Purchasing department use, cannot be nearly as effective as one that integrates both material planning and accounting. The purpose of an MRP II system is to integrate the "islands of automation."

Corporate strategy can best be defined by gathering together the executives or leaders of each major functional area (Marketing and Sales, Engineering, Manufacturing, Finance, Human Resources, Information Systems) for a planning session, using an experienced outside facilitator as a guide. Starting with the corporate mission and objectives, the facilitator leads the group through a series of exercises and workshops designed to help them reach consensus concerning the strengths and weaknesses of their:

Customers

Suppliers

Competitors

Own company, or division

Based on these assessments, the group creates a corporate strategy that utilizes their strengths (and is aware of competitors' capabilities) to assist their customers to succeed. The ultimate goal is to have the customers passionately loyal to "our company" as their favorite supplier. This process is illustrated in Figure 10-2.[1]

Company objectives are those goals that stand by themselves; they do not

[1]Ronald T. Pannesi and Helene J. O'Brien, *Systems and Technologies Certification Review Course Student Guide,* APICS, Falls Church, VA, 1992, Chap 1.

Figure 10-2. Corporate strategy model.

help to accomplish other objectives. In fact, individual company objectives often conflict with other objectives. For example, increasing market share is much easier to accomplish if a company is willing to forgo profitability. Typical company objectives include the following:

Gross profits of *x*% of sales (annual average) by a given date

Market share increase of *y*% by a given date

Excellent community citizen

Return on equity (ROE) of *z*%

Sales volume of $*x* per year (inflation-adjusted)

The company objectives can be graphically displayed by grouping them in a column of boxes on the left side of a page, as shown in Figure 10-3.

Defining Enabling Tasks

After department leaders have tentatively defined the company objectives, they must determine how these objectives can be attained. The immediate *means* of attaining these objectives become the "enabling tasks;" they answer the question, "How are we going to have gross profits of *x*%, or market share increase of *y*%, and so on." An enabling task for the objective of *doubling sales* might be *quicker to market.* Each objective will probably have 3 to 5 enabling tasks, although one objective could have as many as 10 enabling

Double Sales in 3 yrs

Increase ROE 30% in 3 yrs

Figure 10-3. Corporate objectives.

tasks. These enabling tasks can be graphically displayed as shown in Figure 10-4. The lines connect each enabling task to the objective(s) that it supports.

Although these enabling tasks are more detailed than the company objectives, they are still broad, high-level tasks. Each of them can become a major "program" in its own right.

To avoid diffusion of effort, an astute company president will prioritize these enabling tasks, based on how critical each is to the overall mission of the company (the benefits) and on the cost. To assist in visualizing priorities, we encourage the company president to use shades of color on the objectives and enabling tasks, to reflect their priority. The use of color enhances the communication of information, as in Figure 10-4, for example, we might have shaded the most critical items a light red, the moderately critical items in a light blue, and the least critical items in white. Once priorities are clearly defined, the company president will assign responsibility for accomplishing each important task to the most appropriate department leader, clearly stipulating expected results and due dates.

Defining Information Systems Objectives

We define corporate objectives and enabling tasks in order to accurately determine what functions and features are required in our information system. If the feature or function does not assist us to achieve our objectives, we should not have it. Thus, the executives define the objectives of the information system in terms of the features and functions required by the corporate enabling tasks. In turn, the information systems objectives form the skeleton of the information system plan, which culminates in the design and selection of the MRP II system.

Some enabling tasks absolutely require information systems support, such as direct and error-free customer order entry, through Electronic Data Interchange (EDI), or rapid and complete detection of all potential future

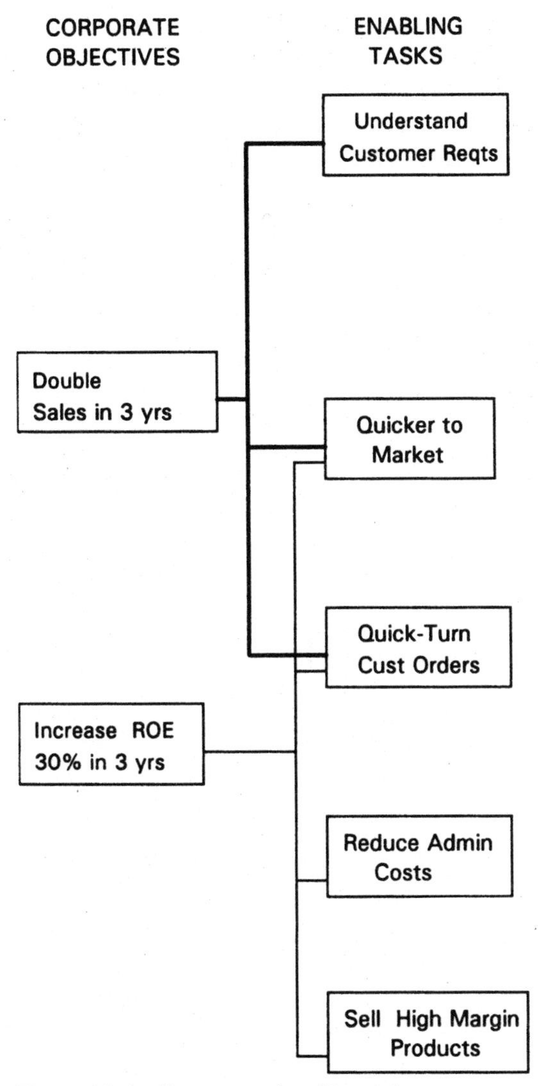

Figure 10-4. Objectives and enabling tasks.

material shortages, through MRP. Some tasks are aided by information systems, such as on-line credit approval. Some tasks have almost no information systems component, such as defining customer requirements.

Each enabling task can be supported by one, or several, information systems functions. These functions can best be conceptually defined by the executive staff in a brainstorming session. A skilled facilitator can encourage a healthy mix of creativity and appropriate realism at this meeting. After the information systems requirements have been defined, the group will want to prioritize them based on their criticality to the organization. These information systems requirements are actually functional objectives for the information system. Figure 10-5, which shows the corporate objectives, the

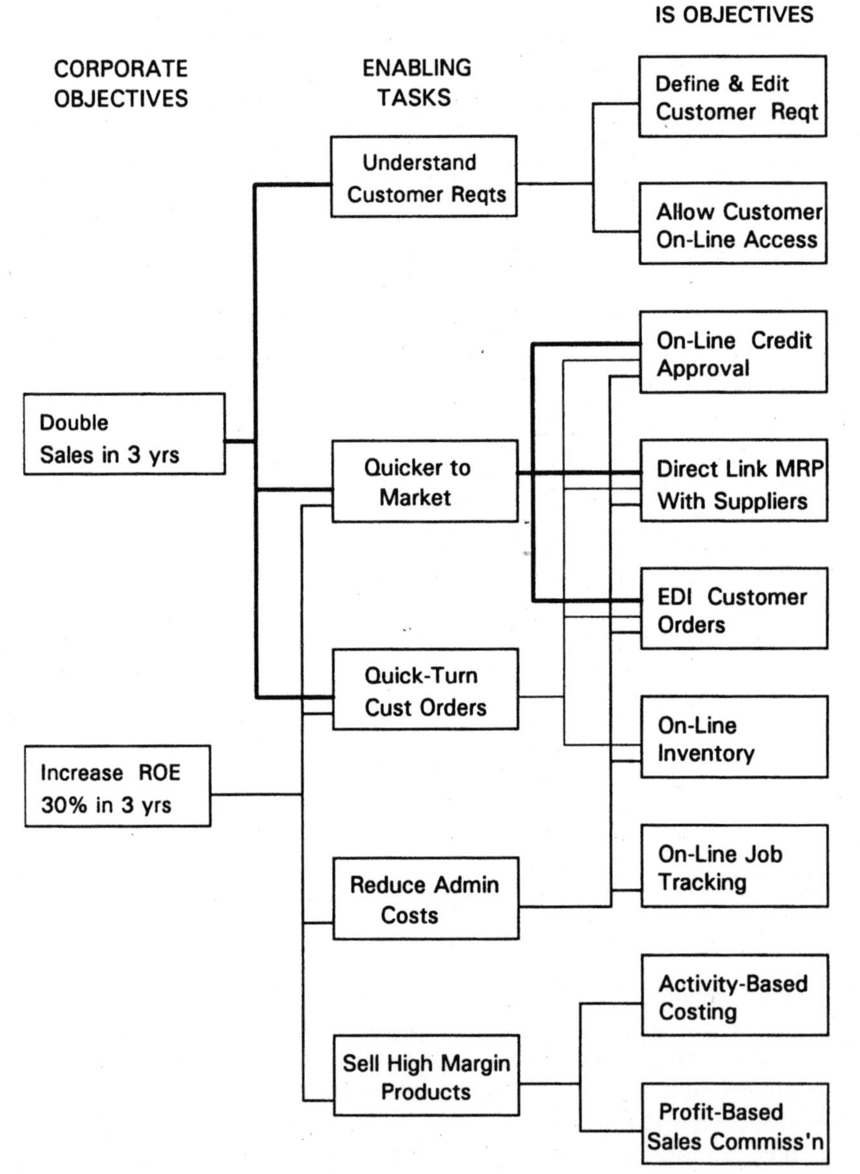

Figure 10-5. Objectives linkage diagram.

enabling tasks, and the information systems objectives, is called an "Objectives Linkage Diagram," because it links corporate objectives to information systems objectives.

These high-level information systems objectives, although vitally important, are not sufficient for selecting and successfully implementing a new system. They are lacking in two key areas:

1. Depth and detail. Although almost all MRP II systems on the market include Bill of Materials functionality, not many support expert system-based options and features for order entry, coupled with serial number effectivity.
2. Commitment of the people who will have to make the new system work. We find that holding face-to-face meetings with the people who have to live with the system on a daily basis is the most effective way of helping each one to prepare for the pain of an implementation.

Defining Information System Functions and Features

One or more people (quite possibly from outside the company, to avoid carrying any political "baggage") should interview the key functional personnel within the company, or division. In a smaller company, this could be as few as 6 to 10 people; in a larger company, this might involve 20 to 30 or more people. Those to be interviewed should be chosen on the basis of the following attributes:

Knowledge of their individual functional area

Knowledge of the company as a whole

Their willingness to try new things, and the probability that they will lead their department in implementing the chosen system

Each interview will require one to two hours. During the interview, the interviewer can take a proactive stance, asking the functional leader if having a specific capability would improve the work environment. Additionally, the interviewer must determine the areas that are currently working well, to ensure that these functions will be performed equally well by the new system. This step is actually the first in the implementation process, and is, perhaps, the most critical step. Performed well, it sets the person's expectations properly, and assists in supporting the upcoming change. Performed poorly, the person being interviewed might feel threatened (with loss of job, loss of prestige or power, or by the fear of failure to perform his or her new tasks), and might consciously or unconsciously hope that the new system will not succeed. We discuss this vital topic in more detail in Chapter

11. Alternatively, the selected individuals can meet as a group, and jointly define the capabilities required.

After the interviews, the interviewers create the first draft of the function and feature specifications of the new system. They then give copies of each appropriate section of these specifications to each of the persons interviewed, and schedule a follow-up meeting for any feedback. *These specifications will form the heart of the Request for Proposal.* It is most important that each individual feels comfortable that he or she can perform all the required functions with the proposed system.

The interviewers must write the specifications to meet the needs of several diverse audiences:

- The key individuals at the company, who must understand that the information system function and feature specifications are a blueprint of how they will perform their daily tasks in the future.
- The system suppliers, who must understand the company's culture and methodologies, as well as the expectations of system performance.
- The software selection team, which must eventually use these requirements as the basis for differentiating between proposed systems, and finally selecting the best.

Thus, the requirements must be technically sound and appropriately detailed, but at the same time, understandable to a person who is not an information systems professional.

This task is complete when the function and feature specifications have been reviewed and approved by all appropriate persons. Although guidelines concerning length are general, we suggest that the specifications should range from 40 to 75 pages. Specifications less than 40 pages in length probably lack sufficient depth to objectively select the best-fitting system. Specifications longer than 100 pages can become so cumbersome that they lose their effectiveness. In addition to the system functions and features, the specifications can include a narrative that describes how the department will operate with the new system. Appendix A contains part of a section from a function and feature specification.

Preparing the Nontechnical Portion of a Request for Proposal (RFP)

A Request for Proposal (RFP) typically contains two major sections:

Nontechnical portion, or boilerplate

Functional specifications

The nontechnical portion of a Request for Proposal contains all the information other than the specifications that helps suppliers to respond. It can include sections on confidentiality, response format, selection criteria, performance expectations, contractual commitments, intended timetable of events, publicity, and other items. Some corporations have standard RFP shells, or "envelopes," that can be adapted to this process. Other organizations prefer a more informal atmosphere, and deliberately minimize the nontechnical portion. At a minimum, we suggest that the nontechnical portion include:

- The name, address, and phone number of the person to be contacted inside the company
- Background information concerning your company (objectives and business strategy, business history, current information systems, objectives of the new system)
- RFP response instructions (conditions of the RFP, submission instructions, proposal format)
- Estimated timetable (proposal due date, decision date, projected implementation schedule)
- Selection process (methodology, evaluation criteria, responsibilities)
- Technical requirements and specifications (throughput requirements, network requirements, security, auditability, recovery, database specifications)
- Request for financial information, so that we can assess long-term financial strength

Whereas the functional requirements are of greatest interest to the various departments, the nontechnical portion provides the system supplier with vital information regarding the framework of the process, and defines several elements that are critical to the eventual success of the system. It must, therefore, be carefully crafted, so that all selected suppliers will invest the resources to provide quality proposals.

Soliciting Proposals

Like all businesses, software and hardware suppliers must allocate their limited resources. Many suppliers who received unsolicited RFPs will choose to not respond, because they assume that the probability of their making the sale is relatively small.

Given this fact, the company that is performing a software search should select 6 to 10 potential suppliers, based on industry reputation. In Purchasing terms, this becomes the basis for the approved supplier list. The importance of an excellent reputation cannot be over-emphasized. You are not buying a stand-alone product; you are buying a long-term professional relationship, on which you are, in some respects, betting the survival of your company. The excellence of the choices included in this initial list directly determines the quality of the eventual selection. Because there are hundreds of potential suppliers, some companies utilize consultants to help them to prepare this initial list. A committee member, or the consultant, then screens the potential suppliers, by telephone, for the critical factors that would remove them from further consideration. These factors directly reflect the corporate objectives and enabling tasks, as previously defined. When the list of potential suppliers is narrowed to 3 to 5, the company asks each of the suppliers if they are interested in proposing. Under these circumstances, most potential suppliers will want to respond.

As a matter of style, we suggest that you refrain from sharing the names of the potential suppliers in the initial stages, so that each supplier will accentuate their own strengths, rather than aiming for a competitor's known weaknesses. In the final stages, suppliers will generally know who their competitors are.

If appropriate, you can send nondisclosure or confidentiality agreements to the suppliers, before the RFP will be ready. These agreements prohibit suppliers from disclosing the contents of your RFP to anyone else, thus protecting your competitive secrets. Most software suppliers have no difficulty in signing and returning confidentiality agreements relatively quickly; some of the major hardware suppliers must have any such agreement reviewed by their legal department, and this can take many weeks. In fact, we have experienced one major hardware supplier who routinely refuses to sign any such agreements.

The RFP should be structured in a manner for easy supplier response, and easy evaluation by your committee. Assuming that you have prepared the RFP on a word processor, you might furnish a disk containing the RFP with the package. Instead, or in addition, you might furnish the response forms (possibly in spreadsheet or PC-based database form) on computer disk, so that suppliers can provide their responses in a form most easily used by your committee.

Suppliers will ask for the opportunity to meet key members of your selection committee, and to tour your plant(s). You can schedule this either shortly before, or after, the RFP becomes available. From a supplier's perspective, if a prospective customer does not agree to a meeting and a plant tour, the chances of winning the business are probably too small to warrant any investment in completing the proposal. Additionally, each supplier is concerned about the possibility that another supplier has an "inside track." To the extent that you can address these issues honestly and directly, the suppliers will respond by working harder to win your business.

During the review process, one person on the committee should keep a log of *all* questions and answers. Some questions will resurface later during the implementation; having ready access to the answers will be invaluable at that time.

One to two weeks after you release the RFP, you need to meet with each supplier, one at a time, to answer the myriad questions they will have. The suppliers will not only ask many questions, they will also try to sway your selection criteria in their favor. For example, one supplier will suggest that a relational database should be a show-stopping criterion (because they are among the few that operate on a relational database); another will suggest that you insist on local technical support (because their local support office is only 5 miles away), and so on.

Selecting Two to Three Finalists

The selection team should consist of a key representative from each major functional area: manufacturing, materials, marketing, accounting and finance, and information systems. The objective of the selection process is to find the best solution as quickly as possible, with the minimum expenditure of corporate resources in the process.

After you receive the supplier proposals on the due date (although one supplier will almost invariably require an extension for a reasonably legitimate reason), the selection team should read all of them from cover to cover. Exceptionally weak proposals should be removed from active consideration.

You can assign point values to each requirement in the RFP, and enter the suppliers' responses into a spreadsheet or database. For example, if the ability to backward schedule *and* forward schedule from a gating or bottleneck work center in CRP is absolutely required, that feature might have a weight of 5. An important feature would be weighted a 3, and a "nice to have" feature would be a 1. A supplier response that indicates that the feature is available as part of the standard package could receive a 5; a response that the feature is available through a report writer could receive a 3; if the feature requires custom modifications, the response could receive a 1, and if it is not available, the response would receive a 0.

In theory, the supplier with the most points is the most attractive. However, this ignores the number of questions in each section. For example, the MRP section might have 20 questions, whereas the general ledger section might have 60. In such a case, the supplier with a somewhat better general ledger could overwhelm the supplier with a better MRP module. Even after weighting the modules, this point valuation technique should be used only as a guide.

Another approach is to gather the entire selection team in the conference room for a forced ranking. Ask each person to rank the proposals in preferred sequence, preferably by secret ballot. Then tally the votes. The potential drawback here is that some committee members might be swayed by outstanding writing style or presentation (e.g., professional hardback bindings with the company logo on the cover). An experienced consultant can add some objectivity to the process, by helping the selection team assess relative strengths and weaknesses.

Many companies use both approaches: If they result in the same finalist, the process moves forward. If not, the selection team needs to examine the reasons for the differences, and to redefine how to decide on the finalist. We suggest that the second method actually yields better long-term results.

Selecting the Winning Proposal

Selecting the winning proposal can be simultaneously energizing and agonizing. Sooner or later, the committee must decide on one favorite solution, which often includes multiple suppliers.

We encourage companies to focus on how well the software fits their current and projected future business operations, as long as the hardware is from a reputable supplier. Most companies do not insist on buying their software from a computer hardware supplier. Having two suppliers generally presents few problems later on, because the software suppliers tend to be very experienced on the hardware as well. The final selection process includes checking references and demonstrations.

Reference Checks

We suggest that the selection team call references before the demonstration, to be better prepared with the "tough" questions. During the reference check, each person on the selection team talks with his or her counterpart at the reference company. Each functional reference check will take 10 to 15 minutes, and can incorporate the following:

- How the reference site runs its business
- How the reference site is using the software (as a planning tool and/or an auditing tool; in place of the informal systems, or beside them)
- The ease of implementation
- Excellence of supplier support
  Local
  National

- Quality of the software
  Functional completeness
  How many "bugs"
- Quality of the documentation
- Quality of supplier education
- Which other packages the reference site considered, and why they chose *X*
- Would they buy the software again, now that they have been through it? Why (not)?
- Would they recommend the software to a friend?

A sample reference checksheet is included as Appendix B.

Generally speaking, if a software package has been in existence for over two years, the supplier should have no difficulty finding 3 to 5 companies who are sufficiently pleased to serve as references. If an outside consultant has been involved in the process, the consultant may also be able to provide other references for the software packages, perhaps companies that are not as ecstatic about the software, or the supplier. Additionally, the company can ask to attend a user's group meeting, if one is being held during the selection process. The same is true for hardware families.

Demonstrations

The selection team schedules each of the remaining 2 to 3 suppliers for a one-half-day demonstration, preferably at the supplier's office. The selection team should require that most of the hands-on demonstration be provided by the technical support person who will be assigned to the company. Before the demonstration, each member of the selection team prepares several questions for their functional area. These questions can serve the following purposes:

- Clarification. "When you answered, 'Yes,' to serial number tracking, what did you mean? How does your software actually handle this function? Please show us on the system."
- Verification. "Did you really mean to answer some modification required for the MRP pegging report? What is involved? Who will perform the modifications?"
- Determining the level of capability of the key supplier personnel. "How many companies have you personally helped to implement the system? What were their greatest challenges?"

Immediately following a supplier visit, each member of the selection team writes a one or two paragraph memo, stating their overall opinion of the supplier's proposal and capabilities. This memo finishes with the answers to four questions:

1. How well does this software fit your function? Could you run your department with it?
2. How well does this software fit the entire company? Can we run the company with it?
3. What are the three greatest strengths?
4. What are the three greatest weaknesses?

The day following the second and each subsequent demonstration, the team meets briefly to establish a consensus concerning which is the best supplier so far, all things considered. This is another forced ranking process. The objective is to narrow the field to two suppliers.

The selection team can invite the final two suppliers to attend separate meetings in order to ask any remaining questions, and to review any critical issues. This provides the company with some additional bargaining leverage. The selection team must remember, however, that they are placing their company's future in the hands and goodwill of the selected supplier. When the chips are down, the goodwill of a committed supplier is worth far more than an extra $5000 wrung out during the final negotiation session. A systems supplier, unlike a new robot, remembers negotiating tactics.

After the final meeting, the selection team meets and finalizes their recommendation to top management. The selection team then notifies the suppliers.

Selection Criteria

Our experience indicates that the following criteria are the most important when selecting a system:

Ease of use

Closeness of fit between the system design and how the company wants to do business

Quality of support by the software supplier

Quality of the applications software (bug-free)

Quality of the hardware

Quality of support by the hardware supplier

Additionally, there can be some financial trade-offs. A company can choose to minimize additional investment (by utilizing existing hardware and software wherever possible), ignoring the potential long-term cost of such a decision. We would, instead, strongly encourage a company to consider information systems as a strategic and tactical tool, to maximize achievement of their business objectives.

Based on all the factors above, the committee selects one solution as the best for the company. After negotiating the financial and support arrangements, the committee can sign a letter of intent or "contingent" PO, subject to:

Successful on-site test

Final approval by the board of directors

Any other appropriate conditions

This letter of intent tells the supplier team that you are indeed serious about implementing their system. Given this assurance, they will generally work very hard to ensure the success of the on-site test, and to remove any final obstacles.

Running the Preselection Pilot

Most companies skip this section owing to (1) the psychological pressure to "move forward," and (2) the general euphoria surrounding the final decision. This testing process is, however, the least expensive career insurance available to the system champion and to the decision maker.

The selected supplier installs a working copy of the software on an appropriate hardware platform located in a conference room within the company. During the course of the pilot run, each member of the selection committee tries to closely simulate a typical day with the system. The committee will:

- Create the fundamental database, including:
  Part masters
  Bills of materials
  Routings
  Customers
  Suppliers
  Chart of accounts
- Enter a few forecasts

- Enter and print some quotations
- Enter some customer orders
- Enter some purchase orders
- Run MPS
- Run MRP
- Release purchase orders to suppliers
- Release shop orders
- Track and report shop order completion
- Answer typical customer questions, including:
  "When will you ship my order?"
  "Can I change the quantity?"
  "Can I change the due date?"
- Receive purchase orders and shop orders
- Ship finished goods to customers, including spares
- Invoice customers for shipments
- Receive customer payments
- Receive and pay supplier invoices
- Close the books for a month, running the following reports:
  P&L
  Balance sheet
  Sales analysis
  Other management reports

This pilot will require 4 to 12 hours from each person on the selection team; we encourage them to bring other key individuals as well. MIS will spend almost full time on this effort. The preselection pilot can last 1 to 4 weeks.

The supplier will probably have technical personnel available during this preselection pilot, to answer questions and to provide general advice. Based on the outcome of each phase of the process, the selection team will probably ask the supplier to revise its proposal, to address areas that were overlooked in the original RFP.

There is no such thing as a "perfect" MRP II system. Even the best system will not fit the needs of most companies 100 percent. Either the company, or the system, or both will have to adjust to specific situations. The benefit of the preselection pilot is that it exposes many, and possibly most, "surprises" early in the cycle, so they can be addressed with a minimum of risk and cost. This process also helps to confirm (or bring into question) many of the assumptions that underlie the RFP and proposal process. It is possible that the selection team will want to reconsider its selection, after the preselection pilot is completed. If so, the preselection pilot was well worth the investment. Otherwise, the preselection pilot costs essentially nothing, because it is the first step in the familiarization process that must occur, in any case.

Based on the information gained in the preselection pilot, the selection team lists each of the selected system's important shortfalls, and the expected solution, including:

- Missing functionality, and how that will be overcome
  Manual workarounds
  In-house customization
  Purchased customization
- Throughput, support, or reliability
- High-level backup and disaster recovery plans

The system supplier should be an integral part of this process.

Creating the Implementation Plan

Based on the preselection pilot, the selection team and the supplier prepare a first-cut implementation plan, such as the one shown as Figure 10-6. This plan outlines the overall flow and interrelationships of the implemen-

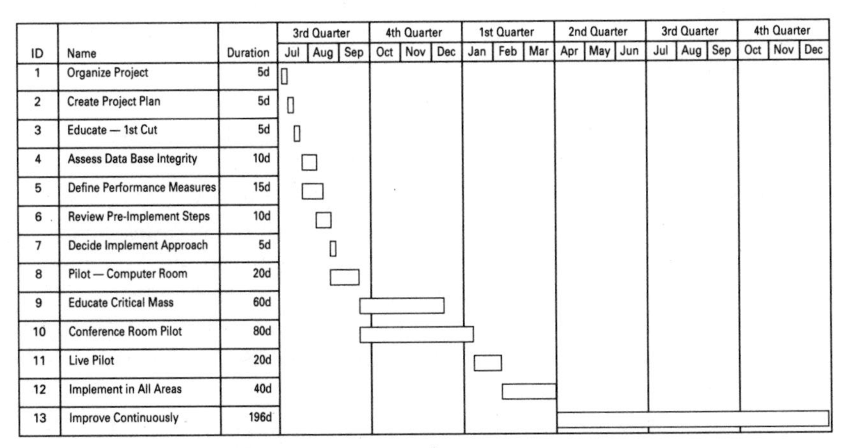

Figure 10-6. Initial MRP II implementation plan.

tation, including module precedence; it also provides the approximate timing of each module's implementation, so that the benefits provided by each module can be time-phased more properly in the system justification.

Justifying an MRP II System

The implementation plan created at this stage is a first draft, primarily intended for budgeting purposes. It should be no more than one page long. It forms the basis of the more detailed implementation plan that is developed, and then continually refined, during the implementation process. Usually, these initial implementation plans turn out to be overly optimistic, because the committee that creates them does not fully comprehend the full scope of work required. For example, many companies assume that their Bill of Material and Routing data are sufficiently accurate to be used in the new system. However, this assumption is probably incorrect more often then it is correct. Thus, it is no uncommon for a Board of Directors to mentally double the proposed costs, and cut the proposed benefits in half as a crude form or risk analysis. Also, a wise Board will mentally increase the projected time requirements, to estimate when the project will probably actually finish.

The actual length of an implementation of the MRP II system usually ranges from 12 to 18 months. If a system implementation takes longer than 18 months, the risk of failure to achieve its objectives increases. Furthermore, the increase in risk is not even linear; it increases at an increasing rate.

An MRP II system should first be justified as the initial search process commences. After the system is selected, the selection team should revalidate the justification, and confirm the costs. If the perceived benefits now appear much smaller, and the costs much larger, management might wisely decide to cut their losses, and stop the system implementation.

MRP II systems usually involve total expenditures of well over $100,000. In previous years, the totals often topped $1,000,000; but with the continuing reductions in computer hardware costs, and the evolution of software suppliers who sell MRP II software for under $100,000, the total costs continue to drop. To justify expenditures at these levels, companies can legitimately expect returns in most of the areas outlined in Appendix C (percentage improvement numbers reflect actual achievements by various companies). However, each company must understand how each saving will actually be achieved. More importantly, the respective department managers and directors must also willingly agree on the proposed savings. The expected benefits become the goal against which progress can be measured periodically, as discussed in Chapter 12. Projects without well-defined goals drift into oblivion. Of equal importance, the benefits themselves are used as the decision criteria for answering the countless questions that are raised during the implementation.

Appendix C contains a chart that includes many of the benefits and costs that typically form the basis of justifying an MRP II system. We highlight the more important ones in the following section.

Tangible Benefits

1. Decreased inventories, because of better planning and lower safety stocks; reductions can exceed 25 percent. A division of Polaroid reduced its raw material inventories 54 percent in two years, as they became "Class A."

   Smaller inventories can cause several costs to decrease. The division of Polaroid reduced off-site storage space requirements, with an attendant reduction of material movement costs, by 80 percent!

2. Increased plant productivity (increases can be from 15 to 30 percent), because of fewer materials shortages, and fewer crises.

3. Increased sales (5 to 10 percent) because of more on-time deliveries.

4. Reduced costs because of reduced overtime and rework.

5. Reduced costs of purchased parts and materials (1 to 5 percent), because of better purchasing visibility of future requirements.

6. Reduced clerical effort (5 to 10 percent), because of better systems integration, with fewer forms to complete and pieces of paper to find and update.

7. Reduced administrative costs as a result of reduced expediting and material chasing, and fewer disagreements about whose numbers are right.

8. Eliminated expenses for old computers and systems, perhaps including the costs of maintaining a climate-controlled computer room.

Intangible Benefits

1. Greater productivity on the part of administrative personnel, because of the availability of better information.

2. Improved communications throughout the company, yielding better and more timely decisions.

3. Substantially reduced "fire-fighting," or reacting to crises. An MRP II system provides a foundation for proactive planning, which eliminates the underlying causes of many crises. However, a company must also change a

culture that rewards "fire-fighting" behavior, in favor of one that rewards proactive planning.

4. Sales personnel receive thanks for on-time deliveries, rather than having to make excuses for late deliveries.

5. Management understands and controls its inventory investment more intelligently.

6. Direct connection to customer's computers for order entry, and to supplier's computers for purchases, thus reducing clerical effort, errors, and mail time.

A properly implemented MRP II system provides intangible returns that are equal to the tangible returns. Companies that have effectively utilized MRP II systems, have achieved returns on investment (ROI) greater than 200 percent.

Elements of Cost

The following cost elements can be considered when a company is creating and refining its MRP II justification.

People are the most important cost element, because they have to understand and support the new system. People costs include:

Time spent on the project by the project leader and project team

Education for all individuals

Training for all individuals

Outside counselor (explained in greater detail in Chapter 11.

Overtime and additional staff

Data costs are often considerably underestimated in initial cost projections. Data costs include:

Inventory record accuracy

BOM accuracy

Routing accuracy

Additional data gathering

Computer costs are also underestimated, because of the tendency to ignore the more mundane categories, such as wiring, or to underestimate modification, conversion, and maintenance costs. Computer costs include:

Hardware (purchase and annual maintenance)

Applications software (license purchase or lease, and annual maintenance)

Systems software (license purchase or lease, and annual maintenance)

Systems and programming

- Bridges between new and old systems to keep the company operating during implementation
- Initial modifications to packages
- Ongoing maintenance of packages
- Complete program life-cycle cost, if using in-house development
- Data conversion (building automated bridges between old and new systems)

Site prep (wiring, and so on)

The most effective justifications are the result of many meetings, and much up-front planning. They reflect the commitment of the Project Team and the Steering Committee to achieve tangible results by certain dates, and to have the detailed plans in place in order to achieve those goals. Justifications and plans dictated from the top down (like some budgets) are rarely effective in an MRP II implementation, because the people who have to make them happen do not "own" the numbers, and do not feel personally accountable for their achievement.

Most companies fail to realize that the total cost for the people involved in an implementation routinely exceeds (sometimes by a large factor) the purchase price of the hardware and software. Thus, a wise selection team will be more concerned about the ease of learning and the ease of use of the new system, rather than the initial purchase price. Some systems also carry much higher maintenance fees than others, and some suppliers bundle more services than others.

Preparing the Final Recommendation for Board Approval

Organizations usually have a formal approval process for a capital expenditure the size of an MRP II system. This routinely involves having the division general manager, the president, and/or the Board of Directors, and all other appropriate persons sign the request for a capital expenditure.

Information Systems Elements: Alternative Strategies

In this section, we turn our attention to several of the more common elements that comprise an information system. Because each of these options affects most of the others, a company selects one or two elements from the following list to serve as a foundation, and then adjusts other variables to fit. The elements include:

1. *Hardware suppliers.* Some companies buy computer hardware only from a single supplier; others view hardware as a commodity. Although the second alternative is gaining wider acceptance, we suggest that hardware be acquired from suppliers who:
   - Provide excellent local support (the cost of downtime is far greater than any potential savings in purchase price)
   - Will be around in 5 to 10 years

2. *Hardware architecture.* Some companies standardize on one type of hardware architecture, such as PC or minicomputer; others mix hardware architectures. In general, mixing architectures creates higher cost because of interface and support requirements. The higher cost can sometimes be justified, however, on the basis of the excellence of the applications available for individual functions.

3. *Operating systems.* Some operating systems run a wide variety of hardware; others run only on a single manufacturer's hardware.
   - The current trend is toward "platform independence," which refers to an operating system (such as MS-DOS on PCs, or UNIX, or Pick on work stations, minicomputers, and mainframes) that operates on hardware from a wide number of manufacturers.
   - "Proprietary" operating systems run only on a single manufacturer's hardware, such as OS/400 (on IBM AS/400 computers), and VAX/VMS (on DEC VAX computers).

In general, applications that are "platform independent" provide better price/performance ratios than those that are "platform specific" (e.g., those applications that run on only one manufacturer's hardware). However, the competitive pressures on price/performance exerted by the platform independent software suppliers can cause the others to become more competitive. Unless operating systems form an integral part of a corporate strategy, we encourage companies to focus, instead, on the excellence of the applications software, and on the suppliers that will support the software and the hardware.

4. *Databases.* We are not attempting to teach computer theory; however, in buying an MRP II system, one should be aware of what types of databases are available. The choices are briefly described below and are described more completely in Chapter 8.
   - *None,* which uses sequential files, with or without indexes, to store data. These methods of filing are the oldest. They are more simple,

and potentially somewhat faster, than the more complex methods, but they often lack the ability to relate data easily to such elements as parts and customer orders.

- *Hierarchical,* which connects data with a vertical structure like a Bill of Materials. This is an older type of database that can be somewhat slow and rigid, because the program has to go up a chain to a common parent, and then down a different chain to reach the desired data.
- *Network,* which uses a single-level vertical structure, such as a Bill of Materials, but which can have several indexes for a given piece of data. This is somewhat more flexible than hierarchical, and can relate data better than sequential files.
- *Relational,* which is the most advanced database currently available (object-oriented databases are still in the future). It is extremely flexible, and is very easy to query. Although earlier versions were usually somewhat slow, the new types of relational database are successfully addressing this problem.

All things being equal, a relational database provides the greatest advantages in flexibility and ease of inquiry. The database type, however, should not be a major determining factor in system selection.

Summary

We have outlined a reliable process for defining the true information systems needs of a company, and for selecting a system that closely meets those needs. This process includes:

Defining corporate mission and objectives

Defining enabling tasks

Defining the information systems objectives that support the corporate objectives

Defining the information system functions and features (including a sample from one section as Appendix A)

Preparing the nontechnical portion of a Request for Proposal

Sending the RFP to selected systems suppliers

Selecting 2 to 3 finalists

Selecting the winning proposal

Running the preselection pilot

Creating an implementation plan

Justifying the new system

Preparing the final recommendation for board approval

Because the system will probably cost well over $100,000 (and some mainframe systems may exceed $1,000,000), we have also included, in Appendix C, a form for developing the justification for an MRP II system.

Although there are many other ways to select systems, including buying the same one that is used by someone else in your industry, we strongly believe that the process outlined in this chapter minimizes your risk and provides the lowest total cost and the greatest possible payback. The system selection process is the first, vital, phase of a successful system implementation, which is where the payback originates. We have seen too many examples of companies that short-cut the selection process (to save time and money), only to bitterly regret that decision during implementation and the years of frustrated use of a system that did not truly meet their needs.

Selected Bibliography

Clark, James T.: "Selling Top Management—Understanding the Financial Impact of Manufacturing Systems," *MCRP Reprints,* APICS, Falls Church, VA, 1991, pp. 112–119.

Gray, Christopher: *The Right Choice: A Complete Guide to Evaluating, Selecting, and Installing MRP II Software,* Oliver Wight, Essex Junction, VT, 1987.

——— and D. Landvater: *MRP II Standard System—A Handbook for Manufacturing Software Survival,* Oliver Wight, Essex Junction, VT, 1988.

Pannesi, Ronald T. and Helene J. O'Brien: *Systems and Technologies Certification Review Course Student Guide,* APICS, Falls Church, VA, 1992, Chap 1.

Smith, Spencer B.: *Computer Based Production and Inventory Control,* Prentice-Hall, Englewood Cliffs, NJ, 1989, Chap 15.

Wight, Oliver W.: *MRP II: Unlocking America's Productivity Potential,* CBI Publishing, Boston, 1981, Chap 16.

11 Implementing MRP II

"It should be borne in mind that there is nothing more difficult to arrange, more doubtful of success, and more dangerous to carry through than initiating changes. The innovator makes enemies of all those who prospered under the old order, and only lukewarm support is forthcoming from those who would prosper under the new. Their support is lukewarm partly from fear of their adversaries, who have the existing laws in their hands, and partly because men are incredulous, never really trusting new things unless they have been tested by experience."

N. MACHIAVELLI, 1514

Introduction

Most MRP II systems implementations are "failures", because they fail to achieve even a modest fraction of their potential benefits. Why? Because most organizations do not understand that the essence of M&CRP is to change the fundamental way in which they operate on a day-to-day basis. They treat the implementation, instead, as a technical project.

Implementing MRP II systems is difficult, time-consuming, frustrating, and full of risk. This difficulty is surprising when we consider that MRP formalizes and integrates the existing informal shortage-driven systems. When implemented properly, the potential results include:

A return on investment of greater than 200 percent

A substantially improved competitive position

In fact, in the near future, just remaining competitive will *require* the effective use of an MRP II system.

This chapter describes what *really* happens in an MRP II implementation, including an approach that maximizes the probability of achieving the potential benefits:

Measures of implementation success

Implementation types

Prerequisites for a successful implementation

Classic implementation steps

Education and training

Additionally, we cover two vital topics:

Why most MRP II implementations fail

Implementing after a previous failure

Measures of Implementation Success

Several organizations, including The Oliver Wight companies, the David Buker organization, and Mike Donovan, have created checklists to determine the relative success of MRP II implementations. The results can be ranked as Class A, B, C, or D, as shown in Figure 11-1.

CLASS	USES OF SYSTEM	WORKS IN	RESULTS
A	Manage the Business	All areas of the business	Outstanding
B	Schedule & Load	Mfg & Matls	Very good
C	Order Launch & Expedite	P&IC Dept	Fair to good
D	"Another computer failure"	MIS only	Disappointment Frustration Wasted resources

Figure 11-1. Class ABCD characteristics.

A Class A company uses MRP II as its complete game plan, including manufacturing, engineering, sales and marketing, finance and accounting, materials, and human resources; in short, as the foundation for running virtually all aspects of the business. Each function of the business is integrated with the others; the MRP II system serves as the foundation for very strong communications throughout the company.

A Class B company uses M&CRP and shop floor control, but has not yet fully integrated purchasing or finance and accounting. Top management does not use the system to directly run the business.

A Class C company uses MRP primarily as an order launching and expediting platform. Shop scheduling is still done from the shortage list. Inventories remain near their traditionally high levels.

A Class D company runs MRP in the computer room, but operations people ignore the reports and, instead, run the company with the informal system.

Chapter 12 contains two separate checksheets that you can use to determine if you are Class A, B, C, or D, as well as a detailed description of various performance measurements for MRP II and M&CRP.

Implementation Types

The difficulty of implementing an MRP II system is directly proportional to the amount of formal and informal change required on the part of the organization. The amount of change is dependent on the type of implementation:

1. Initial (no prior MRP system), first time through. This is difficult, because it generally involves substantial change from the old way of doing business.
2. Initial (no prior MRP system), after one or more failures. This is the most difficult type of implementation, because people have already been "burned" before. Companies in this category will benefit from comparing their effort to the classic MRP II implementation described in this chapter, then organizing their new implementation effort with increased sensitivity to the *people* aspects.
3. Second or subsequent, after one or more successes. In some respects, this is the easiest, because the last implementation was a success. However, considerable organization change might be required to take advantage of the capabilities of the new system.

The terms *failure* and *success* in these categories are completely subjective; they are the *perceptions* of the people involved. For example, one company intended to be a Class A site, but only achieved a high Class C status after a year of struggle. The people viewed MRP II and their efforts as *failures*. Another company decided that they would be Class C after one year, and Class A a year later. They also achieved Class C after their first year of struggle, but viewed their efforts as *successes*, and went on to achieve Class A the second year.

Upgrading an MRP II system with a new release of the same software is not considered an MRP II implementation, unless there is significant change in the functionality that causes moderate or greater change in the organization.

Prerequisites for a Successful Implementation

The major prerequisites for a successful implementation of an MRP II system are summarized in this section and discussed in more detail later in this chapter.

Clear understanding of strategic goals

Supporting change in organizations

Total organization commitment

Data accuracy

Systems

Clear Understanding of Strategic Goals

As mentioned in Chapter 10, most people in organizations do not understand their company's strategic goals, or how their department impacts those goals. Likewise, when implementing an MRP II system, most people do not understand how the system will support the company's strategic goals or competitive edge. People tend to protect their own turf. Thus, when they make the myriad decisions required, both globally and at a detailed level, their decisions do not actually support the company's strategy. Instead, they help the department's performance. For example, a materials planner might deliberately set safety stocks higher than necessary to avoid the possibility of stock-outs, without realizing the true cost of the high safety stocks (that they not only increase cost, but create false priorities, and impede the effort to improve quality).

The best way to overcome this very common failing is to have communications both from the top down and from the bottom up, in which all parties actually listen to what the others are saying. While setting strategic goals is clearly the prerogative of top management, the rest of the organization must understand and agree with these strategies for the organization to be effective.

Supporting Change in Organizations

Implementing an MRP II system changes many social systems, as well as most functions in an organization. The successful company approaches MRP II as an exercise in the leadership of change. Successful change also involves an executive "champion," a person who stakes his or her personal reputation on the success of the new system.

The main goal of MRP II implementation must be to provide a foundation for manufacturing excellence, including:

Empowering people

Improving the competitive edge

Providing superlative customer service (creating fanatical customer loyalty)

Working more closely with suppliers

Therefore, *the most important aspect of an MRP II implementation is the leadership of top management, the Project Leader, and the Project Team.* These leaders must convey an attitude of adventure and enthusiasm, rewarding failures as part of the learning experience. Peters and Waterman popularized learning, as an attribute of excellent companies, in their classic, *In Search of Excellence.*[1]

McAteer has outlined several steps for a learning organization, one which can successfully incorporate new technologies (such as MRP II):[2]

1. Be purpose-centered, rather than leader-centered or personality-centered. This prevents the organization from achieving a major success, then losing it later. The purpose outlives any individual charisma.
2. Be future-focused. The implementation teams should have a sense of creating something with continuing value. Walt Disney has said, "If you can dream it you can do it."
3. Provide active management support for creativity and innovation. Trust permits people to focus on the future, encourages risk taking, and sup-

[1]Thomas Peters and Robert Waterman, *In Search of Excellence,* Harper & Row, New York, 1982.

[2]Peter F. McAteer, "The Learning Organization," *NETworking,* vol 4, no 2 Spring/Summer 1992, Massachusetts Chapter of the American Society for Training and Development.

ports cooperative relationships. As Albert Einstein said, "Imagination is more important than knowledge. For knowledge is limited, whereas imagination embraces the entire world."

4. Encourage self-leadership. Individual initiative and self-leadership are not contrary to a team environment. A team represents a group of individuals with a common sense of purpose, not identical ideas. Collective gains cannot emerge without individual curiosity that creates enthusiasm and commitment. Diversity of opinion, contrasting viewpoints, and creative insights are all critical to the team process.
5. Use team performance as the primary measure of long-term success and recognition. Most corporate reward systems are directed at individual performance and create competitive barriers that inhibit sharing information and the development of a shared sense of purpose. The team becomes the framework for an organization's imagination when groups of individuals can share both responsibilities and rewards.

Senge,[3] suggests that leadership involves several roles that have traditionally been neglected, including the leader as:

- Designer, creating the vision of the future (in concert with colleagues throughout the organization).
- Teacher-coach, assisting those throughout the organization to gain more insightful and empowering views of current reality (rather than in the authoritarian sense, in which a teacher knows the answer and grades students).
- Steward, compelled by the desire to serve others and a sense of commitment to the organization's larger mission (rather than a desire for power or material possession). Several failures have resulted from the Project Leader (or member of top management) subtly using the project to promote personal agendas, rather than the larger vision.

One very effective way to encourage people to change is to lavish education throughout the organization. Education is so important that we have created a separate section devoted to it later in this chapter.

Organization Commitment

A company that wishes to implement MRP II successfully must carefully obtain commitment to the process across all functions and levels of the or-

[3]Peter Senge, "The Leader's New Work: Building Learning Organizations," *Sloan Management Review,* vol 32, no 1, 1990.

ganization. That commitment must emanate from top management, which provides leadership as the Steering Committee. The Steering Committee should commit to specific goals, which should be communicated throughout the organization. The Steering Committee should commit to the change process, knowing that successful change *requires mistakes* and *failures* as the price of learning and developing the new structure.

People must understand how MRP II directly supports the company's objectives and strategies. Sometimes, an MRP II system is the wrong choice for a company. For example, TQM is a much better tool than MRP II to address quality issues. Additionally, an MRP II system must properly integrate with other techniques, such as CAE, JIT, or FMS.

Data Accuracy

As mundane and unexciting as data accuracy seems, it is absolutely required for an MRP II system to function properly. Data includes not only traditional data, such as on-hand inventories, but also the MRP planning parameters, general ledger account numbers, and so on. This requirement for data accuracy is one of an MRP II system's greater weaknesses. Humans, by contrast, tend to be much more forgiving and flexible with respect to data accuracy. However, if the first two prerequisites have not been properly addressed, the humans will more likely use any system inaccuracies as evidence that the system cannot work. They may passively allow the inaccuracies to continue (failure by consent), or deliberately introduce them (failure by sabotage). Either way, the MRP II system will not survive.

Systems

Even though *System* is intentionally last on this list, it is also a requirement, just like highways are a requirement for automotive travel. MRP II is impossible to compute by hand or with a calculator, not because of the complexity of the calculations, but because of the quantity of calculations required. Additionally, given the hundreds of MRP II software packages available on the market today, coding the MRP II system in-house is not a cost-effective or time-sensitive option.

Classic Implementation Steps

The following steps (illustrated on the Gantt chart in Figure 11-2) define the classic approach of truly successful Class A implementations. These steps include not only the activities normally regarded as implementation, but also those that form the preimplementation foundation, because the implementation can only be as solid as the initial foundation.

ID	Name	Duration
1	Organize Project	5d
2	Create Project Plan	5d
3	Educate — 1st Cut	5d
4	Assess Data Base Integrity	10d
5	Define Performance Measures	15d
6	Review Pre-Implement Steps	10d
7	Decide Implement Approach	5d
8	Pilot — Computer Room	20d
9	Educate Critical Mass	60d
10	Conference Room Pilot	80d
11	Live Pilot	20d
12	Implement in All Areas	40d
13	Improve Continuously	196d

Figure 11-2. MRP II implementation steps.

Preimplementation

The preimplementation activities create the foundation for implementation success. These activities assume that you have completed the system selection and justification process, as outlined in Chapter 10. Many companies give these activities only cursory attention. This is like starting to build a house by starting to assemble the concrete, lumber, and other materials, without having a detailed blueprint of the finished house, or obtaining the necessary permits. The work will apparently progress more rapidly at the start, but it cannot yield an excellent result, and has a much greater risk of failing completely. The preimplementation steps include:

Organize the implementation project

Create the project plan

Educate key individuals

Assess database integrity

Define performance measures

We discuss the first four of these in this section; the last is discussed in Chapter 12.

Organize the Project. The classic MRP II implementation project is organized with the following critical elements:

1. *Steering Committee,* composed of top executives plus the Project Leader and the Outside Counselor, who meet at least monthly (weekly during the critical phases), to:

- Provide the *strategic (long-term) vision,* including:
  Internal (virtually all departments and functions)
  Suppliers
  Customers
  Specific anticipated benefits
  Specific anticipated costs
- Intelligently *guide the process,* including *consistently* acting in ways that support the process
- Hold the project in *high priority* (the implementation should be the number two mission of everyone in the company, right behind keeping the customer satisfied)
- Provide *sufficient resources* to enable the implementation to continue on schedule
- Hold the project participants *accountable* for their commitments of results and time schedule; this means understanding that an MRP II system cannot be implemented *for* a company by outsiders; it *must* be done in-house *by* employees
- Approve and monitor implementation *budgets*
- *Resolve* companywide *issues* (e.g., changes in how departments will interact, changes in authority under the new system, allocation of critical path resources)

The Steering Committee must understand MRP II sufficiently to make intelligent decisions that will have far-reaching impact. In smaller companies, the Steering Committee and the Project Team might be the same committee.

2. *Project Team,* composed of one member of each key department (Design Engineering, Industrial Engineering, Materials Planning, Purchasing, Manufacturing, Marketing and Sales, Accounting, MIS, and Human Resources). The Project Team performs most of the implementation work. They meet weekly to:

- Discuss and decide how to use the system on a day-to-day basis
- Resolve issues
- Review progress against the schedule
- Review education and training efforts
- Address interdepartmental issues

The Project Team keeps a log of unresolved issues, and records of decisions and assumptions. Project Team members are responsible for communicating to and from their respective departments, and for leading the implementation within their departments. Thus, a Project Team member must be a person who enjoys learning new things and who is willing to take risks. Co-opting skeptics into the Project Team is a risky strategy; team members who are not fully on board can create problems and can actually sabotage an implementation.

3. *Project Leader* is the most important individual in the entire project. The Project Leader:

- Controls the project on a day-to-day basis
- Manages the project schedule
- Chairs project team meetings
- Communicates (endlessly, it seems) the myriad details required to:
  Keep the project coordinated throughout the entire organization
  Key customers
  Key suppliers
- Participates as a member of the Steering Committee

The Project Leader should be full-time, an experienced insider with an operations background (not MIS), possess excellent people skills, and be the best available person for the job (and probably the least available). The Project Leader should have already participated in a successful implementation. In reality, the Project Leader is the best compromise available. In smaller companies, the Project Leader is not full-time, because the company cannot afford such a critical resource. Alternatively, some companies deliberately hire a new person as the Project Leader because the lack of *political baggage* can outweigh the disadvantage of not knowing that company's products intimately.

4. *Outside Counselor,* who should be part-time, and who must have participated actively in several successful implementations. The Outside Counselor can be external to the organization (a consultant), or from another division or location of the company, but *must be* external to the day-to-day process. The Outside Counselor:

- Coaches the Project Leader
- Provides an objective viewpoint
- Helps create and sustain the unifying vision
- Helps the Project Leader maintain realistic expectations
- Helps the Project Leader and the Project Team avoid various pitfalls, both organizational and technical
- Participates in Steering Committee meetings
- Participates, as appropriate, in Project Team meetings

Companies on a second successful implementation might consider the Outside Counselor to be unnecessary; companies that have failed on a previous implementation attempt, or those that are starting their initial implementation, can usually benefit greatly by using one. They should bypass any potential Outside Counselor who insists on heavy involvement.

Create the Project Plan. The Project Team prepares a first-cut project plan, highlighting activities with responsibilities and due dates. This project plan forms the shell into which the Project Leader inserts additional detail later.

On the surface, implementing an MRP II system looks deceptively simple. One merely replaces the old inventory system with the new, orders materials using MRP instead of the old way, and so on. In reality, the changeover process is as complex a process as most companies undertake, rivaling a move into a new building in its scope and detail.

Many Project Managers turn to PC-based Project Management software. These tools assist with the manipulation of data, and are invaluable when the Project Leader is trying to define the critical resources (both human and material), because the better software contains resource-leveling capabilities. However, the software does not, and cannot, provide the management insight into the actual events; this is still the realm of experienced management.

For the first cut, the project plan should include only high-level items (limited to one page), as shown in the example in Figure 11-3.

The Project Leader then assigns responsibility for each major task to a key individual, asking that individual to create a realistic schedule for the task. Finally, the Project Leader integrates the task subschedules into the top-level schedule and manages accordingly.

On a routine basis (monthly, or more often), the Project Leader reviews the status. After several months of effort, most Project Managers cease using the formal project plan; they now understand intuitively which issues are critical, and they follow up on those issues regularly.

Educate Key Individuals (First Cut). The Steering Committee and the Project Team participate actively in an MRP II executive overview, to gain a fuller understanding of the potential and pitfalls. They then refine their objectives and review the projected costs and benefits with respect to company strategy and the business plan. This step is so vital that we have devoted an entire section to education and training later in this chapter.

Assess Database Integrity. If the required data is not available, or is not sufficiently accurate to support formal planning, the Project Team brings this matter to the Steering Committee's attention, together with a suggested plan of action. Assuming Steering Committee approval, they execute the plan. The team must address, at a minimum:

- Item Master accuracy, including both inventory quantities and planning parameters
- Bill of Materials accuracy
- Routing accuracy

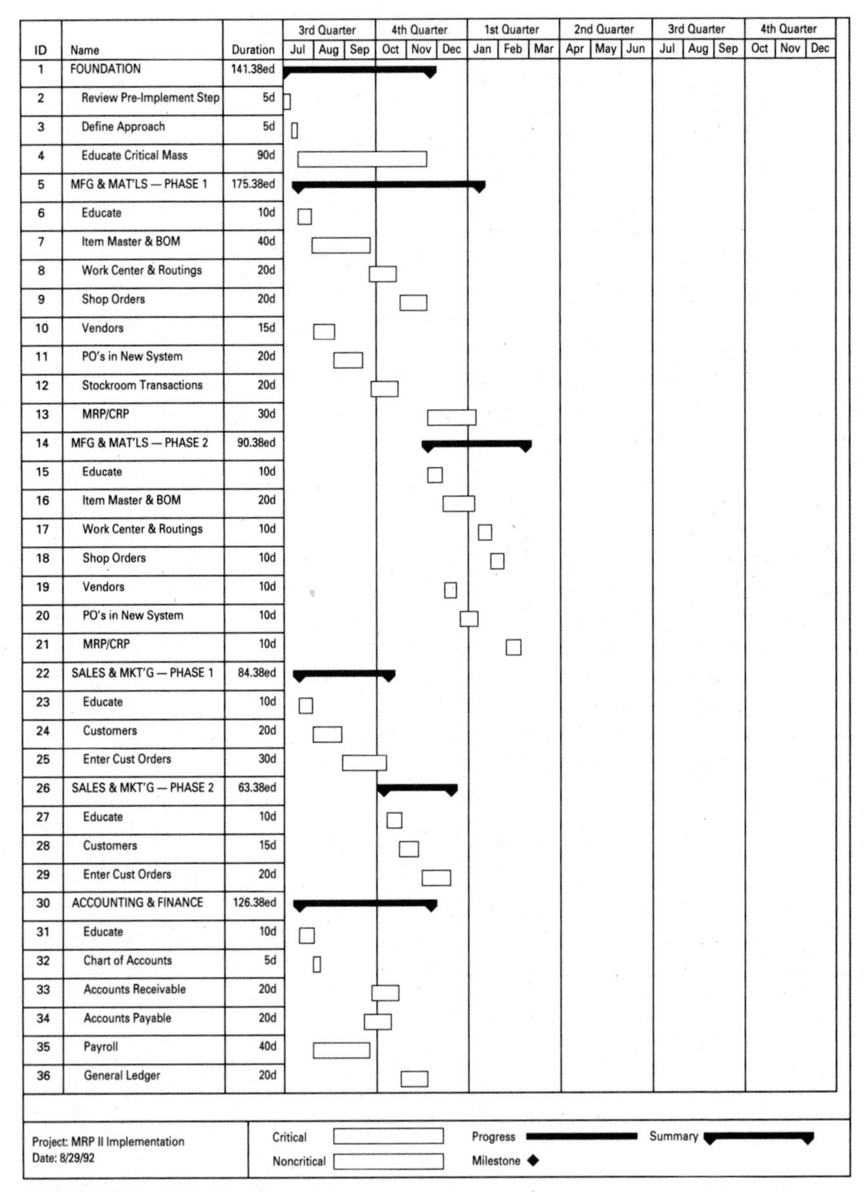

ID	Name	Duration
1	FOUNDATION	141.38ed
2	Review Pre-Implement Step	5d
3	Define Approach	5d
4	Educate Critical Mass	90d
5	MFG & MAT'LS — PHASE 1	175.38ed
6	Educate	10d
7	Item Master & BOM	40d
8	Work Center & Routings	20d
9	Shop Orders	20d
10	Vendors	15d
11	PO's in New System	20d
12	Stockroom Transactions	20d
13	MRP/CRP	30d
14	MFG & MAT'LS — PHASE 2	90.38ed
15	Educate	10d
16	Item Master & BOM	20d
17	Work Center & Routings	10d
18	Shop Orders	10d
19	Vendors	10d
20	PO's in New System	10d
21	MRP/CRP	10d
22	SALES & MKT'G — PHASE 1	84.38ed
23	Educate	10d
24	Customers	20d
25	Enter Cust Orders	30d
26	SALES & MKT'G — PHASE 2	63.38ed
27	Educate	10d
28	Customers	15d
29	Enter Cust Orders	20d
30	ACCOUNTING & FINANCE	126.38ed
31	Educate	10d
32	Chart of Accounts	5d
33	Accounts Receivable	20d
34	Accounts Payable	20d
35	Payroll	40d
36	General Ledger	20d

Figure 11-3. MRP II implementation project plan.

Implementation

Once the preimplementation steps are finished, the company is ready to surge ahead with the implementation. In this section, we discuss the key steps in implementing an MRP II system so that you can achieve the maximum payback:

Review all preimplementation steps

Educate the critical mass

Perform the computer room test

Decide on the implementation approach

Perform the conference room pilot

Perform the live pilot (the actual implementation)

Continually improve

Review all Preimplementation Steps. In the euphoria of finishing a software selection, most companies plan to forge ahead, assuming that all previous steps have been completed properly. Just like an airline pilot who performs each task on the takeoff checksheet, a company must review the progress on the initial steps at this time. Because those who have participated in the activities to this point probably have the normal pride of authorship and accomplishment, performing this kind of objective assessment almost requires an unbiased outsider. After reviewing each of these steps, the outside person should report to both the Project Team and the Steering Committee, helping them create plans to address any important issues.

Perform Computer Room Test. As soon as the software is installed on the computer, the MIS department, assisted by the Project Team, performs the computer room test. This constitutes an initial acceptance test of the software and hardware, and uses the supplier's demo database. The project team must continue to keep the written log of all questions and answers.

The computer room test involves much more than merely loading the software into the company's own computer. Its purpose is to ensure that the system will operate smoothly during the forthcoming Conference Room Pilot. To do this, the MIS department must:

- Ensure that the system operates properly with the company's software environment, including:

  Operating system
  Database and file structures
  Network management system

- Ensure that the system operates properly and compatibly with the company's hardware (both new and existing), including:

  PCs and CRTs
  Network and communications devices
  Printers

- Verify that menus function properly

- Verify that batch job streams execute properly
- Ensure that the test database (for the Conference Room Pilot) is large enough
- Establish sign-on and password security, if appropriate at this time
- Execute every program, screen query, and transaction to assure that each operates properly

None of these activities could be performed during the Preselection Pilot, because that pilot is usually performed on supplier-furnished hardware (frequently a PC), using the supplier's operating system, and so on.

Although the MIS Department is primarily responsible for the computer room test, the Project Leader and (probably the rest of the Project Team) should be involved, especially in trying every program and screen.

Educate the Critical Mass. Not long after the software and hardware arrive, the Project Team educates the critical mass of individuals. This critical mass is large enough to ensure that the company could run with the new system, even if no other persons were educated. This step might well require educating 75 to 80 percent of the salaried workforce, and familiarizing the hourly workforce with the concepts. For each category and function (e.g., order entry clerks, cost accountants, shop supervisors, and so on), the initial project plan must include a rudimentary education plan, which should show:

Number of people

Education modules required

Dates required (to support project plan)

Decide Implementation Approach. *Phased Implementation* is usually the superior approach. The alternative is "Cold Turkey," which we discuss later in this section. Phased MRP involves quickly implementing MRP in a specific section of the business to attain tangible, bottom line results (e.g., improvement of on-time shipments). Because the perception of success or failure of the first attempt at implementation often decides the fate of the entire project, the Project Team must select the first phase with care. We encourage companies to strongly consider this approach, because it minimizes the enthusiasm-sapping delay between the start of the project and the attainment of tangible results.

There are several approaches to phasing:

- By module (first the item masters, then the finished goods stockroom, then the BOM, then the inventory, then purchasing, then accounts payable . . .)
- By product line (first the large machines, then the small machines, then the large consumer products . . .)
- By plant (first Pittsburgh, then Des Moines, then Santa Clara . . .)

Phasing by plants is similar to phasing by product line, because different plants usually produce different products. Even in a plant or product line implementation, most companies additionally phase by module, with some overlap. We suggest that a company consider having multiple efforts, running simultaneously, such as Accounting and Finance, Materials and Manufacturing, and Sales and Marketing, to minimize the actual elapsed time for the first results. Whatever the approach, the implementation plan should emphasize early, tangible successes, which reward the bottom line. This is especially critical if a company:

Needs a financial fix immediately

Needs a quick payback for strategic reasons, perhaps new top management trying to prove itself

Is very large

Has a middle management that understands the need, but is unable to convince top management without tangible results

The disadvantages to phasing include:

- Total elapsed time required. If the total elapsed time for an MRP II implementation exceeds 18 months, there is serious risk that the project will never be completed because of personnel turnover and waning enthusiasm.
- Cost of bridging between the old system(s) and the new. One author used 7 years of programmer-analyst time creating and testing automated (electronic) bridges between the old system and the new, to support a one-year implementation.
- Confusion. When one department is starting to use the new system, but other departments are still on the old system.

An alternative to Phased Implementation is *Parallel Implementation*. Parallel implementation requires running the old system and the new system side by side, comparing the results at a summary level, and then investigating the detail to determine the causes of any discrepancies. Although it is useful for accounting package implementations (e.g., a Payroll department will run the checks on the old and new systems, using the same input, then compare totals and detailed fields), this approach presents major problems for MRP II implementations:

- It assumes that the new system should provide results identical to the old system. However, if the old system is an EOQ system, the objective of the new MRP II system is to provide *different* answers across the board. If the answers are going to be the same, what is the benefit of the change?
- It is extremely detailed and time-consuming. This approach requires:
  Entering all data twice, once into the old system and once into the new
  Investigating every discrepancy to learn the fundamental cause

Most organizations do not have sufficient staff to handle a double workload. The staff spends the bulk of their time on the old system, because that is how they operate the business, thus badly short-changing the new system.

Perform Conference Room Pilot. The Project Team directs the Conference Room Pilot, which runs all the software for a specific phase in the conference room, creating small databases to use for education and training. These databases consist of small representative amounts of real data. (This data should be carefully constructed so that it will challenge people to use and understand the entire system.) As the Conference Room Pilot progresses, and participants are more comfortable with the functioning of the software, a designated person intentionally stresses the participants with "realistic" bad news, such as a late supplier shipment, or a machine down in the plant, or a rush customer order for which there are insufficient parts in stock. The participants then use the system to replan, based on the new information. In this way, the participants prepare in advance for many of the problems they will actually encounter, gaining confidence in the system and their abilities in the process.

The creators of test data need to create formal test plans, which create data specifically to test various conditions, recording the data as well as the expected and the actual results. These plans address, at a minimum, the following characteristics:

- System functionality, for all major functions. This should be tested fully to ensure that good data functions as expected (alpha testing), and that bad data is intercepted before it can affect any other data (beta testing). Good test designs focus on the point of logic decisions. For example, to test the way MRP reorders based on Safety Stock, you can create three situations: (1) stock-on-hand is one unit greater than Safety Stock, (2) equal to Safety Stock, and (3) one unit less than Safety Stock.

- Volume testing, for speed and the ability to handle routine data volumes, then very high and/or very low data volumes. One author experienced an MPS run time approaching 72 hours during a volume test two months before scheduled implementation. The company postponed implementing MPS until the run times were acceptable.

This pilot tests and confirms the fit of the software to the business, including:

Any software modifications

New user procedures

Participant training

Acquisition of new data

Conversion of existing data from the old system

The detailed plan for the live pilot

Again, the Project Team records all questions and answers. Before the Project Team can recommend that the company proceed to use the software in a "live" pilot, they must review all questions to ensure that the answers are still valid.

Comparison of Preselection Pilot, Computer Room Test, and Postselection Conference Room Pilot

The Preselection Pilot, Computer Room Test, and Postselection Conference Room Pilots are important steps in current MRP II selection and implementation procedures. To ensure that you thoroughly understand these three steps, we will summarize and compare them in this section.

The Preselection Pilot is used to select and support procurement of one specific MRP II package from among several candidates by

- Discriminating between multiple MRP II software packages
- Testing supplier's capability claims
- Quantifying cost and benefit estimates
- Determining how well the package satisfies the needs of firm

The Computer Room Test is used to ensure that the selected system functions correctly from a technical standpoint. This includes:

- Ensuring compatibility with the computer's operating system, database, and network management system
- Ensuring compatibility with existing hardware:
  Installed PCs and CRTs
  Installed printers
- Checking job stream execution
- Performing initial database sizing
- Installing security sign-ons and passwords

The Postselection Conference Room Pilot is used to predict and control the impact of implementing a selected MRP II package in a "live pilot" (defined in the next section) by

- Checking detailed software features
- Defining software change requirements
- Predicting organizational and procedural changes
- Developing detailed implementation plans
- Developing data and conducting training for the live pilot

To better understand the differences between these three pilots, they are compared in Figure 11-4.

Live Pilot. This is the actual implementation. The Project Team directs, hands-on, the first use of the new system in a live environment. They monitor the system carefully, tuning and adjusting as appropriate.

A company should select a product line, or group of items, that conforms as well as possible with the following guidelines:

Small to moderate size (200 to 600 items)

Self-contained, with little overlap to nonpilot items

Good cross-section of the company's functions (marketing and sales, engineering, manufacturing, materials, accounting and finance)

The Project Team should decide, preferably unanimously, that the company is ready to try a Live Pilot. Their recommendation should be approved by the Steering Committee.

The perception of success of the first Live Pilot is absolutely important to the willingness of people throughout the company to trust the system. This is where the "rubber hits the road." It is better to delay two weeks, or a month, than to start a Live Pilot with obvious problems. The Project Team must ensure that:

Criteria	Preselection	Computer room	Operation Postselection
Objective	Select software	MIS Technical shakedown	Develop live pilot implementation plan
Orientation	Requirements checklist	Technical	Preparation for live pilot
Length	1 to 2 months	1 to 2 weeks	2 to 4 months
Participants	Key users and MIS personnel	MIS and Project Leader	Project Team and users reps
Data model	Supplied by software company	Supplied by software company	Developed by Project Team
Scenario develop-ment	Software company and users	MIS and Project Leader	Project Team and users
Scenario execution	Software company	MIS and Project Team	Project Team

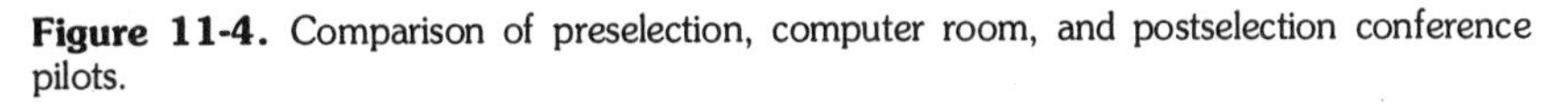

Figure 11-4. Comparison of preselection, computer room, and postselection conference pilots.

- All the demo data has been removed from the live database
- Accurate data is cut over and available to the Live Pilot
- Security is working properly
- All required new forms have been distributed
- (Most important) the people who are involved are committed to succeed

If the first Live Pilot is perceived by the rest of the company as a failure, for whatever reason, you will have a much more difficult time bringing up the system. It is the perception of success or failure that is the key. Given this knowledge, you can set expectations appropriately small at the beginning.

After successfully utilizing the system on the Live Pilot, the Project Team reviews and modifies the schedule for the next Live Pilot, until the company is completely running on the new system. At this point, *the new system is operational.*

As an alternative to the Live Pilot, the *Cold Turkey* approach can also be described as the "bet the farm on one roll of the dice" approach. If a company succeeds, it wins. If not, it could well suffer some serious setbacks. If, as part of the Cold Turkey approach, the Project Team decides to burn their bridges to the old system, as a way of increasing the motivation to use the new system, the penalty for failure may be bankruptcy. The saying "A drowning man does not drown due to lack of motivation" seems particularly applicable here. One author visited a company that had tried the Cold Turkey approach after other approaches had failed. For two months prior to the author's visit, the people in the plant had literally played cards on company time, because nobody knew what they should be making or where any material was!

The Cold Turkey approach involves testing the system extensively in the computer room, training the users as well as possible, then turning on all the modules for the entire company at once, and fixing the myriad little problems as quickly as possible. Although we strongly recommend against this approach for larger companies, we have also noted that the actual cost of keeping two dissimilar systems alive with bridges (either double data entry or a massive design and programming effort for automated bridges) can be very high.

For a smaller company, especially one coming from a manual or simple system, the Cold Turkey approach might make the most sense. In this instance, if the new system did not work, the people could resurrect the old system by feeding it the data that had been transacted in the interim. At a very minimum, a company should keep all transactions that have been fed into the new system for the first week or two, in case they want to revert back to the old system.

Continuous Improvement. Most companies in the United States view an MRP II implementation as a one-time project, somewhat like a sporting event. After the system is implemented, the score is finalized, the game is over, and everyone goes home. This attitude prevents companies from achieving the true benefits of MRP II.

Instead, after the initial data implementation, people can use MRP II to continuously improve their operation. They should design exception reports that illuminate opportunities for improvement. For example, most MRP systems have a report that shows slow-moving or obsolete material. During the first few months of implementation, nobody will probably even look at the report. Then, someone will run the report, defining *slow-moving* as any part that has inventory turns less than 2.0. After they have taken action with those parts, they can rerun the report to show all parts with inventory turns less than 3.0. Another common metric is lead time (both purchased and manufactured), which can be reduced, reduced again, reduced some more, and still reduced some more.... Just as a cycle counter investigates *why* an inventory record became incorrect, a planner can use the MRP system to address the causes of slow-moving items or long lead times. An MRP II system is but one tool, although an excellent one, in the never-ending struggle for competitive superiority.

Education and Training

"If you think education is expensive, try ignorance." Successful education and training share a number of characteristics:

Line accountability for education (rather than for Human Resources or MIS)

Total coverage throughout the company

Continuing reinforcement

Instructor credibility

Peer confirmation

Enthusiasm of all participants

Supportive environment that encourages risks

Nonthreatening environment that encourages removal of departmental barriers

Supreme respect for the dignity and worth of all participants

Education

One key difference between Class A and Class D companies is education: Class A companies spend twice as much on education as do Class D companies[4] (otherwise, the expenditures were essentially the same). For that additional investment in education, the Class A companies achieved an average of 200 percent return on investment, whereas Class D companies achieved between 25 and 50 percent ROI. From a purely financial viewpoint, this additional investment in education appears to ensure the best possible return on investment for the entire project.

Why? Because education is an integral part of the change process. The best education encourages people to start thinking in a *company-wide* mode, rather than viewing their entire world as being bounded by their department. Education, supported by proper leadership, encourages people to take risks, and to fail boldly, in the pursuit of learning and development.

In addition to the aforementioned common characteristics, education should include the following:

Active top management leadership and participation

Total immersion for key people

[4]The Oliver Wight Companies, 1985 Newsletter, Essex Junction, VT.

Education starts early, with the first-cut of MRP II overview for executives and the Project Team. This provides the foundation and the forum for them to set realistic expectations.

Education continues with the total immersion of the Project Team and other key individuals, so they can more fully comprehend the scope and complexity of the project, and learn the interrelationships between functions and modules. In addition, virtually every employee should participate in an MRP II overview prior to the start of their individual module training. Each new employee must also attend MRP II education classes, or view purchased video tapes, to more fully understand the *why* of the company's operating style and values.

Training

Training follows education, when people are ready to go "hands-on" on the system in a test mode, or when new employees have completed their MRP II overview. Training is module and software specific. It teaches an inventory clerk how to access and use the transactions that apply to the inventory function, including receiving, issuing, moving, and adjusting items. Effective training includes the following characteristics:

- It is heavily oriented toward hands-on exposure to the system
- It teaches participants how to use the documentation to answer their current and future questions
- It provides total immersion for all people who will use the system

Each company that attempts to implement MRP II depends totally on its people to make intelligent decisions, based on the information in the new system. The classic style of On-the-Job Training (OJT) (throw them in with the sharks; they'll learn to swim just fine), virtually guarantees that the company's entire investment will be wasted.

Why Most MRP II Systems "Fail"

Professional publications and conferences burgeon with articles on "How I Implemented MRP II Successfully: A Case Study," or "The Ten Sure-Fire Steps to Guaranteed MRP II Implementation Success." The continued popularity of the topic suggests that we, as practitioners, still have a major problem.

In spite of the articles and speeches, over 80 percent of the MRP II systems fail to reach, or even approach, their full potential payback for their companies. A 1985 survey of more than 1000 manufacturers by the Oliver Wight companies indicates that the companies in Classes A through D had spent approximately the same amount on their systems. The Class A companies achieved a weighted average annual return on investment of 200 percent! Yet only 8 percent of those surveyed could consider themselves to be Class A. The "failures," the Class D companies, achieved an average weighted ROI of 30 percent.[5] A 30 percent ROI is good enough for a project to be approved in most companies; however, if a company has a choice between achieving 200 percent or 30 percent on the same investment, they will choose the higher return without any hesitation.

Why, then, do most MRP II systems implementations fail to deliver their potential payback? The answers seem to fall into five major categories:

1. People don't want the new system to succeed
2. People have unrealistic expectations
3. People don't understand the basic concepts of the system
4. The basic data is inaccurate
5. The system has technical difficulties

People Don't Want the New System to Succeed

Why don't people at various levels in the organization really want to make the new system work? Let us examine the following interrelated reasons:

- Change requires a transition
- Sometimes the new system makes the participant's job more difficult
- The new system often reduces the social importance of participants
- The new system might cost participants their jobs
- People are afraid to fail
- Formal MRP II systems require much greater precision
- MRP II systems lack flexibility in key areas
- Top management lacks commitment
- Management does not understand the importance of the new system

[5]Ibid.

Change Requires a Transition. This means that an inner human reorientation is required to make the external change happen. As described by Bill Bridges,[6] it involves:

- Letting go of the old (an *endings* phase)
- A time of uncertainty, during which neither the old way nor the new way function very well (a *neutral zone* phase)
- Embracing the new way and starting to utilize it effectively (a *new beginnings* phase)

Education is one way of helping people to let go of the old, to create a new vision, and to want to embrace the new. One very large company that failed in its MRP II implementation did not educate one person for the first two-and-one-half years after implementation!

Companies that are sensitive to the transition process also help the participants by having a *closing* ceremony to end the old process. The closing ceremony is important, because the ending of any process, however pleasant or unpleasant it may have been, inevitably carries some measure of grief. In fact, some people are not eager to change *because* they have become known as the people who could survive in a hostile setting; therefore, the change dramatically reduces the social rewards they receive from their friends, co-workers, and families. The closing ceremony might be burning the old manual inventory cards in a bonfire, or dismantling the old computer and giving pieces to each participant.

This time of uncertainty or chaos, the *neutral zone* in Bridges' terms, is generally uncomfortable for all concerned. Our culture provides poorly, if at all, for people and organizations in the neutral zone. We expect people and organizations to be at maximum efficiency and effectiveness at all times, just as we expect our cars to always start. The neutral zone is the time of *not* knowing the best way, and provides the "safe space" for:

Exploring new options

Analyzing new and different ways of doing things

Making mistakes (learning the results when we try things)

The neutral zone requires chaos and confusion. But *it is only in the neutral zone, through facing uncertainties and attempting new things, that the participants can create their new vision and lay the foundation for new success.* Before the neutral zone can be effective, the participants must have the courage to leave

[6]William Bridges, Ph.D., *Surviving Corporate Transition,* 1990 William Bridges & Associates, Chap 3, 4, and 5.

the old, familiar structure. As long as participants cling to the old way of doing things, they cannot make much progress toward a truly new structure. Organizations that promote growth, learning, and an entrepreneurial spirit wisely nurture and support these times of uncertainty, knowing that this is the most vulnerable part of the change process. This neutral zone can be compared to a crab that has outgrown, and finally split and discarded its old shell. While the new, larger shell is hardening to protect the creature from its enemies, the crab must take shelter in a safe place. For people undergoing the change, the time spent in the neutral zone seems to move slowly, and they often wonder if they will ever find a way out of this chaotic period.

Finally, a new possibility begins to coalesce. Sometimes this happens just like the dawn of a new day, ever so slowly and unobtrusively, so that the person in the neutral zone is not even aware of the change. Sometimes it hits like a thunderbolt. (Interestingly, the method of arrival of the new possibility has no relationship to its eventual success or "failure.") This new possibility forms the basis of a new detailed understanding of "how I will do my job, and how I and my new function will interact with the rest of the people in the company." The confidence, then the competence returns.

A second paradigm of change is illustrated below:

	Unconscious	Conscious
Incompetent	1	2
Competent	4	3

In this understanding of change, a person begins as an Unconscious Incompetent—they don't know what they don't know. This quadrant applies to the initial stages of MRP II implementation, where people do not yet understand the impact of MRP II on their daily lives.

As people begin to realize what they don't know, they become Conscious Incompetents. They begin to demand training, assistance, a vision, an explanation of how things can be better. They want to define their new functions. This could correspond roughly to the ending phase and the neutral phase.

In the third phase, Conscious Competent, people can now perform their functions with the new system, if they pay close attention. This occurs in the late stage of testing (e.g., in the Conference Room Pilot), and in the early stage of actual use. This could correspond roughly to the new beginnings phase.

In the fourth and final phase, Unconscious Competent, people have now become totally accustomed to the system, and they use it effectively and easily as part of their daily tasks.

Sometimes, the New System Actually Makes a Participant's Job More Difficult. When one first starts to use a new tool, the level of job difficulty increases and efficiency declines. However, in MRP II systems, the increased difficulty may last forever. For example, an MRP II system requires formal, accurate inventory records. In many systems, this requires the stock clerk, and all other individuals, to record *each and every* movement of stock across the stockroom threshold. This is much more difficult, frustrating, and time-consuming than the old informal, inaccurate system. Why, then, should a stock clerk support a new system that will increase the stress and frustration level?

The New System Often Reduces Social Importance of Both Clerks and Managers. It is no longer necessary to rely on a manager or clerk to answer questions, because the system can now provide the answers. For example, a production control clerk, who has been the "person who knows what's happening on the factory floor," will lose the mantle of being the expert. The importance of social standing and interaction has been well demonstrated for decades, starting with the Tavistock studies of miners whose work groups were broken up when they were mechanized. Groves[7] lists the 10 factors people want most in a job today:

1. Work with people who treat me with respect
2. Interesting work
3. Recognition for good work
4. Chance to develop skills
5. Working for people who listen to how to do things better
6. Think for myself rather than just carry out instructions
7. See the end results of my work
8. Work for efficient managers
9. A job that is not too easy
10. Feel well informed about what is going on

Job security, high pay, and good benefits *followed* these first ten factors. MRP II implementations that ignore these important factors, especially items 1, 3, 5, and 6, are likely to face stiff resistance on the part of the participants. MRP II effectively rewrites the social contract between the company and the participants, because it dramatically changes the social environment in which the participants receive their rewards. In the worst cases, MRP II re-

[7]Brenton R. Groves, "The Missing Element In MRP—People," *Production and Inventory Management Journal,* vol 31, no 4, 1990, pp. 60–64.

places rich and rewarding social interaction with lonely terminal usage, eliminating much face-to-face contact.

The New System Might Cost Participants Their Jobs. A manager or a clerk who has been a conduit of information up and down the organization might fear, with ample reason, that the new system will eliminate their old job. Computerization has probably been the single most important enabling technology in the downsizing of American white collar ranks in the late 1980s and early 1990s. Automation reduced the ranks of agricultural workers around the turn of the century, then reduced the requirements for direct labor in factories, and is now reducing the requirements for humans to work as information gatherers and disseminators. One of the methods of MRP II system justification is reduced head count (this is discussed in greater detail in Chapter 10 and Appendix C).

Yet, some companies still insist that a person wholeheartedly support a new system that will eventually eliminate that person's job, and are actually surprised when the cooperation is tepid at best.

People Are Afraid to Fail. Learning a new function requires that we risk "failing." "Failure" can occur in at least three dimensions:

- Technical (for example, locking the computer up, not being able to enter transactions accurately, deleting the wrong record, and never being able to get it back)
- Social (for example, looking foolish to one's peers)
- Hierarchical and/or performance (for example, not being able to supply information to the boss, saying something about the new system that makes the boss look foolish)

Until people learn to embrace "failing" as the price of learning and growth, this fear of failure can cause people to resist the new system, no matter how beneficial the system may eventually be for them and/or the company. The "failing" occurs in the neutral zone and in the new beginnings zone, as participants search for a new way and explore options. When people first learn to ride a bicycle, they must have the courage to leave behind their safer (and faster) tricycle. If peers laugh at them as they wobble on the new bike, they might return to the tricycle.

Western industry lacks tolerance of "failure"; our cultural norm only celebrates "successes." Contrast this to Nissan, USA, whose managers say, "We turn mistakes into learning experiences."[8]

[8]Murray Sayle, "Worker/Manager Program Boosts Productivity at Nissan," *Modern Materials Handling*, March 1985.

Education and hands-on training are keys to reducing the fear of failure, as long as management and the corporate culture are willing to accept failure as the *only* pathway to learning and change.

The Shift from Informal to Formal Systems Requires Much Greater Precision. "I'll ship this week," is no longer adequate. MRP II systems require precision to the day, and may require precision to the hour in the future. Informal systems allow customer orders to be taken with an ASAP date, and shipped within two days, because the shipping clerk knows what the customer really means when they say ASAP. Now, actual dates must be entered.

The MRP II System Lacks Flexibility in Key Areas. MRP acts as if the sky has fallen in when somebody actually uses one unit out of Safety Stock; it plans a replenishment order due TODAY! IMMEDIATELY! RUSH! HOT! By contrast, a human understands that we really don't need the order for a week or two; that's what the Safety Stock is for. In Bills of Material, MRP II systems are totally precise, to the 5th decimal. However, in the real world some recipes in chemical processes are much more variable, depending on technical specifications and qualities of the individual components.

Humans can chuckle among themselves about the limitations of the MRP II system; alternatively, they can use this lack of flexibility as a reason, or method, to kill the new system.

Top Management Lacks Commitment. For years, consultants have been exhorting, "Top management must be committed (to the project)." More than one president has asked, "What does that mean?" We discussed the requirements of top management, acting in the form of a Steering Committee, earlier in this chapter.

When top management does not understand the breadth, depth, and requirements of the changes that must be made in their organization in order to utilize an MRP II system effectively, the implementation process has a low probability of succeeding.

Management Does Not Understand the Importance of the New System. This frequently occurs in one or more of the following forms:

- "We know how to run this business, or we wouldn't be in charge." Top and/or middle management are indeed in charge. However, the statement assumes that continuing past practices will ensure future success. Most companies agree that their competition continues to improve, both domestically and internationally. History is littered with companies that grew complacent and then paid the ultimate price for their unwillingness to change.

- "If it ain't broke, don't fix it." The companies that will survive long term are those that demand continual improvement of every function.
- "We're making good money." Companies that have been relatively profitable have, by far, the greatest difficulty in making changes. They are, therefore, courting the greatest long-term risk.
- "It's up to me to protect my boss from making a real mistake." In this case, middle management will do whatever it takes, including sabotaging the new system, to ensure that the boss is protected from making a mistake.

Top management must understand how the MRP II system will help the company achieve its strategic goals. One of the most prevalent causes of failure of MRP II systems is the lack of overall company objectives and strategic goals.

Finally, all persons must understand how MRP II fits with other techniques that are currently in use or are being planned for future use. This sometimes requires considerable effort. Pronouncements of good intentions from the boardroom do not suffice here; a systems design person or committee must investigate the details.

People Have Unrealistic Expectations

Unrealistic expectations are a second major factor in a failure. The five most common unrealistic expectations are:

Overblown promises

"It's an MIS project"

"Aircraft carrier" syndrome

More work than anticipated

Poor implementation project management

Overblown Promises. In an effort to make a sale, a supplier (software and/or hardware) may inflate expectations of results, and/or understate the effort required to successfully utilize the system. Alternatively, the "internal champion," for whatever the reason, might also inflate expectations. The greatest danger occurs when top management actually believes that a system can be implemented in three months. This attitude might be justified when replacing the fleet of company cars ("If our salespeople can drive a Ford, they can drive a Chevy"). However, MRP II systems require tremendous investment in the organization, so that each person can create his or her new role and learn how to use this new technology effectively.

Even more sad are the cases in which the Project Leader is guilty of grossly underestimating the effort required for a successful implementation. One of the most important functions of a Project Leader is to set realistic expectations, not only for top management, but for the rest of the Project Team and for the entire division or company. The Project Leader must understand and communicate, early on, the complexities that lie ahead. The Project Leader must view the MRP II implementation as a golden opportunity for the organization to become much more skilled at learning, implementing, and changing, all of which are required for long-term survival.

Unrealistic expectations lead to overblown promises that the new system will actually be on line to support operations on a *fixed date* (e.g., fiscal year close, summer plant shutdown, expiration of the lease on the old computer, the move into the new building, and so on). Although management might fervently desire that the new system will run the business on a specific date, even for excellent business reasons, the new system might not be sufficiently ready to do so. Forcing the new system to start on the externally imposed date can cause a completely failed implementation, accompanied by extremely high costs.

"It's an MIS Project." If the rest of the company views MRP II as an MIS project, failure is virtually assured. According to one survey,[9] 30 percent of the companies that have installed MRP II are running it only in their computer rooms (Class D). *All* persons in a company, from the executive staff through the office, the plant, and the field, must realize that MRP II will change how they perform their daily functions. The most successful companies treat the computer as merely a necessary evil that supports their new way of doing business. These most successful companies view MRP II as a completely fresh break from their old procedures, attitudes, and communications inefficiencies.

"Aircraft Carrier" Syndrome. In complex, difficult projects, such as implementing MRP II systems, the actual results do not follow the management anticipated norm, as illustrated in Figure 11-5*a*. Instead, they look more like the profile of a jet catapulting off an aircraft carrier, as shown in

[9]Vinnie Chopra, "How to Develop an Effective MRP II Implementation Plan," *1987 Conference Proceedings*, APICS, Falls Church, VA, 1987.

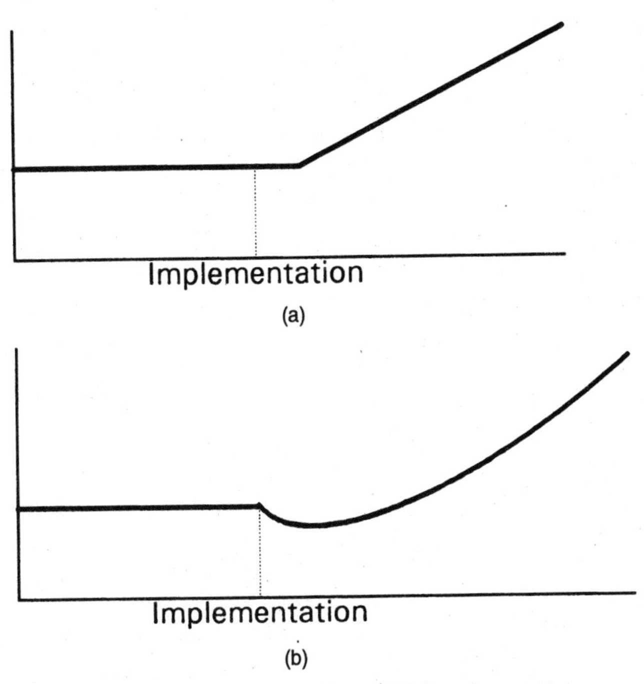

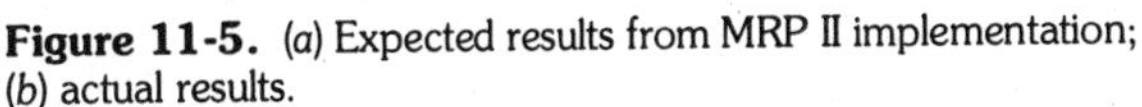

Figure 11-5. (*a*) Expected results from MRP II implementation; (*b*) actual results.

Figure 11-5*b*.[10] This profile is caused by the extreme complexity of the project, in which decisions, actions, and assumptions at any place in the organization can dramatically affect other areas in sometimes unexpected ways. This *confusion factor* increases operating cost, while delaying the payback, causing the production curve to head downward for a short while after implementation. It can be minimized by using some of the implementation techniques discussed later in this chapter.

If management does not understand and expect this phenomenon *in advance,* they can become strongly tempted to cancel the project during this stage. During the critical early stages of implementation, management must also direct the project closely enough to ensure that the production curve, once the new system is turned on, does not continue downward for more than a short period of time. Just as an extended downward trend will put a jet in the water, so will an extended downward trend in production close a business.

[10]W. Bruce Chew, Dorothy Leonard-Barton, and Roger E. Bohn, "Beating Murphy's Law," *Sloan Management Review,* vol 32, no 3, 1991.

More Work Than Anticipated. *Implementing an MRP II system is an exhausting and people-intensive effort* that will affect virtually every department in a company, and most people in those departments.

In complex projects, including major computer systems development projects, overruns of 100 percent are rather common. Most first-time MRP II implementors are amazed at the number of hours required to:

Learn the intricacies of the new system

Develop new procedures

Locate and enter all the required data

Train all appropriate personnel

As the perception of work required increases, the project can at times seem insurmountable, stretching a company's resources (key personnel, budgets, and deadlines) to their limits, and even beyond. It is at these times that the Outside Counselor can refresh sagging spirits, by showing the team how far they have come and objectively assessing how much farther they have to go.

From an operations viewpoint, the implementation of an MRP II system can require a substantial increase in the volume of clerical data entry. Before a company implements MRP II for the first time, its stockroom transactions are usually used only by accountants to value inventories monthly, and shop floor reporting is informal at best. Once the system is implemented, the increase in the number of transactions per day that are required for formal inventory control in the stockroom and on the floor can far exceed the wildest expectations of many people. Management should include this *transaction burden* in the up-front expectations of operations cost.

Poor Implementation Project Management. Most companies, as they start an MRP II implementation, grossly underestimate the eventual scope, size, and difficulty of the project. Thus, they do not initiate detailed Project Management techniques on day one. Only when they realize that the project is out of control do most companies start managing the project properly. Sometimes, that realization comes too late; the project is too disorganized, and people have given up on it.

People Do Not Understand the Basic Concepts of the System

Although M&CRP and MRP II may be intuitive to some people who have been expediting shortages for years in industries with many levels of assemblies and fabrications, people in other industries, who lack that kind of hands-on experience, have greater difficulty in understanding MRP II.

When MRP II is introduced to those companies, the people need to first learn and understand the basic concepts of MRP II before they can use it effectively. This need seems most apparent in the process industries and the true job shops, which were discussed in Chapter 9. Job shops, especially, can have difficulty in converting their thinking from customer orders to parts.

In these situations, management needs to fully appreciate the depth and breadth of learning and change required, and fund sufficient education, training, and pilot testing time, so that the decision makers can become proficient on the new system before they use it to run the company.

The Basic Data Is Inaccurate

Data inaccuracies, the fourth major reason for MRP II failure, reflect the saying, "garbage in, garbage out." Because MRP II outputs need to be accurate to be believed, the data on which the MRP II programs operate must be accurate. More than one company has assumed its data foundation to be accurate as they created their implementation schedule, only to discover that the bills and routings in Engineering did not reflect the way their products are actually made. Reviewing and correcting bills and routings can delay an implementation for six months to a year or more, depending upon the availability of engineers to verify each and every record. If the company has created expectations about implementation dates, this delay can cause critical problems.

Even worse, some companies with admittedly incorrect data forge ahead under the assumption that they will correct the wrong data when they see them. This is dangerous at best, because MRP II uses these wrong bills and routings to create plans for the entire company. Under these conditions, MRP II will quickly lose any credibility it might have had. People will ignore it, and continue to run the company, instead, with the old system. Chapter 12 discusses performance measures that can indicate the basic level of data accuracy. Chapter 8 discusses the requirements of data accuracy in more detail.

Using Old Parameters to Drive the New Software. These can prevent M&CRP from providing the desired improvements. The old parameters include such items as:

Lead time in weeks, rather than days

Putting EOQ lot sizes into MRP

Putting old EOQ-based Safety Stocks into MRP

Using contract lead times for MRP lead times

Adding safety factors to all MRP lead times

Building standard scrap allowances into the Bills of Materials

The System Has Technical Difficulties

Technical difficulties can also cause an implementation to fail, although they are not usually the primary cause. These technical difficulties are discussed in the following paragraphs.

Operations Problems with the Software ("Bugs"). These surprises, also known as "undocumented enhancements" range from annoying to devastating. They are most common in home-grown software; most commercially available software has been tested thoroughly before release. They still occur with surprising frequency in custom modifications to commercial software, however, and sometimes in new releases of commercial software. Many organizations resist being the first to try new computer software for this reason.

Other Operations Problems. Problems pertaining to run times and capabilities do occur. In some older software, on-line functions had to be totally turned off before MRP or other inventory-based batch programs could be run. Some packages still in popular use today permit only one terminal to be in use at a time in each major department. Although these limitations seem to be noncritical during the initial implementation cycle, they can severely hamper the effective utilization of the system on a daily basis.

Interfaces to Other Packages. Interfacing with packages from more than one supplier can cause problems. (For example, MRP from one supplier, Distribution Requirements Planning from another supplier, Shop Floor Control and Finite Scheduling from a third, financials from a fourth, and homegrown order entry). Each interface needs to take into account the design assumptions of each software package. The finite scheduling package, MRP, and DRP might each contain an Item Master file, with each package assuming that it, and it alone, is responsible for the maintenance of that data. Similarly, the separate order entry and accounts receivable systems might each assume that it, and it alone, can add a new customer to the database.

Hardware Difficulties. Although hardware has generally become boringly reliable, hardware problems, especially intermittent ones, can cause considerable difficulty. If a company is system-dependent, when the system becomes unavailable for whatever reason, the company might be forced to send all the employees home for the rest of the day, losing production and perhaps even sales. Most hardware problems currently arise from communications, whether local (inside the same building), or remote (across public phone lines). Because of the relative unreliability of communications, we encourage companies to run the MRP II systems on local computers. This is especially true when the host computer and remote office are multiple time zones apart. When San Francisco has a problem at 3 p.m. local time, it is 6 p.m. at the host computer site in Boston, and the people have already gone home.

Implementing After a Previous "Failure"

If an organization perceives that one or more previous attempts to implement MRP II have failed, the new implementation is substantially more prone to suffer the same fate. The *perception* of failure is much more important than any measurable facts. Many "failures" return 25 percent or better on their investment, and would therefore be considered a success by an outsider. However, the participants may well view the implementation as a failure, because their expectations were not met.

Before proceeding with a second implementation attempt, the Project Team must review the first attempt and answer the following questions:

- What did each person expect?
  Functions
  Features
  Timing
  Effort required
  Results
  Benefits
- Why do people term the implementation effort a "failure"?
- What caused the "failure"?
- What needs to happen, in detail, for people to perceive that the new implementation has succeeded?
- What happened to the people who were associated with the previous effort? (If the participants were fired, demoted, or publicly humiliated, it will be much more difficult for the survivors to support a second try.)
- Have any major changes occurred since the last implementation, to either ameliorate or exacerbate the reasons for failure?

- Why does the company want to try again? Are the reasons valid? Have the reasons for the past failure been examined sufficiently for the Project Team to be able to avoid that same result?

The Project Team should then define the steps that are needed to achieve a successful implementation, using the classic implementation steps (discussed earlier in this chapter) as a guide.

The first results of an initial implementation are critical to the eventual success of that implementation, because people will judge the success of the entire project by that first activity. This is even more true on a second implementation attempt following a "failure." Thus, the Project Team and Steering Committee must choose their first target very carefully, and must make absolutely certain that this activity succeeds in the eyes of the rest of the organization. In some respects, this is not as difficult as it sounds, because the Project Team can set expectations ahead of time. In other words, the Project Team can define what success is, making sure that they can achieve it before they define it to the rest of the world.

Summary

Most MRP II systems deliver only a small fraction of their potential payback, because the organizations do not understand that the essence of a successful implementation requires changing the fundamental way in which the companies function. We discussed how to achieve the success that MRP II can provide, including:

Measures of implementation success

The three types of implementation

Prerequisites for successful implementation

Classic implementation steps

Education and training

We also discussed two critical topics dealing with "failure":

Why most MRP II implementations "fail"

Implementing after a previous "failure"

Paradoxically, those companies that are willing to embrace "failure" as the price of learning are those with the greatest chance of actual success.

We strongly suggested that the success of an implementation rests in top management's understanding of the difficulty of the project, and of the way they support people's attempts to change. We encouraged top management to aim for a 200 percent ROI, rather than a 30 percent ROI, by investing heavily in education and training. Companies that have not yet achieved their first MRP II success, or those that have had a failure, will also benefit by utilizing an Outside Counselor.

Bringing an MRP II system on line is both totally energizing and absolutely draining. MRP II systems seem to demand more data, more decisions, and more details than people ever thought possible. They are also deceptively complex; problems with purchasing exception reports, for example, can be caused by a lack of understanding of the defaults in the Item Master file. Because of this, successful MRP II implementation demands a team effort, with representation from each of the major functional areas.

Selected Bibliography

Anderson, John C. and Robert G. Shroeder: *Material Requirements Planning: A Study of Implementation and Practice,* APICS, Falls Church, VA, 1981.

Berger, Gus: "Ten Ways MRP Can Defeat You," *MCRP Reprints,* APICS, Falls Church, VA, 1991, pp. 128–131.

Bridges, William: *Surviving Corporate Transition,* William Bridges & Associates, Mill Valley, CA, 1990.

Burns, O. Maxie and Walter E. Riggs: "Whatever Happened to the Proven Path?" *APICS 1988 Conference Proceedings,* APICS, Falls Church, VA, 1988, pp. 22–25.

Chew, W. Bruce, Dorothy Leonard-Barton, and Roger E. Bohn: "Beating Murphy's Law," *Sloan Management Review,* MIT, Spring, 1991.

Chopra, Vinnie: "How to Develop an Effective MRP II Implementation Plan," *APICS 1987 Conference Proceedings,* APICS, Falls Church, VA, 1987, pp. 177–180.

Gessner, Robert A.: *Manufacturing Information Systems: Implementation Planning,* Wiley, New York, 1984.

Goddard, Walter E.: "Fast Track to Success," *APICS 1990 Conference Proceedings,* APICS, Falls Church, VA, 1990, pp. 439–442.

Groves, Brenton, R.: "The Missing Element in MRP—People," *Production and Inventory Management Journal,* vol 31, no 4 (1990), pp. 60–64.

Krupp, James A. G.: "MRP Re-Implementation," *Production and Inventory Management*—4th Quarter 1986, APICS, Falls Church, VA, 1986, pp. 73–81.

———: "Why MRP Systems Fail: Traps to Avoid," *MCRP Reprints,* APICS, Falls Church, VA, 1991, pp. 132–133.

McAteer, Peter F.: "The Learning Organization," *NETworking,* vol 4, no 2, Spring/Summer 1992, Massachusetts Chapter of the American Society for Training and Development.

Meredith, Jack R.: "The Implementation of Computer-Based Systems," *Journal of Operations Management,* vol 2, no 1, 1986, pp. 11–21.

Miller, George J.: "The New Conference Room Pilot," *APICS A&D SIG Digest,* April, 1992.

Myczek, Robert M.: "The Conference Room Pilot: Top Gun of MRP II for A&D," *APICS 1990 Conference Proceedings,* APICS, Falls Church, VA, 1990, pp. 623–626.

Pannesi, Ronald T. and Helene J. O'Brien: *Systems and Technologies Certification Review Course Student Guide,* APICS, Falls Church, VA, 1992.

Schultz, Terry: *Business Requirements Planning: The Journey to Excellence,* The Forum Ltd., Milwaukee, WI, 1984, Chap 14.

Smith, Spencer B.: *Computer Based Production and Inventory Control,* Prentice-Hall, Englewood Cliffs, NJ, 1989, Chap 15.

Senge, Peter M.: "The Leader's New Work: Building Learning Organizations," *Sloan Management Review,* MIT, Cambridge, MA, Fall, 1990.

———: *The Fifth Discipline: The Art & Practice of the Learning Organization,* Doubleday Currency, New York, 1990.

Vollmann, Thomas E., W. L. Berry, and D. C. Whybark: *Manufacturing Planning and Control Systems,* 3d ed., Irwin, Homewood, IL, 1992, Chap 10.

Wallace, Thomas F.: *MRP II: Making It Happen, The Implementation Guide to Success with Manufacturing Resource Planning,* 2d ed., Oliver Wight, Essex Junction, VT, 1990.

Wight, Oliver W.: *MRP II: Unlocking America's Productivity Potential,* CBI Publishing, Boston, 1981, Chap 11.

12
Measuring and Evaluating M&CRP System Performance

"Be careful what you ask for; you might get it!"
ANONYMOUS

Introduction

What you reward is what you get and you can only fairly reward what you can measure. In evaluating an M&CRP system, we don't know whether the system is working properly or not until we measure its performance. Neither do we know whether our operations are improving or getting worse until we measure performance.

In this chapter, we discuss why we should measure, what we should measure, and how to develop a system for measurement and evaluation. We then look at two general systems for measuring the performance of the overall MRP II system and the benchmarking approach, to provide a framework for developing detailed performance measures for the M&CRP system. Most of the chapter describes detailed performance measures for the M&CRP system, classified into ten areas. A company can select one or several measures from each area and, thus, tailor the system to its unique needs.

The chapter concludes with a look at some of the problems that plague performance measurement and evaluation systems, and some anticipated future trends.

Why Measure?

Measurement is an important part of the management process. The management process includes:

Setting realistic objectives

Developing action plans

Allocating resources

Assigning responsibilities

Implementing the plans

Measuring individual and organizational performance

Utilizing feedback to take corrective action

The prime purpose of performance measurements is to provide an objective and impersonal basis from which to evaluate performance against stated goals. Measurements provide management with a basis from which we can correct the performance of key areas in the organization. Management establishes a definitive set of operating performance measures for each of the functional areas of the organization, to assess the success or failure in each area. They then monitor the performance measurements and take corrective action when needed.

Which Activities Do We Measure?

Every business activity can be measured, from clerical data entry performance to top management planning. Management determines which facets of the business should be measured. Although a multitude of activities can be measured, operational excellence springs from selecting the critical few that truly improve the performance of the company.

The following are some general guidelines for selecting activities to measure.

- Select an activity that has a *significant effect* on the performance of the overall system being measured. Ensure that the measurements are relevant, that is, that they actually represent the larger reality.
- Select an activity with a high potential for *improvement*. Because management time is itself relatively scarce, the modest amount of discretionary

time available should be spent on solving problems that will provide the greatest benefit to the company.

- Ensure that the selected activity is *measurable.* To be measurable, the measurements must be objective and quantifiable, as we discuss in the next section.
- Ensure that measurement of the selected activity is *productive,* that is, that the benefits gained from measuring exceed the cost of measuring.
- Ensure that the selected activity is actually *controllable* by your organization, and is within your responsibility.
- Be prepared to *take action* based on the results of the measurement.
- Ensure that the measurement process supports *Continuous Improvement.*

Developing a Performance Measurement and Evaluation System

When we admit that we need quantifiable measurements, the need is usually overwhelming and quite urgent. "We must measure this project; we must monitor that project; we must see exactly how we are improving!" In the most perfect of all possible worlds, we would develop performance measurements when the activity is created. Unfortunately, this is usually not the case, and we must often establish measurements to determine the success or failure of a given activity after that activity has begun.

The measurement process itself is more important than when the measurement is established. Good measurement helps to "close the loop" in the management process, and can monitor all areas of the organization. The process of performance measurement and evaluation can be defined as a systematic effort to:

Set clear performance objectives

Select performance measures, standards, and measurement methods

Measure and report performance

Determine if performance meets standards

Take corrective actions, make changes

Review and improve actions

This definition is shown schematically in Figure 12-1.

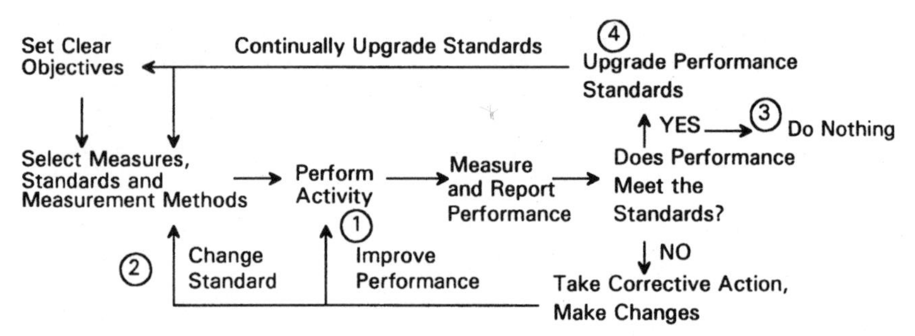

Figure 12-1. The performance measurement and evaluation process.

Set Clear Objectives

The objectives can be individual or group goals, key identifiers for current business activity, or indicators of future business. Some basic questions to ask are:

What will we measure?

What will we gain by this measurement?

What will the measurement tell us about the task in process?

The objectives should be clear and precise to all involved. This phase of the development is the right time to define:

- Realistic objectives. Can we realistically expect to achieve our objectives, or is the current objective too ambitious, or out of reach at this time? Would it serve the company better to begin with a smaller, less challenging goal that, once accomplished, would lead to additional measurements? Would a string of smaller successes be more beneficial in achieving the overall long-term objective?
- Acceptable achievement. Is a performance of 75 to 80 percent acceptable for the stated objective? Will this be initially acceptable, as long as it continues to improve? What is the best we can expect to achieve? Are we successful if several of the measurements exceed 100 percent?
- Acceptable tolerances. We can often state the objective in the form of a tolerance around the desired goal. What are acceptable tolerances around the desired goal? ± 5 percent? ± 2 percent?
- Acceptable timing. When do we expect to achieve the goal? How long should we expect it to take to realize the objective? What percent of the time do we need to hit the goal?

This stage is so critical that it may warrant a second review of the objectives. Is the goal or objective measurable by factual information and not just one person's impression of improvement?

Because "that which gets rewarded gets done," the objectives must encompass the entire spectrum of activities. Most people can skillfully sacrifice unmeasured values to increase measured values. For example, one company manufactures a highly seasonal product on a continuous, year-round basis owing to capacity constraints. The product is very bulky, and low in value compared to the bulk. This company rewards its sales force for moving the product into prospective customers' warehouses during the off-season, so the company itself will not have to acquire extra warehouses. Unfortunately, the company treats the "moved" product as "sold" (although it is indeed returnable later, at no cost to the prospective customer), and pays the sales commissions when the product is shipped. To make matters worse, the company pays sales commissions on a fixed price per unit, regardless of the actual selling price or terms. Because the sales people are rewarded strictly for product movement, they do indeed move the product. However, they give their customers six-month invoice dating and very low prices, costing the company untold millions of dollars in lost profit.

Define the Measurement

The measurements must provide information that is relevant to our objective. As with the objective itself, the measurements should be clear, direct, and precise. They must list not only *what* is to be measured, but also the *how* and *when* (the methods). For example, if we are measuring inventory turns, the *how* could include any combination of production inventories, maintenance inventories, consignment inventories, Material Review Board Inventories, and/or obsolete (tagged for removal and/or disposal) inventories.

The relationship between the measurement objective and the actual measurement criteria is not as simple or obvious as one may think. All measurements intrinsically represent two elements: (1) the integrity of the data used, and (2) the actual performance. When creating a measurement, we first consider:

- The actual source of the data used
- Problems in accumulating the information
- The validity of information gained. If the sources of information (and the information itself) are not clearly defined, the measurement may ultimately be a measure of the quality of information gained (or at best a measure of our ability to gather information) rather than of our objective.

For example, to measure supplier on-time delivery, we would probably compare promised delivery dates to actual deliveries, as shown by our on-line system. However, this comparison may or may not be an accurate reflection of our supplier's ability to deliver. How could this be misleading? (1) Some deliveries might remain on the receiving dock for several days before actually being "received" by our company; (2) our data entry of a timely receipt might be three days late; or (3) our system might automatically assign a system receipt date of three days after delivery, to allow for transportation and receiving processing time. We ask again: Are we actually measuring supplier delivery, receiving and data entry backlog, system ignorance, or all combined?

As another example, inventory accuracy performance is often computed by comparing the inventory record to a recent audit or count. If both numbers agree, we assume the inventory accuracy to be 100 percent. But what if *both* numbers are incorrect? What method, if any, does the organization have to assure they are correct?

The discussions that seek agreement on the measurements and measurement methods can be extremely lively, indeed! Because it is so important to the organization, we state the obvious: all involved persons should agree on both measurements and methods before the company starts to use them.

Measure and Report Performance

Once the performance objective and expectations are clearly developed, we measure current performance, in order to:

- Validate the measurement itself. If the initial measurement falls outside our expectations, we may wish to review the original sources of information or the objectives themselves.
- Validate the current sources of information being used for the measurement. This initial "sanity check" measurement usually produces results close to what would normally be expected, especially if the poor results of the activity are what prompted the measurement objective to be established in the first place.
- Establish the current performance as a benchmark from which we can gauge the success of our efforts.

Measuring the current performance is also beneficial when there are plans to implement a new system, such as MRP II, in the not too distant future. These measurements will provide a benchmark against which the performance of the new system and the performance of the old system can be compared.

Performance measurements help to define our thinking and our goals, but do not, in themselves, attain the goals. It is the effort of responsible people that achieves the goals. To keep their efforts focused and to identify problems, the people should review the results regularly. Problem areas within the organization may need to be reviewed daily or weekly, depending upon the situation.

We communicate all measurements (what we are measuring) and all performance results of the measurements to all the people involved. If the measurements are company-wide, each and every employee should be aware of exactly what we are measuring, how we developed the measurements, and which person or department is responsible for results. The communication can be in the form of a posted department memo, the company newspaper, and/or bulletin boards that are located in key and public areas of the company.

Most people are strongly motivated by a public chart that shows progress toward a goal. In MRP II implementations, companies sometimes post large PERT or Gantt charts, complete with due dates and names of individuals responsible for each step, in public hallways or the cafeteria. JIT-oriented companies deliberately measure critical factors, such as errors and on-time performance, and publicly post performance of each work area. Companies have found that having the persons who are responsible for the results actually post the results provides the best system of rewards; the people feel more personally responsible.

Determine if Performance Meets Standards

Determining if performance meets standards is probably the most straightforward step in the process. If we have already determined *what* to measure, *how* to measure it, and *which* values are acceptable, this step can be relatively easy. We have addressed the measurement complexities in previous steps; now it is a matter of comparing measured results with performance objectives.

As a result of comparing performance to our standards, we can act in any one of the four distinct ways shown in Figure 12-1 and described below:

1. Change and improve the performance of the activity. If the measured performance does not meet the standard, the usual course of action is to change how we perform the activity so that performance improves. We discuss in the following section how to identify and solve problems that prevent us from achieving our performance goals.
2. Change and improve the standard, measure, or method of measurement, which may be in error. We may find, especially in the beginning, that some of our measurement standards are impossible to achieve.

3. Do nothing. If the performance meets the established standard, the traditional response is to do nothing. However, with many companies following a "continuous improvement" program, this response is no longer acceptable.
4. Upgrade the performance objective. If the performance meets the established standard, many companies now choose to upgrade the performance objective as part of their effort to continuously improve performance.

Take Action, Make Changes

Identifying the problem areas that cause the poor performance may not be simple, because the established measurements may be highlighting a symptom caused by some other problem or problems in other parts of the organization. An example of this may be the poor performance of manufacturing in meeting the Master Schedule. Using only a *production versus plan* measurement ignores other factors that contribute to poor manufacturing performance, such as late component delivery from suppliers, overscheduled capacity from the MPS, excessive changes to the MPS within product lead times, rejection of materials at inspection, and so on. These problems are not apparent when viewing the single production measurement.

We suggest using several measurements to pinpoint the exact causes of poor performance. These contrasting measurements can be used to balance the information obtained. Some examples of contrasting measurements include unit production versus on time delivery of work orders, and on-time shipment versus on-hand inventories.

Once we have defined the problems that are contributing to poor performance, we develop action plans to correct the situation. As with the performance measurements themselves, the corrective action plans must have clear guidelines for improvement, including times, dates, and resources required.

Only one person or department should be held accountable for the outcome. We have learned through experience that simply *giving* the responsibility to another person or organization does not necessarily mean that the responsibility has been accepted. The responsible party must actually take ownership of the tasks. Ambiguous responsibility assures the failure of the measurement, and quite possibly the project as well.

We offer the following guidelines for giving performance feedback:

- Be where the performance is actually happening, rather than in the office, removed from the action.

- Seek opportunities for positive feedback. Ideally, positive feedback occurs *four* times more frequently than negative feedback. Even more importantly, positive feedback should be unencumbered by negative feedback. The old technique of "wrapping" negative feedback with positive feedback before and after, leads people to expect negative feedback when someone praises them.
- Give feedback frequently, soon after performance. Annual performance reviews are woefully inadequate from a frequency standpoint.
- Focus on controllable performance. The objective of feedback is to encourage the individual or group to improve performance with respect to those items within their control.
- Keep feedback specific and understandable. "Great job, Pat!" helps the individual feel good, but the individual has to assume why we said the job was great. We praised Pat because the job was on time; Pat thought it was because the label was affixed in a unique position. Much better for us to have said, "Great job, Pat! You shipped this order on time without making any other orders late."
- Provide graphic data feedback where possible. Graphics have much greater impact than numbers.

Review and Improve Actions

Figure 12-1 illustrates that the performance measurement and evaluation process is dynamic. Unless managers see the process through to its conclusion, they are merely monitoring performance rather than evaluating and improving it. The emphasis should be on Continuous Improvement, rather than merely measuring performance and identifying failures.

The essence of Continuous Improvement is to relentlessly devise constructive ways to improve performance. Once the initial target has been reached, and the successful person or teams congratulated, the team mutually agrees to move the target higher. This is *not* to say that the rules of the game continually change once the initial performance goal is about to be reached. On the contrary, when the initial goal is met or exceeded, we reward those involved. We also ensure that each goal, once attained, will continue to be routinely achieved in the future, by reviewing how the person or group achieved the goal. There is a danger that once the original goal is met, the original measurement may become stale and routine. This status quo attitude could actually lead to poor performance against goals previously attained. We can gain the highest level of performance by making continual efforts to upgrade performance objectives in a realistic manner.

Performance review of those people who perform well is one of the pleasures of leadership. Their performance can be honestly and sincerely praised in public (the best way). The rewards should be mutually determined in advance. Appraising performance that does not meet the standards is one of the most difficult activities for management. However, the consequences *must* be delivered. If management allows substandard performance to continue, other people will become less and less interested in performing to standard. The threat of negative consequences, if not actually followed through, loses virtually all effectiveness. Any negative consequences must be applied fairly and consistently throughout an organization.

Systems for MRP II Performance Measurement

In this section, we describe two well-known systems for measuring and evaluating the performance of the MRP II system. Although we are primarily interested in developing a system for M&CRP, these two systems provide useful background.

Class ABCD System

Oliver Wight[1] introduced the first system for measuring and evaluating MRP II systems in 1977. His current system consists of a checklist of 35 questions, shown in Figure 12-2, which classify a company as a Class A, B, C, or D user of MRP II. To use Wight's system, knowledgeable people from a company objectively evaluate their company on each of the 35 questions. To be Class A, you should have 0 to 3 questions answered "No." Class B can have 4 to 7 "No" answers. Class C can have 8 to 10 "No" answers. Class D includes 11 to 14 "No" answers.

The ABCD system is a very general method used to help a company determine how well their MRP II system is operating. This method does enable a company to identify definite deficiencies and to take action to correct these deficiencies.

[1]Walter Goddard, et al., *ABCD Checklist,* Oliver Wight Publications, Essex Junction, VT, 1988.

	Yes	No
1. Management Commitment	____	____
Planning and Control Processes		
2. Strategic Planning	____	____
3. Business Planning	____	____
4. Sales & Operations Planning	____	____
5. Single set of numbers	____	____
6. "What if" Simulations	____	____
7. Forecasts that are measured	____	____
8. Sales Plans	____	____
9. Integrated Customer Order Entry and Promising	____	____
10. Master Production Scheduling	____	____
11. Supplier Planning and Control	____	____
12. Material Planning and Control	____	____
13. Capacity Planning and Control	____	____
14. New Product Development	____	____
15. Engineering integrated	____	____
16. Distribution Resource Planning	____	____
Data Management		
17. Integrated BOM and Routing	____	____
18. Data Accuracy	____	____
BOM 98–100%		
Routing 95–100%		
Inventory records 95–100%		
19. Product change control	____	____
Continuous Improvement		
20. Employee education	____	____
21. Employee involvement	____	____
22. One less at a time	____	____
23. Total Quality Improvement process	____	____
24. Product development strategy	____	____
25. Partner relationship with customers	____	____
Performance Measurements		
Planning and control process measurements		
26. Production Planning performance ± 2%	____	____
27. Master Production Schedule performance 95–100%	____	____
28. Manufacturing schedule performance 95–100%	____	____
29. Engineering schedule performance 95–100%	____	____
30. Supplier delivery performance 95–100%	____	____
Company performance measurements		
31. Customer service delivery to promise 95–100%	____	____
32. Quality performance measured	____	____
33. Cost performance measured	____	____
34. Velocity performance measured	____	____
35. Management uses measurements for improvements	____	____

Figure 12-2. Wight's ABCD checklist.

Manufacturing Performance Measurement System

David Buker has developed a comprehensive Manufacturing Performance Measurement System for MRP II[2] that has a Performance Objective and Key Performance Measurement in each functional area. Figure 12-3 summarizes Buker's system. Using the twelve measures of performance, management can assess which areas of the company are not performing properly. By tracking performance monthly, management can focus on those areas needing attention.

Integrating Performance Measurement with Benchmarking

Benchmarking, or "striving to be the best of the best" is a translation of an old Japanese word, dantotsu, practiced by the legendary Chinese general Sun Tzu, who said, "If you know your enemy and know yourself, you need not fear the result of a hundred battles."

Competitive benchmarking is a program to:

Identify a company's performance standards

Provide insights into how these performance standards can be achieved or exceeded

Develop internal action plans to improve the company's performance

Most importantly, benchmarking is an ongoing learning experience for the company as a whole. In essence, benchmarking involves comparing one company's performance, in selected areas, with the performance of similar companies in the same industry, and with the best companies in the world ("world class companies"). Figure 12-4 provides an example of a convenient form for collecting and comparing data collected during the benchmarking process.

As a result of his extensive experience with Xerox, Robert Camp recommends a four-phased approach to benchmarking:[3]

[2]David W. Buker, "Performance Measurement," *APICS 1980 Conference Proceedings,* APICS, Falls Church, VA, 1980.

[3]Robert Camp, *Benchmarking: The Search for Industry Best Practices That Lead to Superior Performance,* American Society for Quality Control Quality Press, Milwaukee, WI.

Functional area	Responsibility	Performance objective	Performance measurement
Top Management Planning			
Business plan	General manager	Return on investment	Actual ROI / Planned ROI
Sales plan	Sales	Sales performance	Units booked / Units planned
Production plan	Manufacturing	Production performance	Actual production / Planned production
Operations Management Planning			
Master schedule	Materials	MPS performance	Actual MPS / Planned MPS
Materials plan	Materials	Release reliability	On–time orders released / Total orders released
Capacity plan	Manufacturing	Capacity performance	Capacity hours produced / Capacity hours required
Database			
Bills of material	Engineering	Bill of material accuracy	Parts in agreement / Total number of parts
Inventory control	Materials	Inventory accuracy	Number of parts correct / Number of parts counted
Routings	Manufacturing	Routing accuracy	Operations in agreement / Number of operations
Operations Management Execution			
Purchasing	Purchasing	Schedule performance	Parts delivered / Parts scheduled
Shop floor control	Manufacturing	Schedule performance	Parts completed / Parts scheduled
Delivery performance	General manager	Schedule performance	Units delivered / Units promised
Performance summary	Class	Average	Total

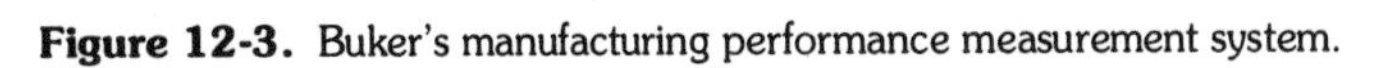

Figure 12-3. Buker's manufacturing performance measurement system.

Performance measure	July 1991			July 1992		
	Ours	Competitor	World class	Ours	Competitor	World class
Delivery lead time (days)	4	3	2	3	2.5	2
Inventory turns	9	14	18	13	16	19
Percent value added	35	50	60	45	55	65
Time: design to production (days)	79	57	48	60	54	42

Figure 12-4. Data collection form for benchmarking.

1. Planning phase
   Identify what you're going to benchmark
   Identify benchmark companies
   Choose data collection method and collect data
2. Analytic phase
   Determine current performance gap
   Project future performance levels
3. Integration phase
   Communicate benchmark findings and gain acceptance
   Establish functional goals
4. Action phase
   Develop action plans
   Take action and monitor results
   Recalibrate benchmarks

Our purpose in discussing benchmarking is to demonstrate its applicability to M&CRP performance measurement, rather than to provide detailed directions for doing it. We strongly encourage companies to incorporate benchmarking as a means of supporting continuous improvement and offer the following general guidelines.

- Expect the gap between your performance and the benchmark's performance to be large. Although this can be demoralizing, it also shows how large the opportunity is. The whole idea is to find out where and how you can improve.

- Start with achievable goals. You don't need to benchmark your whole company to show the value of benchmarking. Pick one area and concentrate on the quality of the effort.
- Go to several companies or industries. No company is best at everything it does. You can learn a lot from looking at the best practices in companies that have different strategies and strengths.
- If you make your comparisons too rigorous, you will spend all your time measuring, adjusting, and compensating, instead of thinking about how to improve. Remember that no other company is exactly like yours, so you will never get completely comparable performance data.

Specific Performance Measures for the M&CRP System

In this section we discuss some performance measures that are specific to M&CRP. They can be utilized in the first step of implementing either a Performance Measurement and Evaluation program, or the more extensive Benchmarking program. Our overall purpose is to maximize the performance of the M&CRP system, so that it will maximize the performance of the entire company. Thus, we are primarily interested in developing measures for those portions of manufacturing that are directly affected by the performance of the M&CRP system. Ten manufacturing performance areas significantly affected by the M&CRP system are:

Response to Customers

MPS Execution

Material Planning

Capacity Planning

Inventory Control

Production Activities

Purchasing Activities

Productivity

Operating Cost

Data Accuracy

Within each of the ten areas, we have developed a number of quantitative and measurable performance measures, and for each we:

- Describe what aspect of performance each criterion will measure
- Indicate how the proper operation of M&CRP will affect the performance that each criterion measures

Each company can tailor its M&CRP performance measurement and evaluation system by selecting appropriate performance measures from the list provided, however, no company should try to use all the performance measures listed. We suggest starting with one or two measures in each area, with perhaps three in a few areas. You can always add more measures as you gain experience.

In a number of areas we have included performance measures that require a comparison with competitors (Benchmarking). Obviously, these cannot be used unless it is possible to obtain valid information on competitors. We prefer these measures, because they provide a means of Benchmarking, as well as measuring performance. Remember that these measures are intended for measuring the performance of the M&CRP system, and not the performance of the entire manufacturing function or the company as a whole.

Measures of Response to Customers

A Material and Capacity Requirements Planning system has a measurable effect on a manufacturing company's ability to respond appropriately to the demands of customers. The following measures can be used to monitor this aspect of the M&CRP system's performance.

Percent Reduction in Delivery Time. Delivery time is calculated by subtracting the date the customer order was received from the date the order was shipped. When this is calculated periodically, we can compare the results and measure the improvement in our manufacturing cycle times for components not Make-to-Stock.

Delivery Lead Time/Competitor's Delivery Lead Time. By itself, delivery time is an excellent performance measure. It is even more meaningful when compared to that of a competitor, especially in time-based competition.

Percent of Items Delivered On Time and Complete. Reliable delivery is the hallmark of a world class manufacturer. When MRP and CRP systems operate properly, delivery percentages will be high.

Units Shipped During Month/Units Scheduled for Shipment During Period. This criterion measures delivery performance. It is easy to calculate, but does not prevent the "end of the month shipping syndrome." To eliminate the end-of-the-month shipping syndrome, use a weekly period rather than a monthly period.

Out-of-Stock Orders/Total Orders. Stock-outs on customer orders are very costly, especially in terms of future business. A well-run MRP II system should ensure that very few finished goods stock-outs occur. Decreasing the manufacturing cycle time will also decrease the stock-outs.

Percent of Past Due Orders. Past due orders can result in the loss of future business. They also confuse priorities on the shop floor and at suppliers. With M&CRP properly implemented, the percentage of past due orders should decrease significantly.

Percent Reduction of Processing Time for Engineering Changes. This measures the responsiveness of the company, including the M&CRP system, in responding to customer technical or design changes.

Time from Design to Production/ Competitor's Time from Design to Production. This measures our capability, including the capability of the M&CRP system, to bring new products to market rapidly. This is an increasingly important capability in time-based competition.

Percent Improvement in Changed Delivery Promises (at our request) Per Period. This measures our ability to make realistic promises and to keep them.

Percent Reduction in Distribution Time. This measures the company's ability to distribute products rapidly. M&CRP can assist by providing stable and reliable schedules for products.

Measures for Assisting MPS Execution

An effective M&CRP system should be able to execute the Master Production Schedule with only a few deviations. Prompt bottom-up planning, based on CRP outputs, should remove any unrealistic requirements in the MPS. The measures of M&CRP effectiveness in assisting MPS execution include the following.

Scheduled MPS Units Produced/Total MPS Units Produced. This is a simple criterion for measuring the "percent compliance with MPS."

Number of MPS Changes Within the Demand Time Fence. This criterion is a measure of the adherence to the time fences in MPS. Changes within the Demand Time Fences can cause lower efficiency and raised costs.

Number of MPS Changes in Production Rate Within the Planning Time Fence. Like the criterion for the Demand Time fence, this criterion measures the adherence to the time fence policy. Changing the production rate, without ensuring that the necessary material and capacity will be available, can be quite disruptive.

Percent Master Schedule Orders Completed on Time Per Period. This is a simple measure to determine how well the M&CRP carries out the MPS.

Percent Past Due Master Schedule Orders. This is a measure of the M&CRP system's ability to comply with the Master Schedule.

Material Planning Measures

Material Planning is the primary function of the M&CRP system. The goal of Material Planning is to develop stable, valid plans and schedules that can be carried out with the resources available. The performance criteria in this area measure various aspects of the M&CRP system's performance in achieving this goal.

Percent Availability of Material to Shop. It is the function of M&CRP to provide the right material, in the exact quantities needed, at the exact time needed for production. Stock-outs to the shop are nearly as bad as stock-outs to the customer.

Percent of Assembly Orders with No Shortages. With proper MRP planning there should be no shortages. Shortages require expediting and disrupt the efficient functioning of the assembly line.

Percent of On-Time Order Releases. Serves as a measurement of the overall on-schedule condition, and the effectiveness of the MRP planner in doing his or her job.

Percent of Order Due Dates Met. This is the essence of material planning, and measures the combined performance of the M&CRP system, the shop, and purchasing.

Percent of Released Orders with Past Due Start Dates. This is a measure of the schedule disruption; it indicates how often orders are likely to be expedited. This criterion measures the performance of the MRP system, and particularly the MRP planner, or MPS changes within the time fences.

Trend in Number of Action Messages. We can compare the number of MRP and CRP nonroutine action messages to last month, last quarter, and last year. Chapter 6 recommends some remedies for M&CRP nervousness.

Percent of Released Orders Halted or Not Worked Due to Material Shortages. This criterion measures the realism of the schedules and the effectiveness of the material allocation that occurs prior to order release.

Percent of Dormant Scheduled Receipts. Count the number of Released Orders that have not been worked on for *X* days. This measure indicates materials shortages, excessive queues, overloaded conditions, or other schedule problems.

Actual versus Planned Stock Issues to Assembly. Compare the actual material issue volume from the stockroom to the shop floor with the planned volume. This is measure of the MRP complying with the MPS in providing material for the assembly process.

Actual versus Planned Lead Time. This comparison measures the accuracy of the planned lead time in MRP and CRP. However, this can be a self-fulfilling prophecy, because orders with longer lead times have lower priorities in the shop, and therefore take longer.

Percent Released Orders Rescheduled. This is a measure of the *nervousness* of the M&CRP system. Chapter 6 suggests actions that can reduce nervousness.

Number of "Emergency" Orders Issued Per Week. This is a measure of the instability of the M&CRP schedule. A properly operating M&CRP will not require emergency orders.

Capacity Planning Measures

Capacity must be planned correctly, and in a timely manner, for a manufacturing company to meet schedule and cost objectives. The Capacity Requirements Planning process translates the material requirements into production capacity requirements, and balances available capacities with required capacities. The measures used to monitor this important aspect of the M&CRP system's performance are:

Percent of Work Centers That Are Overloaded and Underdeveloped. These are two measures of how well load and capacity is matched. Matching load and capacity is the primary function of capacity planning.

Percentage of Work Centers with Past Due Load. This criterion measures the effectiveness of CRP in providing the shop with the capacity to handle the load imposed. It also measures the ability of MRP in providing adequate lead times for shop orders.

Accuracy of Capacity Requirements Forecast. Compare the actual work load issued to the work centers with the work load forecasted by CRP. This is a direct measure of the effectiveness of the CRP system.

Overtime/Available Work Time. Overtime tends to be a costly short-term solution to capacity problems. Better balancing of work load and capacity using other alternatives such as shifting load, realigning shifts, and subcontracting, will reduce overtime and its costs.

Idle Time/Available Work Time. By using CRP to identify and solve load imbalances, idle time resulting from underloaded work centers can be reduced. This measure is not recommended for JIT, because JIT encourages companies to *deliberately* leave slack.

Percentage or Frequency of Work Center Stoppages Owing to Insufficient Work Input. This measure indicates how effectively the load smoothing and work leveling function of CRP is working.

Percent Reduction in Average Queues. Reduction of the average length of the queues through better scheduling will reduce the manufacturing lead time, reduce work-in-process inventory costs, and improve the overall productivity of the manufacturing company.

Percent Reduction in Queue Range. (Range = Highest value – Lowest value.) Erratic queues cause erratic lead times, and hence erratic performance of the shop. Stabilizing the length of the queue through CRP actions will improve the performance of the shop.

Actual versus Planned Work Center Queue Levels. This measure provides an indication of lead time control and scheduling effectiveness of M&CRP.

Actual Work Center Output/Scheduled Work Center Capacity. This measure primarily evaluates the effectiveness of work center supervision in fully utilizing the work time scheduled.

Percent Utilization of Work Centers. This measures the ability of M&CRP to smooth out queues and balance loads, especially at bottleneck activities, to obtain a better use of facilities. Again, however, JIT deliberately underutilizes work centers, in order to maximize throughput.

Work Center Downtime Percentage. Evaluates how effectively preventive maintenance is scheduled and performed.

Percent of Released Orders Halted Due to Tooling Nonavailability. An indicator of prerelease capacity planning effectiveness, and a cause of excessive work center queues.

Inventory Control Measures

Although M&CRP is much more than an inventory control system, it does serve to control inventories. Implementing M&CRP should result in an overall inventory reduction; however, inventories might actually *increase* as an M&CRP system is initially implemented. The criteria in this area will measure the M&CRP system's performance as an inventory control system.

Percent Increase in Inventory Turnover. Inventory Turnover is obtained by dividing the Cost of Goods Sold by the Cost of Average Inventory. This is an easy to use and popular method of measuring the performance of M&CRP in reducing inventories. Because this measure automatically compensates for the sales volume of the company, it is a more meaningful measure of inventory reduction than is the Aggregate Inventory Level.

Inventory Turns/Competitor Inventory Turns. By itself, inventory turns is an excellent performance measure. It is a better measure when compared to a competitor in the same industry, and even better when compared to a world class competitor.

Percent Reduction in Work-in-Process Inventory. The criterion measures the effectiveness of the M&CRP system in scheduling work to reduce queues and the manufacturing lead time.

Percent Reduction in Aggregate Inventory Levels. Levels of inventory should be reduced when using M&CRP simply because orders are time phased to arrive only when needed, rather than having stocks replenished routinely when they are low. Also, most Safety Stock is no longer needed.

Percent Reduction in Inventory Carrying Costs. This is a convenient criterion to measure the effectiveness of M&CRP in reducing inventory. However, the carrying cost percentage rate determined by the Finance department will also directly impact this measurement.

Percent Reduction in Inventory Storage Requirements. A side benefit from inventory reduction is the reduced need for storage space to hold it. We suggest separate measures for finished goods, component stockroom inventory and work-in-process inventory.

Obsolete Inventory/Total Inventory. Excess inventory is bad enough, but obsolete inventory is even worse. When MRP is first implemented, obsolete inventory may rise sharply because MRP identifies inventory as being obsolete if it has no projected future use.

Measures for Assisting Production Activities

The proper operation of the Material and Capacity Requirements Planning system can make activities on the shop floor much more effective. The shop is responsible for producing goods that meet the customer's time and quality requirements. The following measures can be used to monitor production activities, but they also are useful for gauging the M&CRP system's assistance in meeting this goal.

Shop Orders Released On Time Compared to Total Shop Orders Released. Both the late release and early release of shop orders create problems. Late release results in expediting and other costly measures. Early release increases work-in-process in the shop, and dilutes control over shop priorities.

Percent Shop Orders Completed On Time. When the delivery date equals the due date, the Shop Order is completed on time. If the delivery date is later (or earlier), the Shop Order is not on time. Late or early shop order completion may be caused by several problems, including on-time release, material availability, and resource availability.

Percent Reduction of Manufacturing Lead Times. Reduction of lead times is required to support Time-Based Competition. M&CRP makes more effective use of available capacity and reduces erratic lead times by balancing queues and individual operation lead times.

Percent Shop Orders Released Inside Lead Time. Orders released behind schedule will require expediting and other disruptive measures to meet the due date. Effective scheduling by MPS and M&CRP will minimize this problem.

Percent Reduction of Setup Time. Reducing setup time makes the manufacturing system more flexible and better able to respond in time-based competition. If MRP uses reduced lead times in planning, this will provide added incentive for the shop to work on reducing setup times.

Percent Reduction in Product Changeover Times. This is similar to the setup time measure, except that it is more comprehensive. Product changeover requires a number of machine setups for the parts plus a setup of the assembly line.

Percent Reduction in the Lot Size. Reducing the lot size improves the flexibility of the manufacturing system. The lot sizes are controlled by the MRP system.

Percent Past Due Shop Orders. If M&CRP is planning properly, there should be fewer past due shop orders.

Kits with Part Shortages/Total Kits. The M&CRP system should ensure that the parts are available for kitting an order before the order is released.

Percentage of Queue Time in Manufacturing Lead Time. This is measure of the ability of M&CRP and PAC to reduce the queues by better loading and scheduling.

Measures for Assisting Purchasing Activities

A properly operating M&CRP system provides the major inputs to the Purchasing System in an accurate and timely manner. The following measures monitor how well M&CRP supports Purchasing.

Percent On-Time Deliveries. Early delivery is almost as bad as late delivery because it causes increased material handling, increased storage needs, and increased inventory levels. A well-functioning M&CRP system releases purchase requisitions and advance notice of requirements on time, allowing suppliers to plan for sufficient capacity to meet delivery schedules.

Percent Purchase Orders Released On Time. Both late release and early release of purchase orders create problems. Late release results in expediting and other costly measures. Early release results in additional purchase order management costs and, sometimes, storage of material until it is needed. Proper operation of the M&CRP system can reduce both late and early releases.

Percent Reduction of Supplier Lead Times. As M&CRP communicates purchasing requirements accurately and in a timely manner to the suppliers, they can reduce lead times.

Percent of Purchase Orders Placed Inside Lead Time. Late orders tend to be disruptive to our own purchasing organization and our supplier. When M&CRP creates requirements properly, orders have adequate lead times. This, however, also depends on a stable MPS.

Quality Deliveries/Total Deliveries. Poor supplier quality leads to higher levels of Safety Stocks, as well as increased risk of expediting, shipping products of unacceptable quality, and putting our own shop orders on quality hold. With MRP helping Purchasing with the clerical and planning effort, Purchasing can spend more time on selecting suppliers that can produce to the desired quality level.

Percent of Orders Past Due. If M&CRP uses accurate lead times and creates orders properly, the number of past due orders should be minimized.

Percent Reduction in Purchase Costs. By providing advance planning for purchases, M&CRP can reduce the cost of purchasing the required material.

Percent Reduction in Time to Process Purchase Orders. This measures the combined time of MRP and Purchasing to process a purchasing order.

Percent Purchase Orders Requiring Expediting. If M&CRP releases the purchase order correctly and on time, expediting should not be required.

Percent Supplier Delivery to Promise. Proper planning and creation of purchase orders by M&CRP will improve the probability that a supplier will deliver as promised.

Productivity Measures

Productivity = Value of output ÷ Value of input. Productivity is an overall measure of the effectiveness of a manufacturing company. Because productivity is such a broad measure, determining how it is affected by the performance of the M&CRP is rather difficult.

Value Added Time as a Percent of Manufacturing Lead Time. This is a measure of how effectively the M&CRP system makes use of the time available for manufacturing. Run time is the only value-added time in the manufacturing cycle.

Percent Reduction in Floor Space. As M&CRP reduces work-in-process inventories and improves the timing of material flow, floor space will free up.

Rush Orders/Total Orders. Better planning from MRP and CRP reduces the number of unplanned, rush orders to be expedited. Rush orders increase cost through substantial increases in inefficiency.

Total Productivity Measure (TPM). TPM determines the total productivity of a company by considering the value of all outputs and inputs.

$$\text{TPM} = \frac{\text{Products} + \text{Services} + \text{Value added to plant}}{\text{Labor} + \text{Capital} + \text{Energy} + \text{Material} + \text{Other inputs}}$$

Labor Productivity Measure (LPM). LPM measures the productivity of only one input (labor).

$$\text{LPM} = \frac{\text{Products} + \text{Services} + \text{Value added to plant}}{\text{Labor}}$$

Cost of Goods Sold Compared to Number of Employees. This is an easy-to- calculate measure of labor productivity. The employees in the measure can be total employees, direct labor, or some other category. However, this measure penalizes a manufacturer who vertically integrates several operations, compared to one who final assembles purchased components.

Operating Cost Reduction Measures

The M&CRP system is capable of reducing operating costs by reducing inventory, making more effective use of available resources, and reducing waste. The following criteria measure the effectiveness of the M&CRP system in reducing the operating costs of a manufacturing company.

Value Added Cost/Total Product Cost. In the shop, the only true value-added activity is the Run or Processing operation. All the other operations (setup, queue, wait, and move) do not add value to the product. This measure can be used to compare various products, and also to compare a company's products with a competitor's products. M&CRP can affect this measure by recommending the elimination or reduction of nonvalue-added activities.

Percent Reduction in Purchasing Costs. By providing advance planning for purchases, M&CRP can reduce the cost of purchasing the required material.

Percent Reduction in Inventory Carrying Costs. Reducing the amount of inventory in the storeroom and work-in-process is possible with stable and accurate schedules from M&CRP. However, the cost of funds will also directly affect these carrying costs.

Percent Reduction of Setup Costs. Setup does not add value to the product, only to the cost. The reduction of setup costs, by applying group technology and other techniques in the M&CRP system, will result in a more efficient manufacturing system.

Percent Reduction in Material Handling Costs. Material handling does not add value to the product, only to the cost. This criterion measures how effectively the M&CRP systems schedule items to prevent unnecessary material handling.

Percent Reduction In Transportation Costs. The delivery of products on schedule, made possible by good planning from the M&CRP system, will enable a company to minimize transportation costs.

Data Accuracy Measures

Without accurate data the M&CRP system and the manufacturing company cannot operate with a high degree of effectiveness. All actions taken by M&CRP are based on the assumption that all data are accurate. If they are not, the actions may not be correct. The following criteria measure various aspects of data accuracy.

Bill of Material Accuracy. What percentage of the Bills of Material are 100 percent correct, as used to manufacture the product?

Nonscheduled Issues and Receipts from Stores/Scheduled Issues and Receipts from Stores. Nonscheduled requisitions for components, or returns of components to the stockroom, may indicate that certain Bill of Materials are inaccurate and should be corrected.

Routing Sequence Accuracy. What percentage of the routing sequences are 100 percent correct, as measured against how the shop actually manufactures the products?

Routing Standards Accuracy. What percentage of the routing time standards are 100 percent correct, compared to how long each operation actually requires?

Inventory Record Accuracy. What percentage of the inventory records (on hand for each location) are 100 percent correct?

Scheduled Receipt (Open Order) Accuracy (both Manufacturing and Purchasing). What percentage of the manufactured and purchased scheduled receipts are 100 percent correct, including due date and quantity?

Work Center Record Accuracy. What percentage of the work center records are 100 percent correct, including queue times, labor rates, and capacities?

Problems

Even when companies follow the guidelines and implementation steps to introduce performance measurements in their organizations, difficulties will most likely surface quickly. After the measurements have been agreed upon, accountability accepted, and tasks assigned, there will probably still be some initial "slow going." This is because most people, even high achievers, resist change. This topic is discussed in more detail in Chapter 11.

When the measurements are finally accepted and in routine use, management can address additional issues, including:

- Assuring the consistency of the measurement
- Reviewing the source or sources of the information to assure that the correct information continues to be supplied
- Revising the measurements in the face of changes in business policies and procedures and production methods

Trends

Until recently, most performance measurements were based primarily on the accounting and financial aspects of the business. Assuredly, financial aspects are important and should not be dismissed or changed merely for the sake of making a change. However, companies are becoming aware that the financial measurements depend on the underlying operational processes. Companies are starting to also measure these fundamental operational processes.

In the operational areas, a primary focus is on the quality of the goods or services produced. Companies are now developing new measurements in order to highlight the continuous improvement of processes and products. Continuous improvement is just as stated: once milestones are achieved, additional goals are set in order to continuously achieve greater manufacturing excellence. The overwhelming trend is to abandon performance measurements aimed at continuing the status quo, and instead use measurements that support continual improvement.

The main challenge to the future is to develop better performance measurements for customer service, to understand the relationship between quality and operational productivity, and improve time-to-market introductions of new products. Some dissatisfaction with established measurements exists, because traditionally cost based measurements do not match the current emphasis on Just in Time manufacturing, World Class Manufacturing, and Total Quality Management required to compete in a world marketplace.

Summary

We don't know whether the M&CRP system is operating properly or whether its performance is improving or getting worse, unless we measure performance. There are many activities that we could measure, but we should select those that are significant, improvable, controllable, and measurable, in order to obtain the greatest benefit from our performance measurement and evaluation program.

The measurement process itself is more important than the establishment of the measurement. Good measurement helps to close the loop in the management process, and it monitors all areas of the organization. The steps for effectively measuring performance were defined.

We developed significant, improvable, controllable, and measurable criteria in ten areas to measure M&CRP activities. A company can develop a M&CRP performance measurement and evaluation system that is uniquely suited to their company, by selecting appropriate performance measures from the ten areas.

A company can measure operational success as it relates to pleasing the customer in the quality of the products produced, the defects found per items produced, and on time delivery of customer shipments. Benchmarking, which compares our own company to other companies in our industry and to the best-in-class for that activity, provides valuable information on the company's performance.

Selected Bibliography

Abair, Robert A.: "Super Measurements: The Key to World-Class Manufacturing," *APICS 1990 Conference Proceedings,* APICS, Falls Church, VA, 1990, pp. 418–420.

Bechler, Robert: "Performance Measurements: Key to Effective System Utilization," *APICS 1990 Conference Proceedings,* APICS, Falls Church, VA, 1990, pp. 610–612.

Blackstone, John H., Jr.: *Capacity Management,* South-Western, Cincinnati, OH, 1989, Chap 12.

Burke, David W.: "Performance Measurement," *APICS 1980 Conference Proceedings,* APICS, Falls Church, VA, 1980.

Camp, Robert: *Benchmarking: The Search for Industry Best Practices That Lead to Superior Performances,* ASQC Quality Press, Milwaukee, WI, 1988.

Clark, Steven J., James F. Cox, Richard R. Jesse, Jr., and Robert W. Zmud: "How to Evaluate Your Material Requirements Planning System," *Production and Inventory Management,* 3d Qtr, 1982, APICS, Falls Church, VA, 1982, pp. 15–34.

Deis, Paul: *Production and Inventory Management in the Technological Age,* Prentice-Hall, Englewood Aliffs, NJ, 1983, Chap 9.

Erhorn, Craig R.: "How to Audit Your MRP System for Consistent Results," *MCRP Reprints,* APICS, Falls Church, VA, 1991, pp. 230–234.

Goddard, Walter E.: "How to Measure Performance Beyond Class A," *APICS 1987 Conference Proceedings,* APICS, Falls Church, VA, 1987, pp. 488–490.

——— et al.: *The ABCD Checklist,* Oliver Wight Publications, Essex Junction, VT, 1988.

Schultz, Terry: *Business Requirements Planning: The Journey to Excellence,* The Forum Ltd., Milwaukee, WI, 1984, Chap 13.

Smith, Spencer B.: *Computer Based Production and Inventory Control,* Prentice-Hall, Englewood Cliffs, NJ, 1989, Chap 15.

White, Richard W.: "Is It Time for a Physical Exam of Your Materials Requirement Planning System?" *Production and Inventory Management,* 3d Qtr, 1986, APICS, 1986, pp. 24–29.

Wight, Oliver, W.: "Control of the Business," *MCRP Reprints,* APICS, Falls Church, VA, 1991, pp. 125–127.

———: MRP II: Unlocking America's Productivity Potential, CBI Publishing, Plano, TX, 1981, p. 455.

13 The Future of Materials and Capacity Requirements Planning

"I think there is a world market for about five computers."

THOMAS J. WATSON, CHAIRMAN OF THE BOARD, IBM, 1943

"Where a calculator on the ENIAC is equipped with 18,000 vacuum tubes and weighs 30 tons, computers in the future may have only 1,000 vacuum tubes and perhaps weigh only 1½ tons."

POPULAR MECHANICS, MARCH, 1949

Introduction

Fools rush in where angels fear to tread. In this chapter we use several approaches in attempting to predict how M&CRP-MRP II will change in the future.

First, we assess the current performance of M&CRP-MRP II and for each criticism or deficiency, we predict how and when these deficiencies will be corrected.

Second, we look at how M&CRP-MRP II can be adapted so that it can operate more effectively in a number of nontraditional manufacturing environments. In the early days of MRP, software developers believed that the traditional MRP package, which was developed for a traditional Job Shop, would fit all companies. Today, we recognize the fallacy of this position.

Third, we examine the technology that is available for designing M&CRP-MRP II systems, which is essentially the technology of computers and information systems. Today's M&CRP software was designed for the third generation computers of the 1970s. Most existing M&CRP software does not begin to exploit the capability of today's computers, let alone that of the next generation which is coming. For example, M&CRP still assumes infinite capacity as it schedules materials to satisfy demand, then checks the assumption in a later batch program. This sequential approach is an anachronism from the early days when computers did not have sufficient processing power to solve the combined scheduling problems of MRP and CRP.

Finally, we take a somewhat longer and broader view of the situation by looking at how the manufacturing environment and the manufacturing enterprise are changing, and might be expected to change in the future. As we discussed in Chapter 1, the nature of the manufacturing planning and control system is largely determined by the nature of the manufacturing enterprise, which in turn, is mainly determined by the environment in which it must compete. Thus, if we can determine the future trends of the manufacturing environment and manufacturing systems, we have gone a long way toward predicting the necessary changes in the M&CRP-MRP II system.

Assessment of M&CRP-MRP II

In this section, we assess the current performance of M&CRP-MRP II. For each problem or deficiency, we describe actions that are either under way, or probably will be undertaken, to correct the deficiency.

Inflated, Inaccurate, Fixed Lead Times

Deficiency. MRP accepts inflated lead times, causing long planning horizons. MRP systems break the Bill of Materials into main assembly, subassembly, sub-subassembly, fabrication, and purchasing stages, then add the lead time for each stage together to establish the total lead time for ordering. If planners inflate each of the lead times to allow time to recover from surprises caused by unresponsive suppliers, uncertain quality, and undependable shop deliveries, the combined lead time will be even more inflated. This approach also dramatically inflates work-in-process. Unlike JIT, MRP accepts long lead times as fixed, and does not seek to reduce them. JIT focuses on reducing *all* waste, whereas MRP is inclined to accept everything as given, and tell the planner how to live with it. Lead time problems include the following:

1. *The Fallacy of Static Lead Times.* Most MRP systems treat lead time as an independent, static element for each item, ignoring the fact that it is instead dependent on many other factors, especially shop load. This fixed lead time is a major cause of difficulty with shop floor schedules.

2. *Planned Lead Times.* The concept of a planned lead time assumes that it can be planned in advance of the schedule and that it is independent of the batch size and actual shop work load. This is true if the largest component of lead time is the queue time. However, queues are not as static as MRP assumes. Large queue times reflect (and cause) an inefficient flow of work through the manufacturing shop, and suggest a job shop rather than a line layout. Actual queues for a particular batch or job will vary from week to week, reflecting the load on the manufacturing shop and, particularly, on the work center in question. Actual priorities will cause a late-arriving job to move directly to the head of a queue, while lower-priority jobs languish.

Scheduling based on fixed lead times makes MRP essentially insensitive to actual work loads, priorities, and lot sizes. Although actual lead times cannot be accurately predicted, average lead times *can* be used for planning purposes. In a sense, the use of planning lead times reflects the hierarchical nature of planning in MRP II. As long as the planning lead time is consistent with the average actual lead time, then the planning system will work reasonably well. This consistency can be maintained through the application of the Rough-Cut Capacity Planning "what if" analysis and, perhaps, Capacity Requirements Planning.

3. *The Improper Definition of Lead Time.* Some MRP II systems instruct the users simply to load lead time. Others define lead time as consisting of the sum of the following times elements: queue, setup, run, wait, and move. As many companies have discovered, the diligent loading of all lead time elements usually results in total product lead times so excessive that the company becomes noncompetitive in the marketplace. The required element of manufacturing lead time is run time. All the other elements, queue, wait, move, and even setup, are unproductive or wasted time and should be reduced as close to zero as possible.

4. *CRP Assumes Fixed Lead Times.* CRP uses the various factors from the routing file, such as probable actual setup time and run time (based on quantity) to determine lead time. In fact, many sophisticated systems allow

humans to specify operation overlap in the routings, which will be emulated by CRP. However, CRP still must assume fixed queue times and move times, making the resultant profile unrealistic. CRP is best used as a capacity planning device, not a scheduling tool.

Corrective Action. Inaccurate lead times can kill the success of an MRP system. Understated lead times cause expediting. Overstated lead times, which are more prevalent because of the human tendency to be overly cautious, cause early release of orders, higher work in process, and associated control problems. Lead times tend to be self-fulfilling prophecies. At the risk of oversimplification, lead times can be reduced by a planner at almost any time. However, the shop must be informed in advance, so that they understand why the queues are shrinking. For MRP systems with fixed lead times in the Item Master records, planners can set the lead times equal to 3 to 5 times the actual value-added time that was accumulated from the routings for an average lot size.

M&CRP systems do not inherently cause lead times to become so inaccurate and unrealistic. The crux of the problem lies in how lead times are treated by the users of the system. A system will not have all the problems previously described if the users understand the concept and the proper utilization of lead time. A concerted and continuing education program to drive these points home will provide a high return on the investment.

Excessive Levels in Bill of Materials

Deficiency. MRP uses the Bill of Materials to drive fabrication and assembly job orders. Yet, the Bill of Materials often does not represent the way in which the product is actually fabricated and assembled. For example, assume that a product has a four-level Bill of Materials. In practice, however, all the parts are fabricated and sent directly to the assembly line, where the product is completely assembled in one continuous operation. All subassemblies and final assembly require less than fifteen minutes total. In this case using a two-level Bill of Materials would significantly improve the plans.

Corrective Action. An effective MRP system requires that Bills of Material not only be accurate as far as proper part numbers, quantities, and effective dates, but also be structured according to the way a product is really manufactured. An effective and inexpensive way to continuously validate the part numbers, quantities, and effective dates on Bills of Material is to monitor all unplanned stockroom issues and receipts. Another way is to use the feedback from the PAC system. Engineering and Manufacturing can join forces to minimize the number of levels in all Bills of Material, urged on by Finance (because each level causes material movements, stockroom transactions, Safety Stocks, and inventory) and Sales (because each level increases total product lead times, reducing our ability to respond to customers). As a first step in flattening Bills of Materials, engineers can designate intermediate subassemblies and manufactured components as phantoms.

MRP Assumes Infinite Capacity

Deficiency. Assuming that the shop and our suppliers have infinite capacity, MRP schedules orders to meet the MPS without concern for capacity constraints. MRP further assumes that manufacturing orders have materials as their basic bottleneck. In the 1960s and 1970s, this was true, to a large extent, in job shops and plants that had excess machinery. It has never been true in capacity-constrained operations (especially the capital-intensive basic industries, such as steel, paper, petroleum, or other high-volume process industries). When materials are the only real constraint, MRP attacks the basic problem. When capacity, or labor, or EPA permits, or capital is a constraint, MRP alone is not sufficient.

Corrective Action. Because MRP assumes infinite capacity, the first line of defense against overpromising customer deliveries is a realistic Master Production Schedule. The MPS promises only those items that the plant can actually make, based on demonstrated ability in the past, and the pledge of manufacturing, materials, engineering, and purchasing to produce at that level. For Make-to-Stock environments, the Available-to-Promise capability of most Master Production Scheduling software shows how much of each item will be available on a desired ship date. Assemble-to-Order environments might choose to use an approach consisting of a features-and-options modular bill, coupled with Available-to-Promise for each of the base models, features and options, and supported by a very responsive final assembly area. Engineer-to-Order environments actually need software that is not yet in widespread use: integrated, finite constrained M&CRP, in which MRP and finite CRP operate simultaneously.

CRP Assumes Infinite Capacity

Deficiency. Just like MRP, CRP assumes infinite capacity as it backward schedules. Just like MRP, this assumption is often invalid, and can easily cause problems. The CRP system has evolved to be an infinite loading and decision-support planning system, in which the users develop plans and schedules using the "what if" support provided by the CRP system. When manufacturing processes become more predictable, either through process simplification (e.g., JIT) or through engineering and automation, the infinite-loading and decision-support strategy will no longer be appropriate, particularly at the PAC level, and perhaps at the M&CRP level as well. We expect that finite scheduling systems will become widely used at the PAC level. We anticipate that both the infinite-loading and decision-support system and the finite-scheduling system will be used at the CRP level.

Corrective Action. A transparent method of Finite Scheduling (one in which the user can easily follow the internal logic) should be included in CRP. One Finite Scheduling technique, Operation Sequencing, is a simulation technique that models and predicts the sequence that jobs will follow through work centers, based on priority and available capacity. Starting with the highest priority jobs, it forward schedules to each work center's finite capacity, based on priority. It creates its own queue times, based on the projected actual load at each work center.

Finite Scheduling is not intended to replace Infinite Loading. It is intended to complement it by utilizing CRP first, to determine where overloads will occur, based on the infinite capacity assumption, and then utilize the simulation model to determine how best to reschedule the load, redeploy the available capacity, and/or adjust the Master Schedule.

MRP Cannot Accurately Predict When an Order Will Ship

Deficiency. Very few MRP-based systems in existence today can provide quick and accurate answers to requests, from marketing and customers, regarding potential new orders on a daily basis.

Corrective Action. The proper answer to the question, "When can I ship an order?" reviews projected capacity availability, as well as material availability. For any environment except Make-to-Stock, a truly responsive action would require an MRP system with integrated, finite scheduling, which could load the order into the schedule and provide an accurate answer.

Both MRP and CRP Use Backward Planning Almost Exclusively

Deficiency. Backward planning can result in schedules in which the start date of orders is already past due. Typically, a planner-scheduler must modify each of those orders before CRP can be meaningful. This can also impact higher-level assemblies and/or other components and assemblies at the same level.

Corrective Action. The capability for forward scheduling, as well as backward scheduling, should be provided, especially for CRP systems. Backward scheduling could be used for the first pass, then forward scheduling for all orders that have start dates past due, to find a realistic finish date. Some systems already do this. Forward scheduling, in conjunction with finite scheduling, would represent a dramatic increase in the capability of CRP.

No Well-designed Formal Feedback Procedures

Deficiency. When a problem occurs on the shop floor, or a raw material is delayed, the system has no well-defined methodology for recovery. MRP correctly computes vertical dependencies; for example, that each part and subassembly is required on May 15 for a final assembly to ship on May 18. However, if the subassembly cannot arrive until May 27, MRP does not directly assist the planner to identify and reschedule the other components (such as other subassemblies and purchased parts) that are also due on May 18, unless the planner reschedules the Final Assembly order.

Corrective Action. One approach to the problem of improved feedback is Chaining. Chaining is a process that recognizes the dependency between job-to-job relationships, in much the same manner that multiple-level pegging and where-used logic works. Chaining, or job networking, allows instant identification of all job dependencies. This is very useful for customer service inquiries, when each job that is related to a sales order and its status, as simulated by operation sequencing, is available for display. A similar approach would be useful when contemplating a revision to the Master Schedule.

Group Technology Is Not Supported by Most MRP Systems

Deficiency. Few MRP packages have the ability to handle group technology codes. None seem to have the ability to plan and schedule on the basis of groups or families. Group classification codes have been defined and published. However, MRP packages have not incorporated these codes that would permit them to combine requirements or orders on a common schedule.

Corrective Action. M&CRP should have the capability of supporting Group Technology. Group Technology coding and classification of parts allows families of similar parts to be developed. These families can be processed in the same manner, thus permitting the development of Group Technology work cells that convert job shops into flow shops. Group Technology can make the implementation of JIT feasible, where it wasn't feasible before. In short, Group Technology improves the cost, quality, flexibility, and time responsive capability of a company.

MRP Does Not Adequately Support Design-to-Order (DTO) Job Shops

Deficiency. In DTO job shops, the customer order is not definitive enough to use in MPS to drive MRP. The customer order typically consists of specifications or verbal descriptions of what is being ordered. The information that is available in the customer order is sufficient to perform the engineering design, but it is not adequate for manufacturing planning.

Additionally, most MRP systems do not integrate with the engineering design systems; when the engineers change a design in their engineering system, they must remember to maintain MRP. Likewise, when engineers change a process sheet, they must remember to tell MRP.

Corrective Action. DTO job shops can use concurrent engineering to more quickly create the data required for MPS. Some type of decision support, or artificial intelligence system, should be included in the MPS-MRP system, in order to develop a Bill of Material for a design item being ordered, which uses a menu or question and answer session. Otherwise the manufacturing function must wait until the engineering is completed before it is able to load an item into MPS that is capable of driving MRP.

Future MRP systems need to start interfacing more fully with the more popular engineering systems, including CAD and CNC.

Excessive Manual Data Manipulation Required

Deficiency. The day-to-day operation of most systems requires too much manual manipulation of data. In fact, in several companies, MRP has been an acronym for "More Ridiculous Paperwork." This is not a question of data accuracy, but one of systems design and philosophy.

Corrective Action. MRP must become an acronym for "Merciless Reduction of Paperwork." Although most MRP systems are still relatively transaction-intensive, many of the newer ones are becoming less so. The system should be designed so that the computer does as many routine "paperwork" tasks as possible, freeing the human to deal with exceptions and to make key decisions. A stated objective of eliminating all transactions can assist a company to minimize MRP operating overhead.

Difficult to Implement

M&CRP-MRP II systems are notoriously difficult to implement successfully. Chapter 11 addresses this problem in detail.

Changes to Operate in New Environments

The current status of M&CRP-MRP II adaptation for use in a number of nontraditional environments, was described in Chapter 9. However, what has been accomplished to date is only a small part of what can, and will, be done to truly adapt M&CRP-MRP II to many environments in which it can be useful. MRP II software developers are attempting to serve various niches in the market. They are developing, or have developed, modules that extend the usefulness of MRP II to satisfy more specialized needs.

In addition to the necessary technical changes, a philosophical, or attitudinal change, also will occur. Currently, many people believe that M&CRP is only useful for job shops, despite the fact that it is more often used in other types of environments. As M&CRP-MRP II are more widely applied in these environments, this attitude will change.

We now turn our attention to some of the nontraditional environments in which M&CRP-MRP II can be useful, and list some of the changes that will be made to adapt M&CRP-MRP II to these environments.

1. *Repetitive Manufacturing.* Systems for repetitive manufacturing must operate with less paperwork than job shop systems. One technique to reduce paperwork is to produce rate-based schedules, which define the number of units per period, rather than using discrete job orders. In addition, work-in-process inventory is relieved by deducting the number of units completed. Process cost accounting systems should be used. Additional features, such as flexible line and routing definition, optimized production schedules, backflushing, and operation overlapping, are being developed.

2. *Just-in-Time.* In addition to the requirements for repetitive manufacturing, a company using JIT requires the following features: JIT order release, Kanban, supplier management, overlapping, and interfaces to data collection system. The logical approach is to integrate the JIT and MRP II software. This integration is consistent with the practice in many repetitive manufacturing companies of using MRP II as a planning system, while us-

ing JIT as the execution system. Alternatively, the software for MRP II can be extended and enhanced to support JIT.

3. *Process Manufacturing.* In addition to the features required in repetitive manufacturing systems, a process manufacturing company requires M&CRP-MRP II to recognize that capacities are fixed and that they serve as constraints, before material plans are developed. Other required features include the planning of coproducts and byproducts, using recipes or formulas, rather than Bills of Material, theoretical consumption, tank inventories, backflushing, and special routings for off-spec materials. MRP and Process Flow Scheduling (PFS) can coexist in the same planning system. MRP can be used to develop raw material plans by using a two-level Bill of Materials that contains finished products and raw materials. The PFS system can manage work-in-process inventories and create detailed schedules.

4. *Program-Project Management.* The work breakdown structure, specified by MIL-STD-881A for project management, can be used in the same manner that the Bill of Materials is used in MRP. Other concepts, such as lead time, run time, and setup time are also applicable, but must be especially defined for the project environment.

5. *Cost Accounting.* This capability includes the current practice of accumulating actual production costs and reporting variances (labor, material, overhead) from standard. This area requires a major shift in emphasis from job shop processes to repetitive flow processes, Just-in-Time, and Flexible Manufacturing Systems that will require new cost accounting techniques, such as Activity-Based Costing.

6. *Forecasting.* The capability is required to forecast the demand for the company's products for use in Production Planning and Master Production Scheduling. Most MRP II packages have several techniques for forecasting. Forecasting packages that are especially adapted to a particular environment have yet to be developed.

7. *Order Management and Sales Analysis.* This capability has been very difficult for vendors to create, even in similar businesses, because customer requirements are so varied. We foresee considerable interfacing and integration between customers' and suppliers' systems, using some form of electronic communication, such as EDI.

8. *Demand Management.* The module combines forecasting, distribution, and order management to provide a consolidated input to the Master Production Schedule. Future modules in this area can be expected to provide a capability that is better integrated and much more useful.

9. *Quality Tracking and Control and Statistical Process Control.* This capability supports the normal functions of scrap and reject reporting, but more importantly, it also includes statistical techniques for quickly monitoring ac-

ceptable and unacceptable inspection rates and for diagnosing reasons for out-of-tolerance conditions. This module is required for Total Quality Management.

10. *Tool Management and Control.* Companies are beginning to understand that tooling is an important, but often overlooked, manufacturing resource. For a product to be built on time, within quality specifications and cost constraints, management must effectively control, not only manpower and equipment, but also tooling.

11. *Safety and Hazardous Material Management.* The growth of this module reflects the increasing concern of people, companies, and government with the impact that manufacturing has on people and the environment, and symbolizes the movement toward ecologically conscientious manufacturing. It includes managing Material Safety Data Sheets (MSDSs) and tracking all handling, procurement, storage, and disposal of hazardous materials. It can also include tracking worker exposure to such materials.

12. *Human Resources Scheduling.* M&CRP-MRP II systems have routinely ignored a company's most precious resource—its people. Current human resources packages can track an individual's experience and capabilities; finite scheduling packages need to also project a person's availability (e.g., if a key setup person is scheduled for vacation, a company might not be able to run specific products).

13. *Maintenance Planning.* Like manufacturing, maintenance has a schedule of products (maintenance activities) and a bill of resources to produce these products. Adapting the concepts of M&CRP to maintenance is straightforward; carrying out the details of implementation in a cost-effective manner is not so straightforward.

14. *Greater Integration.* Integration with modules outside MRP II is needed. Especially, linkages are required with accounting, computer aided design, engineering, customers, and suppliers. A recent need, especially in the automotive and aerospace industries, is the capability to interface with Electronic Data Interchange (EDI).

Changes Resulting from Changes in Computing Technology

How Will Computer Systems Change in the Future?

At the same time that new approaches to solving manufacturing management's problems are being developed, the computer hardware and software technologies are evolving rapidly. Although mainframe computers have steadily improved their price-performance, minicomputers have improved their price-performance several times faster, and are now capable of supplanting many mainframes in terms of work performed. However, microcomputers and workstations are also improving their price-performance, at breathtaking rates. Micros and workstations can now support reasonably full-featured manufacturing systems, either in:

- The traditional mode of a central computer, sending data to "dumb" terminals (and yes, microcomputers can now support an increasing number of dumb terminals, just like minis or mainframes).
- A Local Area Network, in which the central computer acts as the "file server," which stores and furnishes data, but with the actual processing being performed on the local computers.

Data collection techniques are also improving. Bar coding and process monitoring devices are gaining in sophistication. On the system software side, improvements include relational database management systems (improved reporting), mainframe to micro linking software, new operating systems, and more powerful end user utilities (screen and report writer).

Guidelines for Exploiting Improved Computing Technology

Real advances in utilizing future computer and information system technology to improve M&CRP-MRP II will come only with large doses of creative intelligence. Somehow, over the years, we have contorted the task of manufacturing planning and control by assuming it could be parceled into so many individual activities—inventory planning, production scheduling, capacity planning—as if they were all separable and unrelated functions, each having different objectives. Thus, if splintered techniques are symptomatic of the problem, more stand-alone techniques cannot cure the problem. We can stop expecting Lot Sizing, input/output analysis, priority dispatching, queue control, work-load balancing, and other partial solutions to correct systems failings. They are all band-aids; at best, they address the symptoms, rather than the causes. At worst, they cause, or exacerbate, the problems.

Guidelines for Future M&CRP-MRP II System Development. To guide our search for new methodologies for M&CRP-MRP II, and to ensure that we do not repeat the mistakes of the past, we suggest developing systems that meet the following criteria:

- Supports and substantially improves managerial decision making, instead of reporting on events or accounting for data. Computers have helped to create this sea of data in which most managers drown (and which many leaders ignore). Computer data frequently supports increased analysis without increased understanding of the underlying issues and causes.

- Facilitates the insertion of new knowledge as it is discovered. Our current systems are rigid and are technique-oriented. When we develop a new understanding of manufacturing, our systems are often unable to absorb it. Witness the plethora of articles, for a full decade, concerning whether it is even possible to integrate MRP and JIT!
- Integrates the entire manufacturing enterprise. We are woefully deficient in formally linking engineering, marketing (as opposed to sales), and human resources, for example, to materials and manufacturing.
- Addresses not only the individual manufacturing enterprise, but also its direct environs (e.g., customers, suppliers, community, and regulators). Our current systems are equally deficient in their ability to interface, let alone integrate, to external entities.
- Eliminates work that does not add value in the eyes of the customer. JIT has taken the lead in this most critical area, which is an inherent weakness of most M&CRP systems. The future Manufacturing Planning and Control system must eliminate paper-shuffling, data gathering and entering, and other forms of waste, rather than demanding more of the same. New systems that provide more realistic plans (e.g., computer-controlled shop floor schedules) require exponentially more data in very short time frames.
- Intelligently utilizes the ever-increasing capability of computers. MRP-based planning systems were designed for use on third generation computers of the 1970s, the processing speeds of which were measured in microseconds. Today, computers are easily thousands of times faster, and the next generation of computers, which are just around the corner, promises to be dramatically faster still. In the early 1990s, people were buying computers for less than $2000 (for home use!) that have more power than the largest corporate mainframes did 20 years ago. Yet the most advanced Manufacturing Planning and Control software that is in commercial use does not come close to exploiting this potential.
- Simple to use, yet not simple-minded. One reason the PC-based applications have become so popular is that individuals have been able to understand them. Windowing technology promises to make PCs even easier to use and understand. Most Manufacturing Planning and Control systems have basically ignored PCs and the windowing technology. Most are still designed around a central processor base, with "dumb" terminals as the user interface.

Increased Use of Simulation. We must take advantage of the full power of the computer in Manufacturing Planning and Control systems. Heretofore, most systems have not done that. Specifically, we are not using, to the extent that we should, the ability of the computer to simulate factory operations. This enormous power to simulate plant activities is the single most promising opportunity available for significantly improving the technical capability of the integrated manufacturing system.

The total process, including MPS, MRP, CRP, PAC, and Final Assembly Scheduling, can be done as a single, complete simulator of the total manufacturing plan. This is true manufacturing resource simulation—the ability to imitate, in the computer, the way in which materials and capacity will interact to provide product availability, and in so doing, to see in advance the problems that will interfere with execution of the manufacturing plan.

Manufacturing Resource Simulation—linking of MRP for materials planning and CRP for capacity planning communicating with MPS and PAC—is more powerful than the two techniques used separately, because it shows when capacity will be available to support the materials plan, and when materials will be available to support the capacity plan. And it does this interrelated scheduling without the enormous burden of human effort that is expended in monitoring production orders, tracing material, and guessing probable effects, which is inherent in the standard system.

New Techniques to Utilize Capabilities of Future Computers

Revise MRP II's Sequential, Independent Approach. The current MRP II planning approach first develops a schedule that satisfies demand, and then looks at the resource implications, rather than developing a schedule that satisfies both the demand and the resource constraints simultaneously. This separate modular approach is an anachronism from the early days, when computer systems did not have sufficient processing power and storage capability to solve the combined problem. The result is production schedules that are often found to be infeasible too late in the planning process to allow the company to recover.

One solution to this problem requires a complete rethinking of the planning approach used in MRP II.[1] The MRP II planning process is currently broken down into four levels: (1) PP-RRP, (2) MPS-RCCP, (3) MRP-CRP,

[1]Howard W. Oden, "Designing Production Planning and Control Systems for Flexible Manufacturing Systems," *Proceedings of Manufacturing International 1988,* American Society of Mechanical Engineers, 1988.

and (4) PAC-IO. These decision levels form a natural hierarchy based on the lead time needed to execute decisions, the planning horizons for analyzing and evaluating choices, and the degree of information detail required to make decisions. Each of these levels requires different kinds of models and data to support decision making.

We can envision two phases of the new technology:

1. In Phase I, each level in the hierarchy would have a decision system composed of the following components: a human decision maker, an automatic decision maker, data appropriate for that level, and various models that could be utilized by the human and automatic decision maker to process data. The automatic decision maker would make all the decisions for routine and repetitive situations. For novel and unforeseen situations, for which the automatic decision maker was not programmed, the decision would be made by the human decision maker, assisted by the models appropriate to the situation. The system would perform the material planning, and the capacity planning would be performed simultaneously at each level, with rapid and continuous feedback between all levels.
2. Farther into the future, Phase II might consist of a completely integrated planning tool, which incorporates all four levels of the current hierarchy.

Smaller Distributed Computers in Future. In the past, computer manufacturers advertised systems that were upgradeable to large systems. Today, computer manufacturers recognize the importance of microcomputers in the office and factory. Most now offer some facility to download information, collected through the MRP system, to PCs for use in spreadsheets and reports. Currently the information is not always in usable form when downloaded. In the future, we will see a refining of the uploading and downloading process and an automatic formatting of the information for PC use.

We expect this trend toward smaller, distributed (linked) systems to continue at an ever-increasing pace. We expect the trend of replacing mainframes with minis and minis with micros to increase. PCs can be used independently for word processing or data manipulation, while still being tied into a network with a central file server. Networked systems can expand or shrink to fit the needs of the company.

Electronic Data Interchange (EDI). EDI is the electronic transmission of standard business documents in a standard format. EDI replaces conventional paper documents, such as purchase orders, invoices, and transportation documents, with computer messages in a nonproprietary, noncopyrightable, and publicly accepted format. EDI is rapidly becoming the standard way of doing business, both domestically and internationally. More than 75 percent of the Fortune 100 companies, and almost half of the Fortune 500 companies, had EDI in their business strategies in the early 1990s. These large corporations are forcing their suppliers to use EDI as a condition for continuing to do business, as are many of the large retailers.

EDI can be implemented in two ways.[2] One way is to exchange magnetic tapes, disks, or other storage devices. A courier, in-house or hired, delivers the tape containing the transaction information. The other, and more advantageous, method is transmitting information by direct computer link. There are three ways for a business to transmit and receive directly through their computer.

- Write an in-house program to perform the necessary translation to a standard EDI format
- Purchase a software package that takes your data and translates it into an industry standard that can be transmitted
- Utilize a third party that does all the translation of both your incoming and outgoing data

We recommend the second and third alternatives. Because EDI standards are not consistent and can change, keeping an in-house program up-to-date can present problems. For EDI service and software package providers, that is their business.

EDI can dramatically improve the quality of information in terms of content, accuracy, and timeliness, with a resulting impact on control and operations. Information can be delivered virtually instantaneously. The reduction in the time taken to send, receive, and acknowledge a business document can create a tactical advantage, as fax technology did recently. For example, suppose you normally carry an extra two weeks supply of a critical item in order to allow for purchasing lead time. If you can reduce the time it takes to place an order, you can reduce the amount of Safety Stock carried. The concern about whether your purchase order arrived intact and on time is eliminated. You can know right away if your transmission was received. There are virtually no transcription errors, because the data is transmitted directly from your own computer.

Artificial Intelligence (AI) is a computer technology that is concerned with capturing distinctly human capabilities, such as thinking about prob-

[2]C. R. Banton, "The ABCs of EDI," *APICS 1989 Conference Proceedings*, APICS, 1989.

lems, recognizing patterns, and reacting appropriately to unique situations.[3]

Although AI is used in robotics, speech recognition, and natural language processing, it also includes the decision-making systems known as expert systems. Expert systems use AI technology to provide broader access to the knowledge and reasoning of experts and are simulated on computers that are designed to handle concepts and symbols, as well as numbers. Expert, or knowledge-based, systems have the following distinctive differences from conventional software.

- *Data versus knowledge.* Conventional software deals in absolute data values, whereas expert systems can use incomplete or imprecise data.
- *Quantities versus qualities.* Conventional software manipulates numerical values (1 + 1 = 2), whereas expert systems can produce qualitative as well as quantitative results. They can reason by analogy, discard irrelevant background information, and apply past experience.
- *Passive versus active.* Conventional software responds to user instruction. Expert systems can participate actively, alerting the user to inconsistencies, shortcomings, or overlooked opportunities.

By making use of a combination of artificial intelligence techniques, new advances in expert systems applications are possible. Because expert systems can incorporate qualitative, as well as quantitative, information, the knowledge and experience of seasoned users can be integrated into systems that can guide, interpret, simplify, interact with, and construct relevant models for, and approaches to, real problems.

Expert systems are currently assisting customer service representatives in configuring complex customer orders. AI can potentially be used to support capacity management and to assist the analyst in answering such fundamental questions as, "How can I utilize my existing capacity more effectively?" and "By shifting capacity, can I reduce lead time or minimize work-in-process inventory?" Even though expert systems represent a significant step forward in the science of managing capacity, capacity planning can't stand alone. It depends on many other factors, including good systems for material planning, production planning, and master scheduling.

AI can combine the best traits of human instinct, knowledge, and experience with the power, speed, and consistency of the computer. However, creating a knowledge base is very difficult, time-consuming, and expensive.

[3]Barbara J. Perrier and Mary E. Cross, "Capacity Management and Artificial Intelligence: A New Approach to a Changing World," *APICS 1987 Conference Proceedings*, APICS, Falls Church, VA, 1987.

Even experts quite often disagree on fundamental issues of causality and relationships (for example, economists discussing how to reduce inflation or to stimulate the economy).

Changes in the Heart of M&CRP

As currently used, Material and Capacity Requirements Planning plans for the availability of three primary resources required in manufacturing: material, labor, and machine capacity. As the shop floor becomes more automated and approaches a flexible manufacturing system, we will start employing computer aided planning for resources that are currently being planned on a manual basis, such as, material handling equipment, quality control equipment, tools, fixtures, and NC programs. When this occurs, the process can more appropriately be called *Integrated Resource Planning* (IRP), or *Enterprise Resources Planning* (ERP).

Integrated Resource Planning is defined as the planning that determines the quantity and timing of all production resources needed to produce the end items in the Master Production Schedule. Production resources include raw materials, purchased parts, produced parts, personnel, processing machine capacity, material handling capacity, tools, fixtures, NC programs, and such others as needed to produce the end items.

How Will Manufacturing Change in the Future?

In this section we explore how the manufacturing environment, manufacturing companies, and manufacturing systems will change in future. Thus, we are defining the environment within which M&CRP-MRP II systems will have to operate in the future.

Expected Changes in the Manufacturing Environment

The Unprecedented Nature and Pace of Change. A major distinguishing factor in the new manufacturing environment is the unprecedented pace and character of change in manufacturing, both in its technology and management practices. Although the development of new manufacturing technologies is not unique to the current environment, the pace of technological change and the breadth of products and processes being affected are unprecedented. Developments in microelectronics receive most of the credit for this explosion in technological change. As the cost of computing power has plummeted, a vast array of applications has become economically feasible. The computer revolution not only has allowed the creation of "intelligent" products and processes, but also has greatly enhanced the ability of researchers to improve our understanding of the physical and chemical sciences, which are so important to manufacturing processes. The possibilities for applying new materials and processes to improve the price and performance characteristics of manufactured products have multiplied dramatically.

Emerging Strategies in Inventory, Quality, and Integration. The pace of change in manufacturing strategy has also accelerated. Some newly emerged principles for effective manufacturing practice include such areas as inventory, quality, flexibility, timeliness, and the interdependency of products and processes.

In *inventory,* the emerging principle, simply put, is that inventory is bad. High inventory levels, including finished product, work-in-process, component, and raw materials, are extremely costly. More importantly, they effectively hide operating problems and quality defects, both in supplies and in production. Our profession now understands that minimized inventory is fundamental to effective manufacturing. Practitioners are replacing "insurance" stockpiles of raw and in-process material with quicker processes, higher quality, and greater flexibility.

New approaches to *quality* have also become necessary, as competition has intensified. The traditional practice calculates the cost of quality primarily in terms of after-sales service costs. Poor quality means that our processes are out of control and unpredictable. The real cost of defective parts, including scrap, rework, and service, as well as the opportunity costs of lost sales and lost consumer confidence, must all be included in the cost of quality algorithms.

With the United States' changing position in world competition, *flexibility* is becoming a most important strategic factor. No longer able to compete with low cost mass-produced goods, the United States must be capable of responding rapidly to the changing needs of customers, both in volume and in design. Customers continue to want products, services, and capabilities tailored to their specific requirements, at no or low additional cost in terms of money, time, or quality.

Timeliness is a renewed major competitive battleground. A supplier that can ship within a day has a major competitive advantage over one who requires a week. A supplier that can design a product modification or new product quickly has a major competitive advantage over one who takes longer. The time factor, in this case, is measured by the customer from the date they have the need until the date it is filled, which is far more comprehensive than today's manufacturing order.

The final principle is the realization that effective manufacturing requires *close linkages among materials, processes, and products.* Today, this requirement is perhaps most obvious in the semiconductor industry (new products can be envisioned but the process technology or equipment is not available to make them), but the same principle has applied, at one time or another, across all manufacturing.

All five of these areas relate to manufacturing costs, and are fundamental to the survival of any manufacturing company in the future. They display the need for an approach to manufacturing that does not depend solely on evolutionary improvement, and offer an opportunity for the development of underlying principles that can fuel revolutionary change.

The Changing Basis for International Competition. Another important characteristic of the new manufacturing environment is the changing basis for international competition. Growing international competition in manufacturing is no longer based solely on low prices that stem from cheap labor. The United States no longer has an unchallenged advantage in many other areas—technology, engineering talent, R and D, entrepreneurship—that determine competitiveness. Many countries have developed strong financial, educational, technological, and scientific infrastructures, which allows them to move beyond labor-intensive industries to challenge U.S. manufacturers across the industrial spectrum.

U.S. manufacturers must embrace product customization and production flexibility. Increasingly sophisticated consumers throughout the world are no longer content with identical, mass-produced goods; they are seeking ever more sophisticated products that are tailored to individual needs. Recent developments in computers, communications, and manufacturing technology are also pushing us in this direction.

In the long run, the goal of "total flexibility" (high quality, custom-tailored products at mass production prices) is both feasible and mandatory. We are not quite there yet, but it is clear that customizing products, and being flexible enough to shift production smoothly and efficiently among a broad range of products, are emerging as two of the main competitive arenas for future manufacturing companies.

The latest challenge in manufacturing strategy is *time-based competition.* Aggressive companies are altering their objectives from competitive quality, customization, and cost, to quality, customization, cost, and *time-responsiveness.* To seize the initiative and achieve a competitive advantage, manufacturers must compete in time, as well as in the other elements.

Required Responses to the Changing Environment. The future manufacturing environment will require manufacturers to radically change attitudes toward factory control. Simply increasing throughput rates cannot achieve the rapid product development and production cycles that will be required. Rather, we will need to focus on maximizing flexibility in product output and, at the same time, maintaining control of factory operations, thereby increasing confidence that necessary processes are being performed with high quality and minimum resources.

Expected Changes in Manufacturing Enterprises and Systems

Changes in Corporate Strategy. Criticized for many years for their short-term approach, many U.S. companies operating in the global market have become believers in long-range planning. Manufacturing will play a much larger role in corporate strategy in the future, than it has in the past. The manufacturing capability of an enterprise is an important competitive weapon. It will be far easier for manufacturing managers to master the process of corporate planning, than for the other managers to understand the technical details needed to assess the impact of advanced manufacturing in world competition.

Changes in Enterprise Organization Structure. Customer demand and competition dictate that a greater variety and quality of products be produced with shorter design and manufacturing lead times, which can be accomplished only by a flexible enterprise that is very responsive to the customer. The emerging picture is that of leaner, more flexible companies, with organizational hierarchies of only three or four levels, and with rapid reaction times.

The most used organizational unit will be that of the multifunctional team. The increasing complexity of technology and the press of time will not allow decisions to be made by individuals, in the traditional sequential fashion. Instead, decisions will be made by teams that are composed of people with highly different backgrounds. Such functions as product/process design, quality improvement, and production scheduling will be accomplished by groups. Even routine work will be accomplished in teams, to give members a better opportunity to contribute and exercise autonomy.

With the increasing capability of computer information systems, better communications and data transfer will be possible among top management, marketing, engineering, manufacturing, accounting, and others. Technically, the walls between engineering, manufacturing, and marketing will disappear. However, such capability will be of little value unless the walls are torn down in a management sense. Real integration will occur only when the members of the management team are cooperating to produce quality goods at reasonable prices.

The Factory of the Future. There are two complementary viewpoints concerning the factory of the future; both will be implemented extensively, often in the same industry, and sometimes in the same company:

- Flexible Manufacturing System (FMS) that is extensively automated with a computer-integrated group of computer-controlled machines or workstations, linked together by automated material handling for the completely automatic processing of various product parts, or the assembly of parts into different units. It has an integrated engineering and manufacturing database to automatically design products and processes, estimate and order material, control inventory, program machines, and perform all other activities of the manufacturing process. The purpose of the FMS is to respond accurately and rapidly to the needs of the customer. The FMS can respond rapidly to changes in product design, product volume, or product services. The progress of FMS in achieving maximum flexibility and efficiency is driving batch manufacturing toward cost-effective lot sizes of one.
- Agile Manufacturing System (AMS) enables a company to achieve many of the benefits provided by FMSs, without using extensive automation. AMS is more of a philosophy than a specific set of hardware. In one industry, an AMS will use JIT as the shop floor execution vehicle, because JIT is by far the most appropriate. In another industry, an AMS will use an automated system on the shop floor, because the technology is available and cost-effective. The hallmark of an AMS is its ability to respond to ruthless time-based competition, emphasizing quick response, flexibility, and efficiency. In a more general sense, an AMS is any manufacturing system that has the capability of being completely responsive to the demands of the customer.

Although the details will vary by industry, the factory of the future will challenge our long-held belief that high-volume runs of identical products are required to achieve low cost. It is conceivable that early in the next century, Flexible or Agile Manufacturing Systems will produce virtually all the material goods required by society, except those with high artistic content.

The companies that master this transition will gain nearly unassailable positions in the world market through their ability to produce quality products that are tailored to meet special customer requirements on a very short lead time. For a manufacturing plant to be capable of efficiently producing any and all of its products on demand and with short lead time, while conforming to quality standards, the plant requires:

Well-defined product lines

Tight scheduling, that is, production responsive to customer demand

Efficient, flexible layouts and balance process capabilities

Well-developed processes operating under statistical control

Small lot sizes

High employee involvement

Continuing training and investment in employees throughout their careers

Although the technology and the systems to manage it continue to change, it is apparent that manufacturing is well into a new era. Competition now is worldwide and tough. Obviously, manufacturing equipment, material handling equipment, computers, inventories of all types, technologies, people, and management systems must be capable of rapid, innovative change in order to compete. Having the flexibility to adapt, the ability to integrate the components of manufacturing into an effective system, and the capability to control its functioning is an extremely ambitious undertaking. And yet, it appears that nothing less will be successful.

M&CRP-MRP II Operating in the Future Environment

Customers expect a product to satisfy certain general objectives and to have the specific features it needs in order to perform its basic function. These general objectives are:

Quality

Flexibility and variety

Cost

Time responsiveness

Within specific markets, certain objectives may be more important to the customers than others. Also, if all objectives have been met except one in a specific market, the competition will be based on the unmet objective. In many of tomorrow's world markets, the consumers will expect all producers to automatically satisfy the first three objectives (i.e., they will deliver high quality, customized products at a low price), and thus competition will center on the last objective, time responsiveness. This will result in time-based competition.

The general objectives desired by a particular market will not only affect the design of the product, but will also affect the design of the manufacturing enterprise that produces the product. In the following paragraphs, we will focus on how the processes of a manufacturing company, particularly the processes of the M&CRP-MRP II system, must change when that company engages in time-based competition.

First, let's look at the processes that must be performed for a producer to respond to a customer's order. Response time includes far more than the time required to manufacture (fabricate and assemble) the order. In the customer's eyes, it includes all the time elapsed from when a need first occurs until the need is satisfied. This can involve many or all of the processes shown in Figure 13-1.

Time Consuming Processes

1. Customer Ordering Process
   - The customer detects the need and decides to fill it
   - The customer prepares specs and asks for quotes
   - Producer prepares quotes
   - The customer selects a product and producer to fill need
   - The customer places an order for the product with the producer
   - The producer enters the order in its system
   - The producer acknowledges a ship date to the customer
2. Product-Process Design Process
   - Marketing and R and D Engineer defines customer's needs and product concept
   - Design Engineer develops preliminary design of product
   - Engineering builds and tests a prototype of product
   - Design Engineer develops final design of product
   - Manufacturing Engineer selects process type and major equipment
   - Manufacturing Engineer develops Manufacturing BOM and routings
3. Material and Capacity Procurement Process
   - Marketing prepares a forecast of future sales
   - Manufacturing procures long-range resources (plant, equipment, people, and so on)
   - Manufacturing procures all material needed

     Manufacturing sends material requisition to Purchasing

     Purchasing prepares purchase orders and returns to Manufacturing for validation

     Purchasing reviews possible suppliers and selects supplier

     Purchasing sends purchase order to supplier

     Supplier receives purchase order and enters order into supplier's planning system

     Supplier orders materials and receives them

Figure 13-1. Time-consuming processes in a manufacturing firm.

To compete on the basis of time responsiveness, a manufacturing company must do everything in its power to reduce the time between the instant that the customer discovers the need, until the customer can effectively use the product. In the following paragraphs we describe how M&CRP should change in order to reduce the time of the various processes shown in Figure 13-1. Because the remainder of the book discusses, in detail, how

Supplier fabricates and assembles item
Supplier ships item to producer
Producer receives item and inspects it
Producer moves item to point of use and enters it into inventory system

4. Manufacturing Process
   - MRP develops due dates for all required products
   - MRP planner releases orders for all parts to the shop
   - Part is placed on Dispatch List of *first* work center
     Part is *moved* to the first work center
     Part waits in *queue* until machine is available
     Part waits while the machine is being *setup*
     Part is *run* on the machine to process the part
     The part *waits* to be moved to next work center
   - Part is placed on Dispatch List of *next* work center, and the work center cycle, described in the above five steps, is repeated for each work center on the part's routing.
   - Worker inspects part prior to final assembly
   - Material Handler moves part to final assembly area
   - MPS planner releases assembly orders for all products
   - Assembler assembles part into the first subassembly, the second subassembly, and so on, and finally into the final assembly
   - Inspector inspects product and moves it to finished goods stores
5. Delivery Process
   - Producer picks product from stores and moves it to packaging
   - Producer packages product and moves it to shipping
   - The shipping company transports the product
   - The customer receives the product
   - The customer inspects and/or tests the product
   - The customer moves the product to point of use and informs their system that it is ready for use

Figure 13-1 *(Continued).* Time-consuming processes in a manufacturing firm.

M&CRP can improve the Manufacturing Process, we will restrict ourselves to the other four processes, namely

Customer Ordering Process

Product/Process Design Process

Material and Capacity Procurement Process

Delivery Process

Customer Ordering Process

Although Sales and Marketing is primarily responsible for the Customer Ordering Process, Materials Management can significantly assist in reducing the time needed to carry out these processes. Materials, Sales and Marketing, Finance, and Manufacturing jointly decide which of the company's products will be Make-to-Stock, Assemble-to-Order, and Make-to-Order, based upon the lead time needed in order to be competitive, and the response times of the manufacturing process. Marketing and Materials Management should be prepared to rapidly change the response strategy for products if needed for time competition.

Order entry must be easy, rapid, and error-free, including standard definitions of products currently desired by customers. We should be willing and able to receive orders via Electronic Data Exchange (EDI) from those customers so equipped.

Another possible vision for the system of the future involves having the customer service person sitting in front of a Windows-based screen, with the PC linked directly to the incoming phone. When the phone rings, the computer (based on the incoming phone number) immediately displays the current customer status summary record on the screen for the customer who is calling. If the customer wishes more information on a particular order, the customer service representative points to that order number, clicks the mouse, and windows into the next level of detail.

One final aspect of this vision is that the customer service rep can honestly answer the age-old question, "When can I get my order?" The customer service rep can enter the customer's request, hit the "what if" key, and have the system load the prospective order in a few seconds, based on an on-line version of integrated MRP and Finite Scheduling. Because this reviews other orders in house, as well as current and projected materials availability, the answer is accurate. If the customer likes the promised delivery, the customer service rep hits the "save" key, and the system saves the order and reserves all the resources immediately.

The generation of software after that may actually link directly to the customer's computer, eliminating the need to involve customer service representatives in this particular information dissemination function. If we are willing to link our computer directly to our customer's computer, our customers could potentially be able to inquire on their computer about the status of their order in our plant, without even calling our customer service department. This would increase the need to keep all our shop information truly up-to-date and accurate!

Product/Process Design Process

The speed of the product/process design process, or how fast a company can bring new products to market, is critical in today's time-competitive markets. In the highly competitive automotive industry, Japanese companies, such as Honda, create and sustain an advantage by virtue of the rate at which they can introduce new technology. These companies are not only faster, they are also more efficient. In the 1980s, Japanese automakers, on the average, turned out a new automobile in nearly half the time and with half the engineering hours required by the Big Three automakers.

The key to making new product introduction a formidable time competitive weapon is the adoption of the concurrent approach to product-process design, introduced in Chapter 1. We replace the traditional sequential design and development by concurrent team efforts, with tasks organized on a parallel and cross-functional basis, so team members work in a collaborative, cooperative, and supportive way. The cross-functional team works and lives together as a group on the project. The concurrent process involves simultaneously considering all stages of the product's life cycle during the initial design stage; product cost and performance are engineered to meet the desired objectives. The team considers many product and process alternatives early, in order to select the most competitive (i.e., cost effective) alternative for further development.

In a vision of the future, the design team would be well supported by a completely integrated system, which would not only perform the engineering modeling functions of current systems, but also fully integrate to:

Artificial Intelligence to assist with design trade-offs and considerations

Materials planning and procurement, for capacity and costing issues

Suppliers' design computers, for assistance with purchased parts and processes

Material and Capacity Procurement Process

In addition to utilizing the JIT approach to purchasing, a Supplier Scheduling and Control (SSC) system can reduce the procurement process times. The SSC extends the buying company's MRP system into the supplier's plant. Purchasing sends the MRP schedule directly to the supplier, eliminating the need for purchase requisitions and material releases. The SSC tracks a material requisition, from the time it is initiated to the time it is covered by a released purchase order. Then, in a manner similar to a PAC dispatch report, the SSC system provides each buyer, on a daily basis, a backlog report of all open requisitions in priority sequence. The progress of a purchase order can be tracked in much the same way as a shop order. The purchase routing can be maintained in the computer files in the same manner as a manufacturing routing, and can be automatically assigned to each purchase order. When a single supplier has been designated for an item, much of the purchase order release function can be completely automated. The SSC is described in more detail in Chapter 7.

As we created a vision of possibilities for customer service, we can create similar visions for materials planners. A materials planner might well sit in front of a Windows-based screen (supported by integrated MRP-CRP-Finite Scheduling), using the same drill-down capabilities to browse through information contained anywhere in the central system files. This integrated screen replaces the reams of paper containing MRP action reports, planning reports, and pegging reports. When the materials planner wishes more detail about a given number on the screen, he or she clicks the mouse on that field, and the computer brings up the supporting data. This process can continue for as many levels as data exists. The planner can also access the same powerful scheduling and rescheduling tool that the customer service representative uses, to see the total effect of potential changes of staffing, materials, machine availability, tools, and other necessary resources.

Delivery Processes

Materials Management is responsible for picking the product from the storeroom, packaging it, and moving it to shipping. For Make-to-Stock items, these activities can be a large part of the response time, although they shouldn't be. We can often reduce the shipping response time considerably, by simply making the response time a major goal of the department and by establishing performance measures to provide an incentive for improvement.

The future vision for the shipping department is already a reality. Systems exist today that can guide a material picker to the next location, confirm that the right material has been picked onto the forks of the forklift, and create the paperwork for the trucking, or other carrier. This all functions using bar codes.

Description of Company Engaged in Time Competition

Companies involved in time competition will usually utilize the Make-to-Demand response strategy. This is an extremely flexible strategy that is completely responsive to the customer's order, and is one that delivers the product with a speed needed to beat the competition. It can use any combination of the other response strategies that are appropriate to meet customer demand. Depending on the competitive situation, designs, raw materials, components, assemblies, or finished products may be kept in inventory.

The manufacturing system could be either a Flexible Manufacturing System (FMS) with a Flexible Control System (FCS), or an Agile Manufacturing System (AMS) with an Agile Control System (ACS). The two systems operate in a comparable fashion, the only difference is in the degree of automation. The FMS and FCS is highly automated, whereas the AMS and ACS attempt to achieve the same goals without extensive automation. The goal of both systems is to provide a response to the customer that will beat all competition in regard to quickness, flexibility, and efficiency. Both FMS-FCS and AMS-ACS are more a philosophy than a specific set of hardware.

Relationship between Just-in-Time and Time-Based Competition

From the customer's point of view, what matters is the total time required to deliver the product or service. To compress time, we must significantly change every function that affects the delivery of the product (or service) to the customer. By implementing Just-in-Time, many companies have greatly reduced the times of their manufacturing processes.

Manufacturing is but one of five major processes that must occur in a manufacturing company for a customer to receive a response to an order (see Figure 9-1). To the customer, time saved in manufacturing is no more or less important than time saved in Sales and Marketing or in Shipping and Logistics. Customers want the product as soon as possible after the need arises. If the time saved by time compression in manufacturing is later lost in the delivery process, the customer is unimpressed.

Through the adoption of JIT, many companies have successfully shortened their manufacturing cycles and their material procurement cycles by enforcing frequent deliveries from nearby suppliers. But many of these same companies have snail-like customer ordering and delivery processes. A product may be transformed from raw material to finished product in a matter of hours, but days or even weeks may be required to get the customer's order to manufacturing, and to get the product from the factory floor to the customer. JIT time compression principles apply equally well to *all* processes, administrative and manufacturing alike.

Summary

Survival and prosperity in the future require flexibility and timeliness. Trends toward short production runs, lower inventory levels, and "doing the job right the first time" preclude manufacturing on the basis of historical performance measures. The timeworn alternatives of "more inventory" or "more people" are no longer acceptable, even if they are feasible.

The next decade will see massive changes in the M&CRP-MRP II system. The world of manufacturing is changing in an unprecedented manner, and M&CRP will change with it. The tremendous increase in the power of computers, coupled with their plummeting costs, will enable us to satisfy these requirements in a cost-effective manner. The M&CRP-MRP II systems of the future will be much more fully integrated and much more powerful, able to run mini-MRP with finite scheduling to determine when a customer could receive an order, while the customer waits on the phone!

A M&CRP-MRP II system is an excellent resource for a number of environments, other than the one for which it was originally developed—the job shop. We will continue to develop the adaptations that will make M&CRP more useful in repetitive manufacturing, process industries, and other nontraditional environments.

Finally, companies will learn to more fully utilize the human resources that are available, both inside and outside the company, including educators, system developers, and consultants, in order to make the most progress in advancing the capability of the M&CRP-MRP II system. We sincerely hope that this book provides the framework for this cooperative effort.

Selected Bibliography

Browne, Jimmie, J. Harhen, and J. Shivnan: *Production Management Systems: A CIM Perspective,* Addison-Wesley, Reading, MA, 1988, Chap 10 and 16.

Caruso, David: "Emerging Computer Technologies," *World Class Manufacturing Systems Proceedings,* Oakland University and the Detroit Chapter of APICS, 1992.

Groover, Mikell P.: *Automation, Production Systems, and Computer Integrated Manufacturing,* Prentice-Hall, Englewood Cliffs, NJ, 1987.

——— and Olugbenga Mejabi: "Trends in Manufacturing Systems Design," *1987 IIE Integrated Systems Conference Proceedings,* IIE, 1987, pp. 17–23.

Gunn, Thomas G.: *Manufacturing for Competitive Advantage,* Ballinger, Cambridge, MA, 1987.

Howard, Jeff and Emil Sommerlad: "The Future Direction of Packaged Manufacturing Software," *APICS 1984 Readings in Computers and Software,* APICS, Falls Church, VA, 1984, pp. 48–51.

Kanet, John J.: "MRP 96: Time to Rethink Manufacturing Logistics," *MCRP Reprints,* APICS, Falls Church, VA, 1991, pp. 120–124.

Oden, Howard W.: "Information Systems Architecture for Manufacturing," *1989 SME International Conference Proceedings,* SME, 1989.

Ranky, Paul G.: *Computer Integrated Manufacturing,* Prentice-Hall International, Englewood Cliffs, NJ, 1986.

Sanderson, Gerald A. and Dennis Simmons: "The Last Word in MRP II: A System Interface Using Speech," *APICS 1989 Conference Proceedings,* APICS, Falls Church, VA, 1989, pp. 418–421.

Savage, Edward and Michael Mikurak: "Finite Scheduling: Staging a Comeback?" *CIM Technology,* CASA/SME, Spring 1986, pp. 26–31.

Toye, Charles A.: "Let's Update Capacity Requirements Planning Logic," *APICS 1990 Conference Proceedings,* APICS, Falls Church, VA, 1990, pp. 215–221.

Appendix A
Section of Function-Feature Specification for RFP

Purchase and Manage Inventories

Inventories require careful management. If not managed aggressively, they can be both too high (causing high costs in capital, storage, and potential spoilage) and too low (causing high costs in expediting, poor customer service, and excessively short production times). This section addresses the information systems required to enable XYZ Corporation to minimize inventory levels while maximizing customer service. Successful inventory management has two prerequisites:

- Precisely define each product that is made or purchased. This is normally accomplished by the use of:
  - An inventory Master file, which defines the various characteristics of a given product
  - A Bill of Materials file, which defines all components from which a product is made
  - A routing file, which defines how the product is made. (Purchased parts traditionally have no routings.)
- Ensure that *all* inventory transactions are accurately and quickly recorded in the formal system. Anything that affects the characteristics of any inventoried item, especially its current or anticipated on-hand quantity, must be entered into the formal system. The system needs to assist

this process by editing extensively (for accuracy) and providing exception and action notices (to identify potentially missing transactions).

Transaction entry, editing, and management can require extensive administrative overhead unless carefully implemented. XYZ Corporation must minimize the actual time required for transaction entry and management, because such effort does not directly contribute to profitability. XYZ Corporation is currently envisioning the extensive use of bar code, computer-to-computer communication, and defaults as methods to potentially reduce the time required for data entry. The proposals submitted in response to this RFP need to address this issue directly.

Once these prerequisites are in place, the system computes the materials required, by date, to support the manufacturing plan. Purchasing uses these computed materials requirements, plus other purchase requisitions for maintenance, office supplies, services, and so on, to determine total potential purchase volumes (to be used in vendor negotiations). Purchasing places blanket purchase orders and sends releases against them. In the future, purchasing will send releases electronically, and/or allow the vendor to inquire, with appropriate security, into schedules for the parts they provide.

Finally, the purchased materials are received into stock, and a receiving clerk enters the transaction into the system (using bar code and defaults).

Define Products

Design Engineering, Manufacturing Engineering, and Manufacturing define what a product is, what it is made from, and how it is made. A designated person in the Engineering department maintains that information (on-line, real time) in the Inventory, Bill of Materials, and routing files, respectively.

To assist in retrieving information regarding these items, inquiries are available not only by "Part Number" but by class, type, size, and so on.

Features and Functions

1. Inventory Master, on-line, real-time maintenance
   - 1.1. Part Number
     - 1.1.1. At least 16 characters
     - 1.1.2. Can be restricted to numeric only, at XYZ Corporation option
   - 1.2. Revision level
     - 1.2.1. Obsolete flag

- 1.3. Part description
  - 1.3.1. At least 30 characters on *all* screens and reports
  - 1.3.2. Unlimited description available
- 1.4. Units of measure
  - 1.4.1. Stocking
  - 1.4.2. Vendor
  - 1.4.3. Conversion from stocking to vendor unit
  - 1.4.4. Customer
  - 1.4.5. Conversion from stocking to customer
  - 1.4.6. Vendor pricing
  - 1.4.7. Conversion from stocking to vendor pricing
- 1.5. ABC code
- 1.6. Make-buy code
  - 1.6.1. Manufacture
  - 1.6.2. Buy
  - 1.6.3. Subcontract
  - 1.6.4. Planning
- 1.7. Product line
  - 1.7.1. Primary
  - 1.7.2. Secondary
  - 1.7.3. Other (several)
- 1.8. Inventory codes
  - 1.8.1. Class
  - 1.8.2. Type
  - 1.8.3. Size
  - 1.8.4. Hazardous substance classification
  - 1.8.5. Manufacturing group
  - 1.8.6. Assembly group
  - 1.8.7. Other (user-defined, several)
- 1.9. Responsibilities
  - 1.9.1. Planner
  - 1.9.2. Buyer
  - 1.9.3. Estimator
- 1.10. Stocking-shipping
  - 1.10.1. Weight/unit
  - 1.10.2. Cubes/unit
    - 1.10.2.1. Length
    - 1.10.2.2. Width
    - 1.10.2.3. Height
  - 1.10.3. Normal pallet
    - 1.10.3.1. Configuration pattern
    - 1.10.3.2. Quantity
    - 1.10.3.3. Weight

- 1.10.3.4. Size
  - 1.10.3.4.1. Length
  - 1.10.3.4.2. Width
  - 1.10.3.4.3. Height
- 1.11. Total quantity on hand (summarized from Inventory location file)
- 1.12. Accounting information
  - 1.12.1. General ledger account code
  - 1.12.2. Cost
    - 1.12.2.1. Last purchase, per unit
    - 1.12.2.2. Average purchase, last twelve months, per unit
    - 1.12.2.3. Last manufacture, per unit
    - 1.12.2.4. Average manufacture, last twelve months, per unit
  - 1.12.3. Price
    - 1.12.3.1. Last selling price, per unit
    - 1.12.3.2. Average selling price, last twelve months, per unit
  - 1.12.4. Inventory valuation
    - 1.12.4.1. LIFO
    - 1.12.4.2. Weighted average
- 1.13. Stocking policies
  - 1.13.1. Stock
  - 1.13.2. Nonstock
  - 1.13.3. Floor stock
  - 1.13.4. Bulk
  - 1.13.5. Phantom (transient)
- 1.14. Planning data
  - 1.14.1. Ordering policy
    - 1.14.1.1. MRP
    - 1.14.1.2. ROP
    - 1.14.1.3. Min-max
  - 1.14.2. Order point data
    - 1.14.2.1. Minimum on-hand
    - 1.14.2.2. Maximum on-hand (for exception reports)
    - 1.14.2.3. Reorder point
  - 1.14.3. Safety Stock
    - 1.14.4.1. Order quantity adjustments
    - 1.14.4.2. Order minimum (for exception reports)
    - 1.14.4.3. Order maximum (for exception reports)
    - 1.14.4.4. Order multiple
  - 1.14.5. MRP lot sizes
    - 1.14.5.1. Fixed order quantity
    - 1.14.5.2. Fixed time period
    - 1.14.5.3. Lot-for-lot

- 1.14.6. Scrap factor
  - 1.14.6.1. By operation
  - 1.14.6.2. By component
  - 1.14.6.3. (See waste tables in Section 2.5.4)
- 1.14.7. Manufacturing lead time is derived from the routing file and quantity for an order
- 1.14.8. Material flow rates (for JIT)
  - 1.14.8.1. Per time period (hour, shift, day)
  - 1.14.8.2. Total quantity per time period
  - 1.14.8.3. Minimum quantity per group
- 1.14.9. Actual usage
  - 1.14.9.1. MTD
  - 1.14.9.2. YTD

1.15. Quantity on order (summarized from purchase orders and work orders for this item)

1.16. Quantity allocated (summarized from bucketless requirements records due to explosions for parent work orders)

Appendix **B**

Systems Reference Questionnaire

Interviewer ______________________ Date __________

	Company		Contact
Name		Name	
Address		Title	
Address		Phone	
City, State, ZIP			

Packaged Used ______________________

Module	Revision level	Date installed	Date implemented	Date and reason removed
1.				
2.				
3.				
4.				
5.				
6.				
7.				
8.				
9.				

System Information

System Data:

CPU (model-memory)

Disk (MEG)

Printers (number, speed)

CRTs (number)

PCs (number)

	High	Low	Average
Typical number of users on system			
Response time: Data entry			
Response time: Screen inquiry			
Response time: Batch reports			

Reasons for Fast Response Time:

Reasons for Slow Response Time:

Application Information

Application Strengths:

Weaknesses:

Does It Meet Your Needs?

How Has It Improved Your Productivity?

Does It Do What the Vendor Promised? Why or Why Not?

How Easy Was It to Implement?

Vendor Information

Please Rate the Following:

Category	High	Med	Low	Why?
Pre-sales support				
Implementation support				
Hot line				
On-site support				
User documentation				
Systems documentation				
On-site education				
Vendor location education				

Would You Buy Another Module from This Vendor? Why or Why Not?

What Other Vendors Did You Consider? Why Did You Choose "*X*"?

Would You Recommend that We Buy These Modules from This Vendor? Why or Why Not?

Anything Else We Should Know about the Vendor or the Modules?

May We Visit Your Site if We Select This Vendor?

Appendix C

MRP II System Justification

Benefits—Tangible	Typical %	One-time $	On-going $ per year
Reduced Inventory Levels			
Safety stocks	25–50		
Operating inventories	15–30		
Decreased storage space	15–30		
Improved Plant Productivity			
Reduced overtime	25–75		
Reduced rework	25–50		
Reduced hiring-layoff costs	25–40		
Reduced material shortages	50–90		
Increased Sales			
Increased on time deliveries	50–90		
Reduced product errors	20–40		
Shorter lead times to ship	15–50		
Improved lost business analysis	5–10		
Increased Miscellaneous Revenues			
Improved warranty tracking	30–60		
More accurate costing	10–25		
Reduced Purchased Costs			
Blanket orders, better visibility	2–5		
Reduced expediting	30–75		
Reduced Clerical Effort			
Eliminated manual work orders	100		
Reduced paper chasing	50		
Reduced Old Systems Costs			
Maintenance	50–100		
Total Tangible Benefits			

Benefits—Intangible	Typical %	One-time $	On-going $ per year
Improved administrative productivity	5–10		
Better decisions	5		
Proactive planning (vs. reactive)	5–10		
More effective sales force	5–10		
Better control of inventory investment	5–15		
Direct connection to customer computers	5–10		
Total Intangible Benefits			

Costs—People	One-time $	On-going $ per year
Project leader		
Project team		
Education		
First cut		
Module training		
Refresher		
Outside counselor		
Overtime		
Initial test data load		
Conference room pilot		
Final conversion		
Defining new operating procedures		
Additional staff		
Data entry		
Routine operations (due to learning curve)		
Total People Costs		

Costs—Data		
Inventory Record Accuracy		
Cycle counting		
Correcting environmental causes		
BOM Accuracy		
Verifying each BOM		
Correcting inaccuracies		
Defining missing BOMs		
Revising BOM structure to best use new capabilities (e.g., features-options)		
Maintaining BOMs after implementation		
Routing Accuracy		
Verifying each routing		
Correcting inaccuracies		
Defining missing routings		
Revising routing structure to best use new capabilities (e.g., alternate routings)		
Maintaining routings after implementation		
Conversion		
Converting existing data		
Gathering and entering new data		
Total Data Costs		

Costs—Computer Software	One-time $	On-going $ per year
Applications Software—Purchased		
M&CRP programs		
Finance and Accounting programs		
DRP programs		
Other		
Applications Software—Custom Developed		
Conceptual design		
Detail design		
Code		
Test		
Document		
Systems Software		
Operating system		
Language compilers-interpreters		
Database		
Report writers—query tools		
Communications and network		
EDI (Electronic Data Interchange)		
PC—Mainframe link		
Terminal—Mainframe link		
Systems and Programming		
Interfaces between new systems		
Bridges between new and old systems during implementation		
Modifications to packages		
Total Software Costs		

Costs—Hardware	**One-time $**	**On-going $ per year**
Computer Room		
CPU (Central Processing Unit)		
Disks		
Tape backup		
Main printer		
Communications boxes		
Power conditioner		
HVAC (climate control)		
Wiring—power		
Wiring—data		
Wiring—communications		
User Devices		
PC's		
Work stations		
Dumb terminals		
Local printers		
Remote Support		
Communications boxes		
Other Communications		
Network hubs		
Modems		
Multiplexors		
Total Hardware Costs		

Index

About the Authors

The authors of this handbook are all Certified Production and Inventory Control Managers, and are active in the Worcester, Massachusetts, chapter of APICS. HOWARD W. ODEN, DBA, CFPIM, is a professor and coordinator of the materials management program at Nichols College in Dudley, Massachusetts. A Navy veteran, he has also taught at the University of Bridgeport and Worcester Polytechnical Institute.

GARY LANGENWALTER, MBA, CFPIM, is president of Langenwalter & Associates, a Stow, Massachusetts-based information systems consultancy which specializes in JIT and TQM implementations for manufacturing firms.

RAYMOND LUCIER, MBA, CPIM, is materials manager for Artel Communication Company. He teaches MRP II, M&CRP, and Inventory Management in colleges and review courses.